TEXTILES
for the Consumer

TEXTILES
for the Consumer

Nancy Belck
Central Michigan University

with

Sara M. Butler
Miami University

Marlene Wamhoff
Fabric Coordinator
Michigan State University

MICHIGAN STATE UNIVERSITY PRESS
East Lansing, Michigan
1990

Library of Congress Catalog Card Number: 84-60623

ISBN: 0-87013-237-7

Manufactured in the United States of America

Michigan State University Press

East Lansing, Michigan 48823-5202

In memory of Marjory L. Joseph

Acknowledging our long-term professional and personal friendship
and her contributions to the previous edition of this book.

CONTENTS

Preface

Purpose of this Approach to Basic Textiles

Based on conversations with and surveys of textile professionals, instructors, and students interested in textiles, this text was designed:

1. To present a *consumer-oriented* approach to textiles in which the details of textile technology serve to increase consumer awareness, rather than simply developing technological sophistication.
2. To be *relevant* to the needs and desires of students preparing for professional roles in retailing, design or education; and to provide an overview for the nontextile, business professional working in the industry.
3. To provide an integrated *system* of instruction which ties a specially prepared set of sample swatches to learning about textiles in terms of consumer-oriented *end use,* the most critical dimension of this area of study.

Development

This new and greatly revised second edition incorporates critical and constructive suggestions received from many who used the first edition. The first edition was developed in the College of Human Ecology, Department of Human Environment and Design, Michigan State University, where basic decisions were made concerning course philosophy and specific content which should be incorporated into the text. Information gained from textile literature, visits to sites in the industry, and the direct contributions of university educators and industry leaders are reflected throughout this revision.

Since this particular approach to textiles is *consumer-oriented,* it includes textile information and actual fabric samples relevant to consumer needs. Textile chemistry and physics concepts are included only to the extent that they interpret consumer problems. If you desire a career in textile chemistry, textile technology, or some phase of textile manufacturing, then you should plan to take additional courses emphasizing those aspects of textiles in which you are most interested.

This approach to textiles is designed to provide an *overview of the entire textile field.* If you wish to concentrate on some technical aspect of textiles, or if you come from a nontextile discipline (such as marketing, management, or economics) and find yourself working in the textile industry, you should find through this approach a good overview and solid perspective of the entire textile field. You will be able to appreciate the many facets of the textile industry and how they work together to affect the consumer. Given such a base, further specialization can be increasingly meaningful. Each unit builds on the previous one, beginning in Unit I with a discussion of basic textile concepts and terms that will be used throughout the text. Following this orientation to the textile field, units follow detailing information on *Fibers, Yarns, Structures, Finishes,* and *Color and Design.* Last is a summary unit on Textile Decisions, which provides decisionmaking situations that offer the user an opportunity to apply what was presented throughout.

All units have an introduction, including *objectives,* plus a *pretest* at the beginning and *posttest* at the end. By working through the pretest, you will be able to determine the extent of your present textile knowledge. More importantly, however, you will get some perspective on what will be presented in each unit.

If you do not wish to try the pretests, you may immediately begin the new unit and use the pretest as a second posttest to help prepare you for the course examinations. Also, in each unit there are *Quick Quizzes* and *Review* sections. You will find it useful to note the page numbers of these sections for study at a later date. You can record these page numbers on the matrixes that are included at the end of the text for your use as work sheets.

Acknowledgments

Contributing much to the efficiency of this text is the ''hands on'' learning possible because the text comes with a packet of textile samples, each of which is used in several ways to demonstrate the various characteristics of woven, nonwoven and knitted goods. A summary of the characteristics of each fabric is provided in the Review Summary.

Since the swatch packet would have been far too expensive if its samples had been bought in the marketplace, the authors worked hard to obtain as much yardage as possible from textile industry donors. Fortunately, the industry, despite its own serious financial problems at the time, came through with almost every request made.

Because of the generosity of the following contributors, they must share a great deal of the credit for making this text what the authors hope will be a continuing success story for all in the textile field, whether in terms of its professionals or those who learn about it in the classroom.

Appreciation goes to the management of these companies for the fabric samples listed which they contributed; these are identified by their sample numbers:

American Silk Mills Corp.
Shantung, 100% silk (3)

Burlington Industries, Inc.
Stretch Denim, 75% cotton, 25% spun polyester (8)
Nexus®, spunlaced, 100% polyester (33)
Nexus®, spunlaced, backing-100% acrylic, crushed foam (34)

Cannon Mills Co.
Terry cloth, 88% cotton, 12% polyester (15)
Brocade, jacquard, 56% olefin, 44% nylon, Scotchgard® (18)

Cohama Riverdale Decorative Fabrics
Division of United Merchants
Sateen, 100% cotton, print, Scotchgard®, durable press (11)
Matelassé, 35% olefin, 35% polyester, 30% rayon, (19)
Drapery fabric, 100% acrylic (43)

Cone Mills Corp.
Pinwale Corduroy, 83% cotton, 17% polyester (12)

Crompton Company, Inc.
Velveteen, 100% combed cotton (13)

Culp, Inc.
Upholstery, 100% olefin (7)
Tufted, 100% rayon (38)

Cyrus Clark Co. Inc.
Chintz, 100% cotton, Everglaze® (1)

Dan River Inc.
Oxford, 60% cotton, 40% polyester, Permaset, DanCrest® (6)

Dover Textiles
Yarn samples (1, 2, 3, 4, 6, 7, 8)

Fab Industries, Inc.
Single knit jersey, 100% spun Fortrel® polyester (22)
Warp knit crepe, 75% Arnel triacetate, 25% polyester (23)
Interlock, 100% polyester (24)
Textured Raschel, 100% polyester (29)
Raschel lace, 100% nylon (30)

Galey & Lord
Seersucker, 50% Kodel® polyester, 50% combed cotton (21)

Guilford Mills, Inc.
Textured pile knit, 65% acetate by AVTEX fibers, 35% nylon (26)
Tricot, 100% Zefran® nylon (27)
Tricot, Flannel II, 100% Celanese Fortrel® polyester (28)

Guil Suede Plus, 100% polyester (28A)
Polyester Chamois (36A)
Guil-Tuf® luggage 100% Caprolon nylon (27A)
Leather-look 100% polyester (25A)

Klopman Mills
Ultressa®, Stretch Woven, 100% Dacron® polyester (9)

LaFrance Ind.
Division of Riegel Textile Corp.
Velvet Upholstery, 65% cotton, 35% rayon, Scotchgard® (14)

Lida Manufacturing Co.
Textured transfer print, interlock, 100% polyester (42)

Masland Duran
Supported film (36)

Milliken
Crepe, Textured, 100% polyester (20)

Old Mill Yarn
Yarn samples (5, 9, 10)

Pellon Corp.
Phun Phelt™, 100% Kodel® polyester (31)

Phillips Fiber Corporation
Needle punch industrial backing, 100% polypropylene (35)

Polylok Corporation
Malimo Stitch Bond, 70% cotton, 19% polyester, 11% rayon (37)

Spring Industries, Inc.
Shadow Voile, 65% Kodel® polyester, 35% combed cotton (2)
Racquet poplin, 65% Kodel® polyester, 35% cotton (4)
Patrician Fancies (Dobby), 75% polyester, 25% combed cotton (16)
Waffle cloth, 65% Kodel® polyester, 35% cotton (17)
Flocked Ringer, 65% Dacron® polyester, 35% Avril® rayon (40)
Plissé, 65% Kodel® polyester, 35% combed cotton (41)

Travis Knits, Inc.
Double knit, 100% polyester (25)

V.I.P.
Quilted, 50% Kodel® polyester, 50% cotton face, 100% polyester filling, 100% nylon backing (39)

UNIT ONE

Textile Concepts and Terminology

The primary goal of this unit is to give you a perspective of the complex field of textiles and help you understand the overall relationships of the many details of textiles. To give you an idea of textile complexity and the resulting selection, use and care problems of the myriad of textile products, consider that:

1. There are several generic classifications of man-made fibers in addition to the variety of natural fibers used in making textile products.
2. There are about 500 trade names for the variety of fibers within the generic classes, and for many of these trade names there are further special fiber properties which are identified by the addition of a number or a letter to the trade name.
3. There are approximately 25 different methods by which fabrics are constructed and further variations within these operations.
4. There are a variety of types of yarns which add complexity to the fabric and ultimate textile product.
5. There are myriads of different types of finishes that can be applied to fabrics in various combinations.
6. There are a variety of different types of dyestuffs which can be used on the textile products to provide different degrees of colorfastness as well as esthetic characteristics.

Given these factors alone, there are at least "n" possible combinations of textile characteristics that contribute to the ultimate textile product. Thus, the science of textiles is complicated and careful decisions are required for consumers to wisely evaluate textile products.

Regardless of the role you will fill as a consumer, a consumer consultant, a retail representative, interior designer, apparel designer, merchandiser, or entrepreneur, it should be apparent that it will be necessary to know how to *use* information rather than just to memorize a set of facts. You will need background information that will provide a consistent rationale for identifying textile alternatives, for matching availability with demands, and for creating availabilities and demand. That is what the study of textiles is all about.

Before you can talk about textiles with any degree of professionalism, you must understand some of the jargon or terminology of textiles. Once you have these basic tools, you will then advance to the smallest element of fabric, *fibers,* before moving forward to the final textile product. As stated, the *fiber* is the smallest unit or building block from which textiles are made. Fibers are either spun into *yarns,* or placed in some special arrangement that will enable *fabrics* to be constructed. The fabrics will be subjected to the necessary *finishing* processes, and *color* and *design* may be added.

Now refer to Matrix 1 and notice the items numbered one through four on the left side of the matrix under "Serviceability Concepts": *Durability, Comfort, Care* or Maintenance, *Esthetic Appearance.*

These are the four basic serviceability concepts that will be discussed throughout the text. All of the

MATRIX I
OVERVIEW OF CONSUMER TEXTILES CONCEPTS

Terms	Fibers	Yarns	Structures	Finishes	Color & Design
	Mineral		Nonwoven		
	Man-made noncellulosic		Multi component		Structuring
	Man-made cellulosic	Novelty	Complex weaves	Chemical	Finishing
	Protein	Ply	Basic weaves	Mechanical	Printing
	Natural cellulosic	Singles	Knits	General	Dyeing

Serviceability Concepts	Developmental Processes: Terms	Fibers	Yarns	Structures	Finishes	Color & Design
1 **Durability**						
Properties:[1] strength or tenacity ......; abrasion resistance ....; cohesiveness or spinning quality ....; elongation ....; elastic recovery ....; flexibility or pliability ......; dimensional stability						
2 **Comfort**						
Properties: absorbency ....; hydrophobic ....; hydrophilic ....; hygroscopic ....; wicking ....; electric conductivity ....; allergenic potential ....; heat or thermal conductivity; heat or thermal retention ..; density or specific gravity ..						
3 **Care or Maintenance**						
Properties: resiliency ....; dimensional stability ......; flammability ....; chemical reactivity ....; heat tolerance ....; biological resistance ....; light resistance ....; age resistance						
4 **Esthetic Appearance**						
Properties: color ....; dye affinity ....; luster ....; translucence ....; drape ....; texture ....; body or hand ....; loft or bulk						

[1]For review you may wish to record property definitions on this matrix to the right of the term. Compare this matrix with the contents to see how the logic of the book is developed.

NOTE: All matrixes are repeated in the back of the book; you will find it useful to remove the appropriate matrix and fill it in as you work through the program. So, remove Matrix I and write in the serviceability concepts and developmental processes you will be studying in Unit 1.

details of textiles will be presented in terms of these concepts. Now locate the column headings under "Developmental Processes": *Terms, Fibers, Yarns, Structures, Finishes*[1], *Color & Design.*

[1]A "finish" is a chemical or mechanical treatment applied to a fabric to achieve a particular purpose. Various finishes are discussed in detail in Unit 5.

The serviceability properties of durability, comfort, care and esthetic appearance will be related to each of these developmental processes. Matrix 1 presents the overall framework that will be used in this study of textiles as applied to serviceability and general consumer satisfaction.

You should now understand that your study of

textiles amounts to expanding the information represented in the spaces of the matrix just described. You will start with some fundamental terminology which defines the four basic concepts of durability, comfort, care or maintenance, plus esthetic appearance. Then you will proceed through the developmental processes, which will be studied with reference to the above four basic concepts. As you study the *developmental processes* beginning with the study of textile fibers, your basic vocabulary will become larger, more refined, and more precise; the processes of logical thinking and decision making related to textiles will become more accurate at each step in the total process.

Objectives for Unit 1

1. You will understand the four subconcepts—durability, comfort, care or maintenance, esthetic appearance—underlying the concept of serviceability.
2. You will be able to list and define the specific properties which define each of the four serviceability subconcepts.

Pretest

Answers are at the end of each test.

Mark the one most appropriate answer which completes each multiple-choice question or match listings as directed.

1. Whether a fabric is *serviceable* to the consumer depends on
   a. how well it will wear.
   b. how well it serves its end use.
   c. how easy it is to care for.
   d. how durable it is when subjected to abrasive forces.

2. If a fabric has high *durability* features, it can be considered serviceable.
   a. true
   b. false

3. A fiber with high *tenacity* will be able to
   a. absorb a large amount of moisture.
   b. conduct a high amount of electricity.
   c. withstand a heavy pulling force.
   d. retain its shape after elongation.

4. *Abrasion resistance* refers to a fiber's resistance to
   a. flexing.
   b. elongation.
   c. duress.
   d. friction.

5. Fibers with rough surfaces will have more ____________ than those with smooth surfaces.
   a. flexibility
   b. elongation
   c. cohesion
   d. tenacity

6. If a fabric has high *elongation potential,* it is made of fibers that will
   a. stretch and return to their original state.
   b. withstand a great amount of stress before breaking.
   c. withstand a small amount of stress before breaking.
   d. stretch a large amount without breaking.

7. *Elasticity* refers to a fiber's ability to
   a. stretch and partially recover its shape.
   b. stretch and finally break.
   c. stretch and return to original size.
   d. elongate slowly, without returning to size.

8. A fabric which can tolerate considerable bending without breaking is made of ____________ fibers.
   a. abrasion resistant
   b. flexible
   c. elastic
   d. cohesive

9. If a fabric is dimensionally stable, it will
   a. lose its shape through stretching.
   b. have a low amount of elasticity.
   c. retain its shape over time.
   d. have a high amount of elongation.

10. A fabric that is comfortable to wear must have high
    a. cohesiveness.
    b. abrasion resistance.
    c. tenacity.
    d. absorbency.

11. An absorbent fabric will ______________ the evaporation of body moisture.
    a. assist
    b. retard

12. A hydrophobic fiber
    a. absorbs moisture but doesn't feel wet.
    b. does not absorb moisture.
    c. absorbs moisture readily.
    d. aids in body cooling.

13. *Hygroscopicity* refers to the ability of a fiber to
    a. absorb moisture and aid in moisture evaporation.
    b. resist the absorption of moisture.
    c. absorb moisture internally without making the fabric feel wet.
    d. absorb only a small amount of moisture.

14. *Wicking* is a property certain fibers have that
    a. aids in body cooling.
    b. retards body cooling.
    c. does not affect body cooling.
    d. does not permit movement of moisture.

15. High *electrical conductivity* is associated with fibers which are
    a. hydrophobic.
    b. hygroscopic.
    c. hydrophilic.

16. *Heat conductivity* of a fabric is determined primarily by
    a. fiber generic class.
    b. the thickness of its fibers.
    c. yarn and fabric structure.
    d. the length of its fibers.

17. A fabric has high *thermal retention* when
    a. heat passes from the body through the fabric.
    b. it is constructed of thin, firm yarns.
    c. heat is not conducted away from the body.
    d. the structure is open.

18. *Density* and *specific gravity* are measures of fiber
    a. tenacity in relation to length.
    b. weight in relation to volume.
    c. weight in relation to width.
    d. strength in relation to width.

19. When fabrics made of acidic fibers are laundered in basic (alkaline) solutions, you can expect that the fabric will
    a. be well cleaned.
    b. not react with the solution.
    c. be damaged.

20. "Built" detergents have a higher ____________ content than "unbuilt" detergents.
    a. acidic
    b. alkaline
    c. salt
    d. soap

21. The use of different cleaning agents is influenced by the chemical reactivity of the
    a. fiber content.
    b. yarn structure.
    c. fabric structure.

22. Moths, mildew and silverfish are likely to attack certain fabrics under what conditions?
    a. dark, cool, dry
    b. light, warm, dry
    c. dim light, warm, humid
    d. light, cool, humid

23. Fabrics which soften, melt or shrink in the presence of heat are made of ____________________ fibers.
    a. low strength
    b. heat resistant
    c. heat sensitive
    d. absorbent

24. A fabric that returns to its original thickness after compression has high
    a. hand.
    b. loft.
    c. body.
    d. drape.

25. A fabric with a firm hand could most accurately be described as having a rigid
    a. loft.
    b. texture.
    c. body.

26. A fabric with a soft hand would have a ________________ drape.
    a. flowing
    b. loose
    c. heavy

27. The degree to which fabrics accept dyes or color pigments depends primarily on fiber
    a. absorbency and molecular structure.
    b. resiliency and tenacity.
    c. molecular structure and abrasion resistance.
    d. absorbency and elasticity.

Answers for Unit 1 Pretest:

1. b	15. c
2. b	16. c
3. c	17. c
4. d	18. b
5. c	19. c
6. d	20. b
7. c	21. a
8. b	22. c
9. c	23. c
10. d	24. b
11. a	25. c
12. b	26. a
13. c	27. a
14. a	

UNIT ONE

Part I

Durability and the Concept of Serviceability

The concepts of durability, comfort, care or maintenance, and esthetic appearance undoubtedly have meaning for you now. However, those meanings are probably general rather than precise or specific in terms of textiles. The purpose of this unit is to refine your concepts of durability, comfort, care or maintenance, and esthetic appearance so that you have the same understanding of these terms as professionals in the textile field. One criterion of professionalism is a firm and clear understanding of the terms and symbols used by the profession and the ability to use them in decision making processes.

In this part of Unit I, you will concentrate on developing a more precise understanding of the overall concept of serviceability in terms of general consumer satisfaction. For a textile product to be a wise choice for a particular use, it must be able to meet the desired criteria for the purpose and serve properly while providing consumer satisfaction. One of the criteria for being serviceable is that a product must tolerate the use and care to which it will be subjected during its intended period of application; that is, the product must provide the durability expected by the consumer. However, as explained below, the terms *serviceable* and *durable* are not completely synonymous.

Durability is a concept related to how long a fabric will last; *laboratory tests* can help measure this. Serviceability is a somewhat more general concept and applies to those considerations and decisions concerning the intended *end use* for the textile product. For example, a suit of armor is certainly durable, but it would not be especially serviceable if intended for a business suit!

You have to consider the durability of a product before you can evaluate its serviceability. One of the more obvious elements of a product's durability is its *strength*. The strength of a fabric depends on the ability of its textile fibers to withstand a *pulling force* without breaking. This quality can be measured mechanically in the laboratory without regard to serviceability considerations or concern about the end use to which the fiber might be put.

For example, some cotton will withstand 30,000 to 60,000 pounds per square inch of pull before breaking. However, this durability measurement is meaningless to the consumer unless the end-use requirement or serviceability is considered.

Strength or Tenacity

The term *strength* is usually used as a measure of yarns and fabrics, while the term *tenacity* is used as a measure for fibers. Both have precise meanings and both can be determined by mechanical test instruments. If a particular fiber withstands a heavy pulling force, it is determined as having high tenacity; while a fabric that withstands heavy pulling forces is said to have high strength. Regardless of whether fibers, yarns or fabrics are being measured for tenacity or strength, test results do not actually indicate how satisfactory the product will be for a specific end use.

At this point the various concepts concerning serviceability are discussed in somewhat general terms. As you continue studying textiles, you will see that many factors combine to determine the properties of the final textile item. Thus, strength or tenacity is only one measure of the durability concept of serviceability.

Abrasion Resistance

In addition to being subjected to pulling stresses or forces, a fiber can be damaged by abrasion; this is the technical term for rubbing or friction applied to the surface of a fiber, yarn or fabric. Abrasion resistance can be measured to some degree in the laboratory; however, since actual end use (wearing or using) conditions cannot be duplicated by mechanical testing, abrasion resistance cannot be exactly predicted by laboratory test. Further, the amount of abrasion resistance necessary to make a particular textile product serviceable will be determined by the durability requirements for the end use.

Friction can break fibers and cause strong fiber ends to ball up into fuzz and cling to the fabric surface. The result is called *pilling* and may be seen in Figure 1.1. When a fabric is subjected to abrasion, fiber balls can be created on the fabric surface if the fibers are strong enough to cling. Weak fibers do not form balls because they break and fall away from the fabric surface.

Cohesiveness or Spinning Quality

Another measure of textile durability is fiber *cohesiveness* or *spinning quality*. This property identifies the ability of fibers to cling to one another in the formation of yarns and fabrics. Fibers that have rough surfaces, such as wool, tend to be more cohesive and spinnable than smooth fibers such as silk or man-made filaments. The term *filament* refers to long continuous fibers that can be measured in meters or yards, as silk; or in kilometers or miles, for man-made fibers like nylon.

For smooth fibers and filaments that lack natural cohesiveness, and are therefore difficult to spin into yarn, manufacturers have developed techniques that improve spinning quality. These techniques include making the fiber surface rougher by adding crimp to it or using one of many available texturizing processes.

Filament fibers usually are crimped or texturized to improve spinning quality before being processed into yarns except when high luster and smoothness are required in the end product. Even when texture is not added to these smooth filaments, they still can be spun into yarn because of their long length. However, short fibers like cotton and wool must have a minimum level of cohesiveness or spinning quality to compensate for their short length in order to be successfully made into yarns and fibers.

Elongation and Elastic Recovery

Almost all fibers can be stretched at least a small degree without breaking. This ability to stretch without breaking is called the *elongation* property of the fiber. Depending on the end use of the textile, a certain amount of fiber stretch or elongation is essential for good serviceability. Along with the ability to stretch, the ability to return to size is highly desired; this is called the *elastic recovery* or *elasticity* of the fiber. When a stretched fiber returns to its orig-

---

QUICK QUIZ 1

1. Check the end use items in the following list which would need the durability requirement of high strength to be serviceable:
   - ☐ evening dress ☐ child's playclothes ☐ carpet
   - ☐ sports uniform ☐ infant's garment ☐ rope
2. Check the products which must have high abrasion resistance to be considered serviceable:
   - ☐ carpet ☐ umbrella ☐ hat
   - ☐ coat lining ☐ drapery ☐ upholstery

****

ANSWERS QUICK QUIZ 1

1. sports uniform, child's playclothes, rope
2. carpet, coat lining, upholstery, & possibly drapery (depending on where it is hung, and what it might rub against—floor, window sill, furniture, etc.)

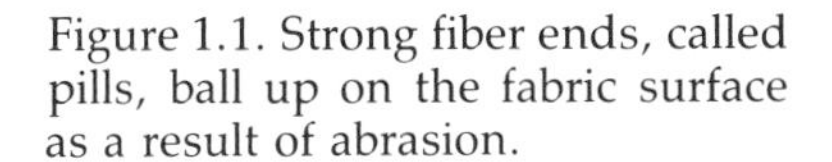
Figure 1.1. Strong fiber ends, called pills, ball up on the fabric surface as a result of abrasion.

inal shape after elongation, it is said to have good or high elastic recovery. If the fiber does not return to its original size, elastic recovery is poor. The speed with which a fiber returns to size is also important. Elastic recovery may be immediate, occurring within a one to five minute interval, or may be delayed, occurring over a period of time.

A fiber that has low elongation and high elastic recovery would not be suitable for active sportswear. If the fiber does not elongate, the elastic recovery is of no value for end uses where stretch is important.

While elastic recovery and elongation are related durability properties which can be evaluated with precise testing equipment, the amount necessary for a product would depend on the end use planned.

In addition to fiber-related elasticity and elongation properties, it is also possible to introduce elongation or stretch into the end use product through fabric construction procedures. Thus, the total elongation and elastic recovery of a textile item is dependent on more than just fiber content.

Flexibility and Pliability

In addition to the stresses already discussed, a fiber may be subjected to considerable bending or creasing in the course of normal use and care. A fiber which can tolerate a reasonable amount of bending or flexing without breaking is said to be *pliable* or *flexible.*

Fabrics which resist splitting, cracking or breaking when folded or creased repeatedly are made from fibers with a high level of flexibility or pliability. Flexibility is essential in textile products that are repeatedly bent or creased in the same place, such as shirts, tableclothes and durable press slacks.

Fibers with low levels of elongation frequently have poor flexibility, which means the consumer must follow special care procedures. For example, tableclothes made of flax, known for its low flexibility, should be folded and refolded in different ways for storage. Or better yet, roll them in order to eliminate folding. Leaving a linen tablecloth on the table for long periods is undesirable since the continuous creasing in the same place is destructive to the fibers.

Dimensional Stability

Dimensional stability is the ability of a fiber, yarn or fabric to retain a given size and shape during its lifetime, using acceptable care procedures. For various reasons to be discussed later, many fibers, yarns or fabrics tend to lose their shape through stretching or shrinking over time.

QUICK QUIZ 2

1. Which of the following factors would be important in determining serviceability of athletic clothing?
   - ☐ low elongation ☐ low elastic recovery
   - ☐ high elongation ☐ high elastic recovery
2. Of the following three end uses, which would demand the highest level of dimensional stability?
   - ☐ swimsuit ☐ necktie ☐ pillowcase

   Why?____________________

****

ANSWERS QUICK QUIZ 2

1. high elongation, high elastic recovery
2. swimsuit, because it must continue to conform to the wearer's body

Plainly, for a garment with high elongation to be dimensionally stable, it would have to have high elastic recovery or elasticity. For it to fit properly over time, it must be cared for properly and have good dimensional stability.

A fiber which has high levels of tenacity, abrasion resistance, cohesiveness or spinning quality, elongation and elastic recovery, flexibility or pliability, and dimensional stability will have high durability. However, it is not desirable or possible in most situations to have all these factors present in high levels at the same time in the same textile product.

The concept of *serviceability* implies that there must be sufficient amounts of certain durability properties to satisfy the end use requirements of the specific item. For example, "white-tie-and-tails" or evening gowns do not require the same amounts and kinds of durability factors as do work or school clothes. In dress apparel, a far lower level of durability properties will be found to provide adequate serviceability for most consumers.

It is important to remember that durability factors of fibers are harder to evaluate when fibers are converted into yarns and fabrics. Thus, while fiber properties are important, the consumer must recognize that converting fibers into textile end products creates additional properties and modifies existing ones. These complexities of textile products make the study both interesting and important for consumers.

REVIEW—Answers are at end of each Review.

1. *Fiber tenacity, fabric strength, elongation* and *elastic recovery* are measures of __________ which can be determined by mechanical tests in a __________.
2. A product (does, does not) need high measures of all durability properties to be serviceable in the planned end use.
3. The *durability* of a product relates to how long it will last. Durability requirements __________ from item to item.
4. *Serviceability* is a relative concept. It cannot be determined by laboratory measurements but depends on subjective judgments related to the planned __________.

5. If a product is intended to be disposable, high ____________ is not essential since the item could still be serviceable for the limited end use.

6. Indicate if the following terms are laboratory measures of durability (X) or relative and subject to interpretation according to end use (0).

____ tenacity	____ economy	____ elasticity
____ flexibility	____ abrasion resistance	____ dimensional stability
____ disposability	____ utility	____ attractiveness

7. Durability implies an objective standard because it can be evaluated to an extent by ____________.

8. Serviceability must be evaluated by subjective judgment based on ____________ ____________.

9. A product may have properties which make it very ____________, but unless the end use has been taken into account, it may not be ____________.

10. List seven measures of *durability* and define them. Do not proceed until you can do this easily, since these concepts are important to the future units in this program.
    1.) ____________
    2.) ____________
    3.) ____________
    4.) ____________
    5.) ____________
    6.) ____________
    7.) ____________

****

REVIEW ANSWERS:

1. durability; laboratory
2. does not
3. will vary or be different
4. end use
5. durability
6. X = tenacity, flexibility, abrasion resistance, elasticity, dimensional stability
   O = disposability, economy, utility, attractiveness
7. laboratory tests (mechanical measurement)
8. the planned end use of the product
9. durable; serviceable

10. *Strength or tenacity*—ability of a fiber, yarn or fabric to withstand pulling without breaking.
    Abrasion resistance—ability of a fiber, yarn or fabric to tolerate rubbing or friction without damage.
    Cohesiveness or spinning quality—ability of fibers to cling together during conversion into yarns and fabrics.
    Elongation potential—ability of a fiber or fabric to stretch without breaking.
    Elastic Recovery—ability of a fiber, yarn or fabric to return to its original size following release of stress that causes elongation.
    Flexibility or pliability—ability of a fiber, yarn or fabric to withstand repeated or long-term bending without breaking.
    Dimensional stability—ability of fiber or fabric to maintain its original size during use and care.

Because these terms will be used through at the text, it is important that you master them.

UNIT ONE

Part II

Comfort and The Concept of Serviceability

If the concept of durability were the only consideration in the overall serviceability of textile products, decisions would be easy! A glass fabric undergarment might be the last word in strength, but wearing it would raise some itchy problems. Therefore, *comfort* is an equally important factor in the selection of apparel and furnishings which clothe or contact the human body. As with durability, the definition of comfort is arrived at through an understanding of selected terms. The importance of comfort in a textile product can be evaluated in terms of how serviceable the item is for a specified end use.

For many items, you have to consider both durability and comfort in order to obtain the serviceability you prefer. In fact, all the textile serviceability concepts should be weighed in choosing textile products, with the values desired for a specific end use helping in the decision making process.

Several properties are involved in the evaluation of comfort. Those included in this discussion are: *absorbency, wicking, electrical conductivity, allergenic potential, thermal conductivity, thermal retention* and *density*.

Absorbency

One important aspect of comfort relates to the ability of fibers, yarns or fabrics to absorb moisture. A characteristic of the human body is that its sweat glands give off moisture through skin pores. This temperature-lowering process is important not only for the comfort of the body but for survival. In terms of clothing, then, comfort is related to how well the fabric absorbs moisture from the body. On a dry day your body's moisture can evaporate easily into the air, as it is given off through the skin and cools the body. You feel comfortable. On a hot, damp day, however, there is already too much moisture in the air; so the air does not pick up the perspiration from the skin as quickly. Thus, the body feels uncomfortable.

Since clothing intercedes between body and atmosphere, the ability of the fabric to assist or retard the evaporation of perspiration into the air is an important factor in body comfort. A fabric that takes up moisture and transmits it to the air will be more comfortable than one that does not. Fibers, yarns and the fabric structure all play a critical role in the ultimate comfort of textile products.

A fabric which takes up moisture is said to be *absorbent*. This property assists the body in giving off moisture for cooling purposes, provided that this moisture is transmitted to the air.

If a fiber has an affinity for water and absorbs moisture readily, it is said to be *hydrophilic*. *Hydro* is a Greek derivative meaning "water," and *philic* is from the Greek word *philein*, which means "to love." A hydrophilic fiber is, therefore, "water-loving." Such fibers when woven into fabrics contribute to absorbent products.

QUICK QUIZ 3

1. Put a "d" in front of the items you would select primarily for durability properties and "c" in front of those where comfort would be more important.

☐ tablecloth	☐ handkerchief	☐ leisure-lounge wear
☐ dish towel	☐ draperies	☐ bath towels

2. Check below the items for which durability and comfort are both equally important to *most* consumers.

☐ underwear	☐ carpet	☐ evening dress
☐ upholstery	☐ sheets	☐ work clothes

****

ANSWERS QUICK QUIZ 3

1. c = handkerchief, leisure-lounge wear, bath towels; d = tablecloth, dish towel, draperies
2. underwear, upholstery, sheets and work clothes

Another term closely associated with absorbency is *moisture regain*. This term indicates the amount of moisture that a fiber will actually hold under specified laboratory conditions (i.e. controlled temperature and humidity). *Saturation moisture regain* is the total amount of moisture a fiber is capable of absorbing when subjected to quantities of moisture. Saturation regain is an important consideration in evaluating fabric absorbency for a given end use.

Fibers which are hydrophilic are those with good to high saturation regain or absorbency. Fibers that have little or no regain are called *hydrophobic*. *Phobic* is derived from the Greek word *phobos*, meaning *fear*. So, fabric made of hydrophobic fibers has low moisture regain and is considered "water fearing" or nonabsorbent.

When a hydrophilic fabric or fiber absorbs moisture, it usually feels damp to the touch. Yet, a few fibers such as wool will absorb a high amount of moisture without feeling damp to the touch. This is a special property of hydrophilic fibers called *hygroscopicity*.

Wool absorbs moisture from the air as well as from liquids, but its surface does not feel damp until the product is totally saturated. Wool, therefore, is a hygroscopic fiber which is also hydrophilic because it absorbs moisture.

The particular structure of a hygroscopic fiber that can absorb moisture but feels dry on the surface will be discussed in the fiber unit. However, in general, hygroscopic fibers do not produce comfortable apparel fabric for hot, humid weather because the high level of moisture retained within the fiber prevents the passage of body moisture into the air.

Wicking

In some fibers which are nonabsorbent or hydrophobic, moisture does not pass through the internal structure of the fiber as it does in hydrophilic fibers, but the moisture may travel along the fiber surface. This property, called *wicking* in the textile trade, is similar to capillary action. Both absorbency and wicking contribute to comfort by allowing moisture to be carried away from the body.

A fiber which has the property of wicking rather than the property of absorbency will assist the body in maintaining a normal body temperature by passing perspiration to the outside air for evaporation.

A fiber may wick moisture either vertically along the length of the fiber or horizontally around the width of the fiber. Refer to Figure 1.2. Horizontal wicking promotes body comfort by passing the moisture around the fiber to the fabric surface for evaporation. For example, the disposable liners in diapers are generally made of nonabsorbent fibers that transmit the moisture away from the baby by horizontal wicking. However, when a rain wets apparel from the outside, the wicking process is reversed. The water is carried toward the body rather than away, which reduces comfort.

Electric Conductivity and Static Charge

A frequent problem with certain fibers is that they generate static charges. You have no doubt experienced these "static shocks" and know they can be annoying and uncomfortable. Static charges are caused by the lack of conductivity for electrical

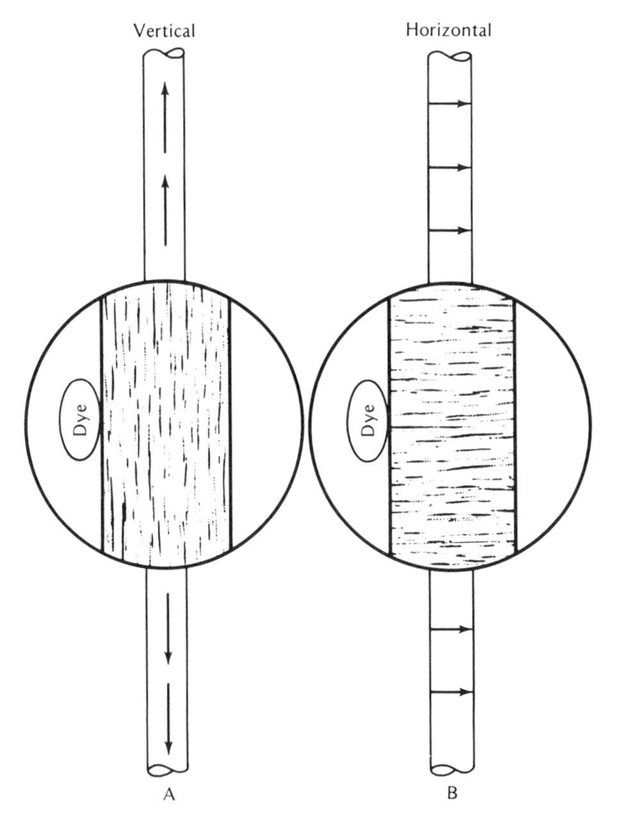

Figure 1.2. In "A" the moisture is "traveling" the fiber length, in "B" the moisture is spreading horizontally around the fiber.

charges that build up on fabrics as a result of friction. In such cases the electrical charges increase until discharged when the wearer comes in contact with a good conducting surface.

Fibers which are hydrophilic conduct electrostatic charges along their surface, thus preventing build up and decreasing the chance of static shock. Fibers that are hydrophobic and have low moisture regain do not provide a conductive path; therefore, nonabsorbent fibers are subject to static build up and resulting static shock. This is because electrostatic charges generated by friction cannot be transmitted to the ground without a shock to the user because there is not a conductive path made possible by moisture.

Thus, an electrostatic charge will build up on a fabric made of hydrophobic or nonabsorbent fibers because there is no way for free electrons to be neutralized until the wearer touches a neutral or oppositely charged object. Then the static charge suddenly jumps to the ground or through the person, resulting in a static shock to the wearer. It should be noted, however, that highly absorbent fibers also may produce static shocks when it is extremely dry and there is little or no moisture in the air to serve as a conductive agent.

Allergenic Potential

In addition to the aspects of comfort resulting from fiber absorbency and conductivity, some people have a body chemistry which makes them allergic to certain fibers, especially wool and other animal hair fibers. Typical signs of an allergic reaction include itching, swelling, a rash, hives or a burning sensation.

Some fibers produce simple skin irritation because they have rough or abrasive surfaces. These create the same types of reactions and discomfort as an allergenic fiber.

QUICK QUIZ 4

1. Check the items which would be most serviceable if made from hydrophilic fibers.
   - ☐ raincoat
   - ☐ awning
   - ☐ swimsuit
   - ☐ diaper
   - ☐ shirt
   - ☐ mop
   - ☐ bath towel
   - ☐ dish towel
2. From your own experience or observation list fibers which might cause allergenic types of reactions. ____________________

****

ANSWERS QUICK QUIZ 4

1. diaper, bath towel, shirt, dish towel, mop (It would not be desirable for a swimsuit to absorb moisture as the added weight would prove uncomfortable.)
2. wool, mohair, nylon, glass fibers are possible examples

Heat Conductivity and Heat Retention

Yarn types and fabric structures largely determine heat conductivity or heat retention, also called *thermal conductivity* and *thermal retention*, which affects the wearer's warmth or coolness. Physical characteristics of fibers likewise can influence thermal properties in determining how the fibers pack when converted into yarns and fabrics.

Since heat comes from the body, fibers, yarns or fabrics can be structured so that body heat escapes—resulting in textile products that have high thermal conductivity and help keep the wearer cool. If fibers, yarns or fabrics do not allow heat to be conducted away from the body and the heat is held in by thermal retention—the resulting textile product will tend to keep the wearer warm.

Fibers with irregular surfaces hold air, and still air is one of the best body insulators. Thus, yarns and fabrics made of fibers that do not pack but that do trap air tend to be characterized by high thermal or heat retention. So the thicker the fabric and the bulkier the yarn, the greater the heat retention.

Body heat also is retained when nonabsorbent fibers like polyester or nylon are made into tightly structured fabrics which do not allow air or heat transfer through the garment. For example, lightweight windbreaker jackets can be quite warm to the wearer, since body heat is retained inside the jacket. Conversely, when apparel is structured to allow air to pass through the fabric, resulting in increased heat conductivity, the wearer will have reduced body heat and feel cooler.

Density or Specific Gravity

Density or *specific gravity* is a measure of the weight of a fiber. Fabrics made from lightweight fibers tend to be lighter per unit volume than fabrics made of heavier fibers. When comparing weight of fabrics made from different types of fibers, it is essential that the fabrics under comparison be identical in structure. Acrylic fibers, for example, have a lower density than wool fibers; so in fabrics of identical structure, the acrylics would be lighter than the wool.

Fibers of low density allow for lightweight yet warm textile products. A ski jacket would be most serviceable if made of a fabric composed of low density fibers which contribute to warmth without weight. Fibers with low density tend to be bulky and are easily made into high-bulk, lightweight fabrics.

REVIEW

1. When selecting a fabric for a hot, humid day, what two fiber properties are important for comfort? ______________________________
____

2. If a fiber does not absorb moisture, it is described as being ______________.
In this fiber, you would also expect (high, low) electrical conductivity. Give two reasons why such a fabric would not be comfortable. ______________________________
____
____

3. If a fabric has high ______________, it will more readily conduct moisture away from the body; however, if the fiber abrades the skin, the fabric still will not be ______________.

4. Terminology Check: List and define, from memory, the ten properties or considerations which bear on the comfort factors of a finished product.
   1.) ______________________________
   ____
   2.) ______________________________
   ____
   3.) ______________________________
   ____
   4.) ______________________________
   ____
   5.) ______________________________
   ____
   6.) ______________________________
   ____
   7.) ______________________________
   ____
   8.) ______________________________
   ____
   9.) ______________________________
   ____
   10.) ______________________________
   ____

****

REVIEW ANSWERS:

1. hydrophilic properties or high absorbency; high thermal conductivity
2. hydrophobic; low; body perspiration would not be absorbed or static shocks could result.
3. absorbency; comfortable
4. *Absorbency*—ability of a fiber to take up moisture.
   Hydrophilic—property of an absorbent fiber; having an affinity for water.
   Hydrophobic—property of a nonabsorbent fiber; having little or no moisture regain.
   Hygroscopic—ability of a fiber to absorb water vapor without feeling wet to the touch.
   Wicking—ability of a fiber to pass water along its length or circumference without absorbing the moisture.
   Electric conductivity—ability of a fiber to conduct electricity to the ground so that static electricity is not created.
   Heat conductivity—ability of a fiber to transmit heat from the body.
   Heat retention—ability of a fiber to hold heat close to the body.
   Allergenic potential—property of some fibers that causes irritation to the skin.
   Density or specific gravity—weight per unit volume of a fiber.

UNIT ONE

Part III

Care and The Concept of Serviceability

A fabric must receive proper care if it is to give continued service; and, as the cliché says, "The better the care, the longer the wear!" *Care* can be defined as any activity that helps preserve the "new" look of a garment or textile product. The term "maintenance" is frequently used as a synonym for "care."

The procedures followed during use, care or maintenance, and storage of textile products are all important to retaining or *maintaining* the desired appearance and performance of a textile fabric. The actual care procedures that should be followed are determined by the chemical structure and composition of the fibers, the yarn and fabric structures, and the finishing and coloring agents used in fabric manufacture.

The chemical sensitivity of the fibers, finishes and dyestuffs all influence the type of detergent, bleach, and other laundry agents used, or dictate the selection of dry cleaning treatments. Also, the heat sensitivity of the fiber and the effect of heat on fabric finishes determine temperatures of laundering, drying, and pressing. Thus, the consumer must match the chemical composition and heat reactions of the fiber, special finishes and coloring agents with the appropriate care procedures to maintain the desired appearance and size of the product. If you have ever laundered a wool or cashmere sweater in hot water with strong detergent and a lot of agitation, you recall the unfortunate shrinking that occurred.

In addition to the chemical composition and heat sensitivity of fibers, finishes and coloring materials, the manner in which the fibers are spun into yarns and manufactured into fabrics also affects care procedures. For example, a loosely woven fabric made of loosely twisted yarns tends to have low dimensional stability, and thus requires special precautions during use and care to preserve the product's shape. Also, the stitching and seam construction procedures used in manufacturing a textile item and the combinations of components used—such as interfacing, thread and trim—will add complexity to the choice of care procedures.

You cannot determine the appropriate care procedures for a textile product simply by looking at it. Because finishes are frequently invisible, and fiber content and colorfastness are impossible to determine by observation, you must rely on information given on labels or hangtags to determine fiber content, fabric finishes and coloring methods. Federal law requires that textile fabrics and items of apparel be labeled to identify fiber content and recommended methods of care. However, there is no requirement that any information be given concerning finishes or colorfastness of the dyestuffs used.

The information in this unit is designed to help provide the basics required for you to determine the most appropriate care and maintenance needed to retain the desired properties of a textile product. Remember that in addition to fiber content, care procedures are influenced by yarn and fabric construction, by the combination of components in the product, by product construction techniques, and by the types of finishes and dyestuffs or coloring materials used on the fabric. Thus, these are other considerations you need to take into account that are not required by law to be on the label.

The following discussion identifies selected groups of properties that influence care during *use, maintenance* and *storage* of textile products.

Selected Properties Influencing Care During Wear or Use of Textile Products

Resiliency

The degree to which a fabric wrinkles during wear and use depends on its *resiliency,* the property that causes a fiber or fabric to return to its original shape after compression, bending, twisting or pulling forces have been applied and removed. While resiliency is identified as both a fiber and a fabric property, it is influenced by *fiber* tenacity, elongation and elastic recovery, and flexibility or pliability. (These terms were discussed in a prior section.)

Thus, a garment that is wrinkle resistant is generally made of fibers that possess high resiliency. Fibers or fabrics that are resilient tend to be flexible and generally recover from wrinkling because they spring back into shape following bending or creasing.

Resiliency and dimensional stability, while related properties, are not the same. Recall that dimensional stability is the ability of a fiber or fabric to retain its shape and size over time. It refers to either change through shrinkage or stretching. Resiliency relates to the ability of a fiber or fabric to recover its shape after wrinkling, bending or twisting.

There is also a close relationship between resiliency and elastic recovery. Fabrics with low elastic recovery tend to wrinkle and exhibit poor resiliency. Thus, they are more difficult to care for than those with good resiliency because wrinkles formed do not readily hang out. The loss of fabric appearance and serviceability through stretching or shrinkage is due to poor fabric dimensional stability, and wrinkling is due to poor fabric resiliency.

Flammability

Flame retardency influences consumer care during wear or use of selected items, and end use determines the importance of having *flame resistant* or *flameproof* fabrics. There are obvious safety advantages for certain home furnishings such as draperies, carpeting and mattresses being flame resistant. There are also advantages to flame resistance in some apparel, particularly children's wear.

QUICK QUIZ 5

1. Resilient fibers and fabrics recover from wrinkling or compression during continued use. What household textile items must have good resiliency to be serviceable over time?
   ____
   ____
2. Would a sofa slipcover also require good dimensional stability? (yes, no) Why or why not?
   ____
   ____
3. The durability properties of strength, elongation, elastic recovery, flexibility and dimensional stability all contribute to the ease of fabric care during use as a result of their contribution to the property of ______________.

****

ANSWERS QUICK QUIZ 5

1. upholstery, slipcovers, carpeting
2. yes; slipcovers not only need to recover from wrinkling, but must retain shape and size to continue to fit the sofa.
3. resiliency

Federal law has set the level of flame resistance required for children's sleepwear, mattress covers, carpeting and selected other textiles, in order that they meet minimum standards required for sale on the consumer market. To produce the necessary level of flame retardancy, manufacturers use either *flame resistant* or *flameproof fibers* or *finishes.* Care instructions included on product labels must be followed to maintain flame retardant properties over time.

A *flameproof* fiber or fabric differs from a *flame resistant* one in that it will not burn even in the presence of direct flame. Only glass and asbestos fibers are flameproof. Other fibers have to be modified so that they are flame resistant, or finishes can be applied to the fabric to produce flame resistance.

Part of the law regarding the flame resistance of selected textile items requires that labels be clearly attached to provide you with information on flammability and indicate conformance to the law. Knowledge of fiber content and care helps determine the flame resistance or heat tolerance of a textile item. However, the only sure way to know that a product is flame resistant and to know how to care for it is to read attached labels required by law.

Selected Properties Influencing Care Treatments For Textile Products

Chemical Reactivity

The chemical reactivity of fibers, finishes and dyestuffs influences the care methods that must be used with textile products. These chemical properties determine the types of detergents, bleaches or dry cleaning solvents that will be recommended for use in care procedures. The damage that certain detergents or bleaches may cause in one type of fiber may have no affect on another; the same is true for finishes and dyes. For example, care procedures that include highly alkaline detergents may actually improve the strength of fabrics made from cotton, but they can seriously damage wool products.

The detailed study of the chemical composition of fibers, finishes and dyestuffs comprises the field of textile chemistry. Since an understanding of textile chemistry requires a knowledge of organic chemistry, you will not be expected to master the complexities of fiber chemistry at this level. However, it is hoped that you will obtain sufficient information so that you can appreciate and understand the care implications involved in the selection and use of certain fibers for given end uses.

A chemical reaction occurs when two combined chemicals are sufficiently different or reactive so that the elements or compounds of one chemical affect the other. For example, when an acid and a base (such as vinegar and baking soda) are combined, a chemical reaction occurs. Sometimes this reaction may completely alter, perhaps even destroy, the products involved. In other cases only minor changes occur.

A potential problem in textile care relates to the damaging reactions that can take place when acidic and alkaline or basic substances come in contact with each other. Some fibers are acidic in nature, while most detergents are alkaline. If a fabric made of fibers that are acidic is laundered using detergents that are strongly alkaline, negative reactions can occur that will damage the textile. Because these reactions are accelerated in the presence of heat, the use of hot water and a strong alkaline detergent to launder a fiber with acidic properties tends to increase the potential for serious damage.

Thus, for fabrics made from acidic fibers, *dry cleaning* is frequently a preferred care procedure. However, laundering can be acceptable for acidic products if warm (not hot) or cool water and a mild detergent with low alkaline content are used.

In addition to acidic-basic interactions, other types of chemical reactions occur between textile products and substances used in their care. Thus, it is important to select the most appropriate type of either laundering or dry cleaning care procedure for a specific textile item. Proper care is difficult without considerable background knowledge about fibers, finishes and dyestuffs; therefore, care labels are required by law to help consumers. So, be sure to read these labels and follow directions carefully if you want the most service from a product.

The degree to which textiles react with other chemical substances is known as *chemical reactivity.* Each textile fiber has identified chemical properties and, thus, identifiable chemical reactions. Knowledge of these reactions for the various types of textile fibers will help you determine the preferred care procedures. However, it is essential to recognize that finishes and dyes used in preparing the textile fabric for consumer use are also involved in care choices.

What Does "Washable" Really Mean?

Difficulties arise in classifying fibers and fabrics as strictly washable or dry cleanable because there are degrees of each process. Some fibers are washable in hot or cold water, with any type of detergent or soap; other fibers are washable only in warm or cool water with a mild detergent.

Some fibers can be safely laundered or dry cleaned; others do not respond to dry cleaning, while still other fibers usually should be dry cleaned. In addition, the different types of cleaning solvents used in dry cleaning processes may react differently with different fibers.

Detergents have varying amounts of alkalinity, and may be described as built or unbuilt. "Built" detergents, strong and highly alkaline in nature, are important as cleaning agents for fabrics that have been heavily soiled. Yet, built detergents cannot safely be used on acidic fibers that tend to react negatively with alkaline or basic substances. "Unbuilt" detergents are mild and relatively neutral.

Detergents and Soaps

Technically, the term detergent encompasses all types of cleaning agents used in laundering or wet cleaning processes. Even soaps are detergents. However, in general use the term *detergent* has become associated with those cleaning agents that have been synthesized from chemicals rather than composed of the natural substances used in making soap. Therefore, in subsequent discussions the term soap will be used for the product made from a fatty acid and lye.

Detergents, usually made from petroleum products and various other chemicals, account for about 90 percent of the laundry agents on the consumer market. They are preferred in areas where the water is hard because no scum or deposit forms on textiles during cleaning. Soap may be used in *soft* water with favorable results and works well as a mild, low alkaline cleaning agent. But, in hard water the scum or curd which forms from soap discolors fabrics.

In the 1960s there was considerable controversy over the environmental impact of synthetic detergents, especially those containing phosphate. In fact, some states completely ban the sale of phosphate detergents, while other states limit the amount of phosphate allowed in detergents.

Another aspect of the 1960s environmental controversy led to all detergents on the market now being biodegradable. Thus, detergents break down through bacterial action to reduce or eliminate the problem of foam formation in lakes or streams.

Today there are several types of synthetic detergents on the market including phosphate, carbonate, citrate, and nonionic. The latter you probably won't see on the consumer market since it is used primarily by industry. Yet, despite the various detergents available and the restrictions on them, phosphate detergents still perform best in cleaning textile products.

QUICK QUIZ 6

1. If you were laundering a fabric made of fibers that were acidic, such as wool, which procedure would be the most damaging to the product? Laundering in:
   a. mild, low alkaline detergent and hot water.
   b. mild, low alkaline detergent and cold water.
   c. alkaline or basic detergent and hot water.
   d. alkaline or basic detergent and cold water.
   e. neutral detergent or substance.
2. If you decided to launder a wool sweater instead of dry cleaning it, what procedures should you use? ______________________________

____

****

ANSWERS QUICK QUIZ 6

1. c
2. cool or warm water and mild detergent

Bleaches

Another important laundering aid is *bleach,* frequently used to whiten fabrics and remove spots and stains. Bleach is designed for use in conjunction with a detergent or soap; it is not meant to be used alone for general cleaning.

Two basic types of bleach are available to consumers: *oxygen* and *chlorine*. The decision as to which to use will depend on the chemical reactivity of the fibers in the fabric and the types of spots or stains to be removed. Chlorine bleach is stronger than oxygen and has more brightening power, but it can damage certain fibers and may react negatively with some finishes or dyestuffs. Oxygen bleaches, on the other hand, are mild and safe for all fibers and fabrics.

Therefore, in deciding whether to use a chlorine or oxygen bleach, you need to be familiar with the chemical reactivity of the finish and/or dyestuff. Both types of bleaches will remove some stains from fabrics and will whiten and/or brighten them. However chlorine bleach is more effective than oxygen bleach for removing difficult stains, if the fiber is resistant to chlorine damage.

Even if a detergent and bleach appropriate to the fiber content are used for a given product, it still can shrink in the presence of moisture and heat due to other fabric properties. Thus, dry cleaning was invented to deal with such products.

Dry Cleaning

There are two basic dry cleaning methods. For general use, a solvent with cleaning substances similar to those found in household detergents is satisfactory. However, products that are heavily soiled may need to be wet cleaned as well. This involves the use of water, detergents and cleaning solvents. The solvents reduce or prevent shrinkage from the water; the water and detergents remove the soil. If you have the knowledge, it's a good idea to tell the cleaner about the fiber content and any special finishes that might be affected by the cleaning solvents.

Absorbency

Absorbency and moisture regain are important properties that need to be considered in the care of fabrics. Earlier you learned about absorbency as an important determinant of fabric comfort during wear.

Fibers that are absorbent and have high saturation moisture regain will take up moisture during care. This added moisture increases required drying time.

QUICK QUIZ 7

1. Given the issues concerning detergents, you have to make careful decisions based on where you live and the type of cleaning required for specific textile products. Identify two considerations that should be a part of the decisionmaking process relative to the selection of a detergent:
   ____
   ____
2. If you were to select a wardrobe made exclusively of fabrics composed of low absorbent synthetic fibers, you would be choosing them primarily for ease of ______________.
3. The choice of fabric for maximum serviceability depends on the ______________. The selection of appropriate fabrics usually represents a ______________ between durability, comfort and care considerations.

****

ANSWERS QUICK QUIZ 7

1. any two: hardness or softness of laundering water; type of fiber and finish in the textile product; degree of soiling of the textile item; local or state restrictions on types of detergents
2. care or packing
3. end use; compromise

Fibers that have low moisture regain and do not absorb much moisture will consequently dry quickly. Fabric structure and finishes also influence absorbency of textile items because spaces between fibers in yarns, or between yarns in fabric structures, increase the ability of the textile to absorb moisture. Certain finishes may be applied to fabrics to increase absorbency.

If you travel frequently, your wardrobe should be a compromise between high absorbency fabrics for comfort reasons, and low absorbency fabrics for ease of laundering and rapid drying. Again, it is important that you read whatever label information is provided and understand properties of fibers and fabrics.

Heat Tolerance

Another property which affects cleaning procedures for textile products is the heat tolerance level of fibers in the fabric, and any special heat-sensitive finishes used. Fibers differ greatly in their resistance to or tolerance of heat. Some fibers will withstand very high temperatures, while others are damaged by iron temperatures at the lowest heat setting. In the unit on fibers, the specific heat tolerances for the various fibers will be described.

The heat tolerance or heat resistance of a textile is a relative concept, since most fibers used in consumer textile products can be damaged by heat if temperatures are high enough. For purposes of this discussion, a fiber will be considered heat resistant if it can be laundered in hot water, dried at high temperatures and ironed at the highest temperature setting provided. On the other hand, heat sensitive fibers require laundering in cool or warm water and ironing at low to moderate temperatures.

Thermoplastic fibers are man-made fibers which soften and eventually melt when exposed to temperatures of 400° F or less. Frequently they cannot even withstand temperatures over 300° F without some weakening or softening; therefore, they are considered heat sensitive. Their heat sensitivity can be used to advantage, however, in that thermoplastic fibers may be "heat-set" into a prescribed shape—as in a pleated skirt or slacks which has its creases set by heat.

Heat setting of thermoplastic fabrics is a finishing process which affects fabric performance in terms of both appearance and care. Temperatures used for industrial heat settings are higher than those available for normal care procedures, so most thermoplastic fabrics set into pleats or creases remain dimensionally stable during consumer use and care. Further, they resist wrinkling and creasing during wear or care.

If care procedures should accidently involve temperatures similar to, or higher than, those used in the heat setting manufacturing process, a new heat set shape can occur, thereby altering the end product. As with most textile decisions, the relative advantages and disadvantages of thermoplastic fibers can be evaluated only in terms of serviceability requirements based on planned end use.

While thermoplastic products have an advantage of being heat set for ease of care, they can melt if exposed to too high a temperature. Whereas nonthermoplastic fibers, like cotton and linen, scorch rather than melt when exposed to excessive heat, they can withstand higher temperatures than thermoplastics before any damage results.

REVIEW

1. The chemical reactions of textile fibers are important aspects of care and must be considered in the selection of appropriate agents used in the care process. In addition, these reactions help determine which of the two care methods should be used, which are: ____________________ and ____________________.

2. The best source of information on fabric care is the ____________________, which must be permanently attached to the textile item as required by law.

3. Indicate the major uses of the detergent types listed below. Keep in mind that the use remains the same regardless of what compounds have been used in the detergent.
   Built detergents: ____________________
   ____
   Unbuilt detergents: ____________________
   ____
   ____

4. Name the two major types of bleaches available for home use and tell when you would select each.
   ____
   ____

5. What care procedures should be followed in laundering textile products made from thermoplastic fibers?____________________
   ____

****

REVIEW ANSWERS:
1. laundering; dry cleaning
2. care label
3. built for heavy duty laundering and cleaning; unbuilt for cleaning lightly soiled articles and delicate products that require gentle handling
4. chlorine bleaches for heavy duty bleaching and stain removal on durable fabrics; oxygen bleaches for whitening of delicate fabrics
5. use warm (not hot) water; dry at low dryer temperature settings; iron at warm, not hot, temperatures

---

Selected Properties Influencing Care During Storage of Textile Products

The third category of care considerations covers the precautions which should be taken when storing textile products. Factors discussed in this section that influence *care during storage* include the *biological resistance,* and *age resistance* of the fibers and finishes used.

Biological Resistance

Biological resistance refers to a fiber's level of resistance to insects and microorganisms. The major organisms that attack fibers or finishes include moth larvae, carpet beetles and silverfish; and mildew, fungi and other microorganisms. In general, damage from these organisms can be prevented if textile products are stored clean in dry atmospheres, with proper protection in relation to the amount of light. Humidity, warmth and darkness tend to encourage insect and microorganism growth.

Moths and carpet beetles are attracted to food particles and soil remaining in textiles. Also, wool contains a high proportion of protein or amino acids, particularly *keratin,* which is an attractive food for insects. Since moth larvae are especially damaging to wool when it is improperly stored, it is important that wool products be kept clean and in a dry atmosphere with as much light as feasible.

Another aid to protect wool and animal hair from moth larvae is to include in the storage area mothballs, crystals, powders or sprays which emit a chemical that is poisonous to the insects. Further the biological resistance can be increased by applying chemical finishes to the textile product. The only way the consumer can be certain such a finish has been applied is to read the information on the label or hangtag.

Silverfish, another insect which damages certain textile fibers, are particularly attracted to cellulosic products and to fibers which have starch base finishes applied. This means leaving starch out of garments going into storage even if it is a normal part of care during continued use. Mildew likewise attacks cellulosic fibers in warm, moist, dark conditions. While the storage environment is the most important in preventing the growth of microorganisms like mildew, certain biological resistant finishes are available to retard attack.

Light Resistance

An important aspect of storage, and to some extent use, is the resistance of textile fibers to light. Fabrics with low light resistance must be kept away from direct sunlight or light with ultraviolet rays. In addition to damage to the actual fiber, sunlight can cause a loss of color due to its effect on the dyestuffs used. Thus, fibers and fabrics with low light resistance should not be used in products such as draperies and window curtains which will be exposed to the sun for long periods of time.

Age Resistance

The ability of a fiber to last over long periods of time is said to be *resistance to aging*. Some fibers tend to have a high resistance to age, while others deteriorate quickly. These fiber characteristics will be discussed in Unit 2.

Certain finishing processes to be explained in Unit 5 may also increase the speed of deterioration over time. Factors most affecting the life of fibers and fabrics include resistance to cleaning agents and care procedures, sensitivity to light and biological organisms, and other environmental conditions such as temperature and humidity.

REVIEW

1. Mildew and moths are likely to attach fibers with (high, low) ______________ under what two conditions? ______________

   ____

   ____

2. One of the most important properties of fabrics used in draperies and curtains is resistance to ______________.

3. A fiber is ______________ if it burns; it is ______________ if it does not burn; and ______________ if it does not support combustion.

4. A detergent which contains weak alkaline compounds is called ______________; one with strong alkaline compounds is ______________.

5. Detergents, bleaches, cleaning solvents and other laundry or cleaning additives must be compatible with both the ______________ and ______________.

6. A fabric which has the ability to maintain its original appearance under various conditions has good ______________.

7. The addition of finishes to fabrics can alter the biological resistance of a fiber, its resistance to light, heat and flame properties, and chemical behavior. Thus, it is important for consumers to have information about fibers and finishes on a securely attached label in order to determine ______________.

8. As in the case of durability, ______________ applications are influential in the determination of care procedures.

****

REVIEW ANSWERS:

1. low, biological resistance; any two: warm or hot temperatures; darkness; dampness; high humidity; when soil is present
2. sunlight
3. flammable; flameproof or nonflammable; fire or flame resistant
4. unbuilt; built
5. fiber content; finishes or dyestuffs
6. age resistance
7. care procedures
8. end use

UNIT ONE

Part IV

Esthetic Appearance and The Concept of Serviceability

Properties which contribute to the esthetic appearance of a textile product will be discussed in this section. Of all of the concepts related to serviceability, this one is the most subjective—what is esthetically pleasing to some people is not to others. There are often major disagreements as to whether or not an item is attractive because such judgments depend primarily on individual perspective or interpretation rather than on absolute laboratory measurements. An exception involves changes in fabric appearance due to color or finish loss, which can be measured scientifically. However, the influence of color or finish loss on the consumer is still a subjective matter.

Under a given condition, a large group of individuals with similar orientations has a high probability of judging an object as being acceptable. Under different conditions, the same group might rate the object as unacceptable. Individuals with backgrounds in art and design are more apt to evaluate appearances with similar results. In addition, individuals may evaluate appearance from a *psychological* point of view, and results on this basis tend to vary widely among consumers.

The purpose for evaluating textiles according to esthetic considerations is to provide a background that will enable you to make reliable decisions concerning the esthetics of a textile item. It is important to understand how fiber and fabric properties contribute to esthetic criteria used in selection of products for specified end uses. A textile item which is perceived as unattractive or unacceptable will not be considered serviceable regardless of its other properties related to durability, comfort and care.

In spite of the amount of subjectivity or lack of precision when evaluating the esthetic aspects of a textile, much similarity in evaluation can be achieved through shared experiences. The experience can be visual; it can be tactile; or it can be a combination of both. If a consumer finds a textile product esthetically unattractive for a given end use, the product will not be serviceable regardless of other properties.

In order to develop a common base for discussing the esthetic appearance of textile fabrics, a set of properties representing both visual and tactile characteristics will be used. These properties include color, luster and translucence, which are evaluated through the visual sense. Texture and drape may combine the visual and tactile; while body, hand, loft or bulk tend to be primarily tactile properties. To maintain appearance it is important that proper care procedures be identified and followed, as they are extremely important to the retention of appearance.

Color is probably the first esthetic property that consumers notice in a fabric. Often rated as the most important quality in the selection process, color is a visual characteristic whose purity or fidelity is determined by a combination of the fiber and dyestuff used. Dye affinity describes the degree to which fibers and fabrics accept color through the application of dyes or pigments.

Luster, another property that is primarily visual, is the shine, gloss or sheen of a fiber or fabric caused by the light reflecting from the surface. Luster may be controlled in manufactured fibers, but in natural fibers the amount of luster depends on the fiber type. The smoother a fiber and the fewer the compounds that break up light rays, the higher the level of luster.

If a fabric has a high shine, light is reflected directly. If a fabric has a subdued shine, light is diffused or reflected indirectly. Therefore, a fiber that tends to reflect light directly rather than indirectly tends to have a high luster.

Translucence is the ability of a fabric to pass light through, and the degree of translucence affects the appearance and use of a fabric. For example, the selection of fabric for window curtains where you want light to come through differs from choosing draperies when you want to limit light. Fabrics which allow no light to pass through the structure are not translucent; they are described as *opaque.*

The translucence of a textile item is related primarily to the method of construction used in forming the fabric; however, light reflection or luster is basically a property of the fiber. (How this property can be modified is discussed in a later section.)

The manner in which fabric drapes is primarily dependent on the method of construction used in forming the fabric, and to some degree on the type of yarn structure used. A fabric with a stiff hand tends to have less drapability than a fabric with a soft hand. *Body* and *hand* relate to the lightness or heaviness of a fabric, and to its firmness and looseness. A fabric that appears to have a heavy, stiff draping quality also tends to have a stiff body or hand.

Although the properties of body or hand and drape can be evaluated through both visual and tactile senses, drape is more often evaluated or determined by the visual senses, and body and hand by the tactile senses.

Texture is an important property used in describing esthetic characteristics of textiles. Fabric texture refers to the surface smoothness or roughness and involves both the visual and tactile senses. When a fabric is described as having a smooth texture, most consumers perceive the fabric as being smooth to both of the senses.

A fabric texture can look shiny or dull, smooth or rough, just as it can feel smooth or rough. In the comfort section, the texture of a fabric was related to how the fabric would feel next to the skin. The smooth feel of certain fabric textures would contribute to serviceability through both comfort and appearance.

Texture is often considered second only to color as a most important property for evaluating esthetic appearance. The external surface of the fiber, methods of yarn and fabric construction, and the presence of special finishes or surface designs all can contribute to the texture as well as luster of a fabric.

Some fabrics have the ability to spring back to their original thickness after being compressed. This special kind of resiliency is referred to as *loft.* Among the household furnishings which must have considerable loft to retain their esthetic appearance over time are carpeting and certain types of upholstery.

After you have handled a variety of fabrics, you will develop subjective judgments in describing fabric *hand.* Words like soft, warm, silk-like, and bulky are frequently used to describe how a fabric feels. Such descriptions lack the precision obtained with laboratory testing equipment but do allow different people to arrive at a similar system of describing textile fabrics. This system involves the use of shared experiences and is a subjective system of evaluation.

REVIEW

1. Which esthetic property is described in the phrases below?
   a. has a silky feel ______________
   b. has a shiny appearance ______________
   c. has considerable weight and firmness ______________
   d. hangs loosely and appears soft and flowing ______________
   e. has good pigment or dye ______________
   f. feels rough and uneven ______________

2. Since the entire field of esthetic evaluation is very subjective, this presentation is just one perspective; another description could serve as well. However, experienced textile professionals have little difficulty in communicating when using the terminology just discussed. See how successfully you can match the following properties with their definitions:

____	1. hand	a. weight or firmness of a fabric
____	2. color	b. manner in which a fabric hangs
____	3. translucence	c. degree of light reflection
____	4. loft	d. degree of light transmission
____	5. luster	e. the way a fabric feels or handles
____	6. drape	f. ability of a fabric to return to its original thickness after compression
____	7. texture	g. a visual fabric quality that contributes greatly to esthetic appeal
____	8. body	h. an all-encompassing fabric property that involves visual and tactile perception

****

REVIEW ANSWERS:

1. a-hand or body; b-luster; c-hand or body; d-drape; e-color; f-texture
2. 1-e; 2-g; 3-d; 4-f; 5-c; 6-b; 7-h; 8-a

Posttest

1. A fabric suited to a particular end use is said to be
   a. serviceable.
   b. durable.
   c. flexible.
   d. elastic.

2. Durability is determined by mechanical measurement. A high measure of durability assures serviceability of a textile product.
   a. true
   b. false

3. For carpet and upholstery fabrics to be durable, they must have high
   a. absorbency.
   b. cohesiveness.
   c. abrasion resistance.
   d. electrical conductivity.

4. What fiber property can increase spinning ease?
   a. wicking ability
   b. hygroscopicity
   c. cohesiveness
   d. elongation potential

5. A fabric with low flexibility is very susceptible to ____________ forces.
   a. pulling
   b. bending
   c. abrasive

6. How does fiber elasticity differ from fiber elongation?
   a. Elasticity refers to stretching and eventual fiber breakage.
   b. Elasticity refers to stretching and return to shape after release of stretching forces.
   c. Elasticity allows for unrecovered fiber stretch.
   d. Elasticity refers to different degrees of elongation and stretch potential

7. A fabric which has high elongation potential and low elasticity is quite serviceable for apparel.
   a. true
   b. false

8. A fabric which retains its shape over time has high
   a. cohesiveness.
   b. resiliency.
   c. dimensional stability.
   d. tenacity.

9. The comfort of a fabric is ________________ concept.
   a. an absolute
   b. a relative

10. An absorbent fabric is made of ____________ fibers.
    a. hydrophilic
    b. hydrophobic

11. A hydrophobic fiber would be a wise choice for which end use?
    a. blouse
    b. towel
    c. raincoat
    d. mop

12. Wicking is a fiber property that increases the comfort of fabrics made of ____________ fibers.
    a. hydrophobic
    b. hydrophilic
    c. hydroscopic

13. A fabric with good electrical conductivity will
    a. glow in the dark.
    b. create sparks or shocks.
    c. allow electrostatic charges to build up.
    d. pass electrostatic charges to the ground.

14. A fabric with which characteristic would most likely have high thermal conductivity?
    a. thick
    b. bulky
    c. textured
    d. thin

15. Thermal retention is high in a fabric made of ____________ fibers.
    a. bulky, textured
    b. highly flexible
    c. smooth filament
    d. high tenacity

16. Comparing two fabrics of equal thickness, the one made of ____________ fibers weighs more.
    a. high-density
    b. low-density

17. ____________________ detergents used in concentrated amounts are likely to damage acidic fibers and delicate fibers.
    a. Built
    b. Unbuilt

18. Soaps made from fatty acids will
    a. form more suds in hard water than a detergent.
    b. form in hard water a deposit which grays clothes.
    c. form in soft water fewer suds than a detergent.
    d. not form in hard water a deposit as detergents do.

19. Oxidizing bleaches are used to
    a. remove fabric color.
    b. brighten and whiten fabric.
    c. assist in fabric cleaning.
    d. do all the above.

20. Textile care procedures are primarily determined by
    a. detergent chemical composition.
    b. yarn chemical structure.
    c. fabric structure.
    d. fiber chemical composition.

21. Which description best defines a thermoplastic fiber?
    a. plastic
    b. heat-sensitive
    c. heat-resistant
    d. thermal

22. A highly flammable fabric is
    a. self-extinguishing.
    b. not self-extinguishing.

23. Which of the following environmental factors can accelerate a fabric's aging or degradation?
    a. chemicals used in laundering
    b. dry-cleaning solvents
    c. insects and biological organisms
    d. sunlight and heat
    e. all of the above

24. Which property below would be least likely related to fabric resiliency?
    a. absorbency
    b. elastic recovery
    c. dimensional stability
    d. flexibility

25. Match the definition with the fabric property:
    ____ 1. resiliency
    ____ 2. dimensional stability
    a. retention of fabric shape
    b. recovery from wrinkling

26. Since "hand" is primarily understood through the tactile sense, which of the following properties contributes most directly to fabric hand?
    a. luster and color
    b. drape and translucence
    c. body and loft
    d. color and texture

27. Fabric loft is closely related to fiber
    a. tenacity.
    b. resiliency.
    c. elongation.
    d. cohesiveness.

28. Hand, loft and body are all related properties because they can be understood primarily through the ______________ senses.
    a. tactile
    b. visual

29. Which of the following properties directly affects the drape of a fabric?
    a. color
    b. hand
    c. luster
    d. translucence

30. Fabric texture contributes the most to which of the following textile concepts?
    a. comfort and esthetic appearance
    b. comfort and durability
    c. durability and care
    d. esthetic appearance and durability

31. Fabric luster and texture are related. A fabric with a ______________ texture generally has a ______________ luster.
    a. rough, high
    b. smooth, low
    c. rough, low

Answers for Unit 1 Posttest:

1. a	17. a
2. b	18. b
3. c	19. b
4. c	20. d
5. b	21. b
6. b	22. b
7. b	23. e
8. c	24. a
9. b	25. l-b; 2-a
10. a	26. c
11. c	27. b
12. a	28. a
13. d	29. b
14. d	30. a
15. a	31. c
16. a	

UNIT TWO

Fibers

Fibers are the smallest units in textile products, and the fiber content in a fabric is one of the major determinants of its ultimate serviceability. The fibers selected for a particular fabric are important in determining the finishes, color, and design procedures which will be applied. As the basic building block in the manufacture of textile products, fibers are converted into yarns and/or fabrics by a variety of processes which will be presented in later units.

By the end of the Fibers Unit, one of the most demanding and important ones, valuable progress toward the mastery of the complexities of textiles will have been achieved. This unit is divided into five parts, including a general description of the various fiber categories followed by a discussion of each category's basic serviceability concepts.

Serviceability concepts and specific fiber information should be remembered. The activities included throughout, such as fill-in matrixes and reviews, will help you master the content. What is learned in this unit should be linked with information provided in the first unit; all material will be used later to provide a background on the conversion of fibers into ultimate textile products.

Fibers have specific properties and characteristics. While they may be modified by subsequent operations, the basic fiber properties are still important in determining the performance of the final product. Fiber properties, yarn and fabric characteristics, and finishes and design applications are all important in determining the final characteristics of textile products.

Objectives for Unit 2

Upon completion of this unit the student will be able to:

1. Describe the serviceability characteristics of each of the following fiber categories: *natural cellulosic, protein, man-made cellulosic, man-made noncellulosic* or *synthetic,* and *mineral.*
2. Recall at least three specific fiber examples from each fiber category.
3. Discuss each fiber category in terms of potential serviceability.
4. State the appropriate care procedures for each of the fiber categories and special subgroups.
5. List and justify potential end uses for which a fiber category is suited; identify advantages and disadvantages that each category offers for particular end uses.
6. Apply the information concerning fiber properties to the serviceability concepts.

Pretest

1. Natural fibers are generally ______________ length.
   a. filament
   b. staple
2. If the internal structure of a fiber has amorphous arrangement, the molecular chains are
   a. parallel to each other and to the fiber's longitudinal axis.
   b. unaligned to each other and to the fiber's longitudinal axis.
   c. crystalline and not parallel to the fiber's longitudinal axis.
   d. crystalline and aligned with the fiber's longitudinal axis.

3. When molecular bundles of a particular fiber have high internal orientation but are not parallel to the fiber longitudinal axis, the fiber molecular structure is described as
   a. highly oriented.
   b. nonoriented.
   c. amorphous with crystalline arrangements.
   d. completely amorphous.
   e. highly crystalline.

4. What is the special bonding which occurs across highly oriented molecules in certain fibers?
   a. cross linking
   b. electron bonding
   c. high orientation bonding
   d. copolymerization

5. The hollow center canal in a cotton or flax fiber is called the
   a. epidermis.
   b. medulla.
   c. convolution.
   d. lumen.

6. The longitudinal axis of cotton fiber is characterized by
   a. convolutions.
   b. scales.
   c. nodes.
   d. striations.

7. The epidermis of wool is composed of
   a. scales.
   b. convolutions.
   c. nodes.
   d. lumen.

8. The lumen is to cellulosic fibers what the ________ is to protein fibers.
   a. epidermis
   b. cortex
   c. medulla
   d. node

9. Both cotton and flax are classified as natural cellulosic fibers, but the cotton fiber surface has convolutions while the flax fiber surface has
   a. scales.
   b. epidermis.
   c. striations.
   d. nodes.

10. Striations are characteristic markings on ________ fibers.
    a. wool
    b. rayon
    c. cotton
    d. flax

11. Regenerated cellulosic describes the basic component of certain ________ fibers.
    a. natural cellulosic
    b. man-made noncellulosic or synthetic
    c. man-made cellulosic
    d. man-made mineral

12. How are manufactured fibers classified according to the TFPIA?
    a. trademarks
    b. generic names
    c. trade names
    d. manufacturer's name

13. What is the spinning process in which evaporation of a chemical from the fiber solution causes the fiber to solidify?
    a. wet spinning
    b. dry spinning
    c. melt spinning

14. Fibers solidify upon exposure to a chemical solution in the ________ spinning process.
    a. melt
    b. dry
    c. wet

15. Alteration of a synthetic fiber's cross-sectional shape would not necessarily improve which of the following properties?
    a. bulk
    b. loft
    c. hand
    d. tenacity

16. Which fiber listed below decreases in tenacity when wet?
    a. silk
    b. cotton
    c. wool
    d. flax

17. Which category is known for having fibers with high tenacity?
    a. man-made noncellulosic or synthetic
    b. natural cellulosic
    c. man-made cellulosic
    d. protein

18. Abrasion resistance is lowest in the ________ fiber category.
    a. protein
    b. man-made cellulosic
    c. natural cellulosic
    d. man-made noncellulosic

19. Which of the natural fibers listed below typically has the highest abrasion resistance?
    a. flax
    b. jute
    c. wool
    d. cotton

20. In order to be spun into yarns, staple fibers must have
    a. flexibility or pliability.
    b. tenacity or strength.
    c. elastic recovery.
    d. cohesiveness or spinning quality.

21. Fiber surface textures contribute to cohesiveness.
    a. true
    b. false

22. Generally, high elongation potential with low elasticity is found in fibers with ____________ molecular arrangements.
    a. amorphous
    b. oriented
    c. crystalline

23. The majority of man-made noncellulosic fibers have ____________ elasticity when compared to natural fibers.
    a. higher
    b. lower
    c. the same

24. Which of the molecular arrangements listed below does not contribute to high elastic recovery?
    a. crimped molecular chains
    b. cross-linked molecular chains
    c. weak bonds across molecular chains
    d. strong bonds across molecular chains

25. Low elasticity and low elongation contribute to low fiber
    a. absorbency.
    b. resiliency.
    c. cohesiveness.
    d. density.

26. Which fiber group contributes the highest elastic recovery to textile products?
    a. cellulose
    b. mineral
    c. protein
    d. elastomeric

27. Which fiber category listed below would be most suitable for a permanently pleated skirt fabric?
    a. protein (wool)
    b. man-made noncellulosic (polyester)
    c. natural cellulosic (cotton)
    d. man-made cellulosic (rayon)

28. Which fiber category has the poorest dimensional stability?
    a. protein
    b. man-made cellulosic
    c. natural cellulosic
    d. man-made noncellulosic

29. Many of the man-made noncellulosic fibers have good dimensional stability because they are
    a. nonabsorbent.
    b. thermoplastic or heat settable.
    c. flexible or pliable.
    d. good electrical conductors.

30. In general, man-made noncellulosic fibers have ________ absorbency and ________ wicking ability.
    a. high, high
    b. low, high
    c. high, low
    d. low, low

31. The high absorbency of natural cellulosics indicates that they are likely to have ________ electrical conductivity.
    a. high
    b. medium
    c. low
    d. no

32. The amorphous molecular arrangement of cotton contributes to its
    a. high thermal retention.
    b. low electrical conductivity.
    c. high absorbency.
    d. high allergenic potential.

33. If all properties related to yarn and fabric structure are held constant, the ________________ the fabric density, the ________________ the fabric weight.
    a. higher, lower
    b. higher, greater
    c. lower, greater

34. Which man-made cellulosic fiber has properties most like those of natural cellulosic fibers?
    a. acetate
    b. rayon
    c. triacetate

35. The protein fibers are known for their
    a. hygroscopicity.
    b. hydrophobicity.
    c. hydrophilicity.

36. Wool fibers have
    a. high heat conductivity.
    b. low thermal retention.
    c. high thermal retention.
    d. high heat transfer.

37. Which factor below determines how well protein fibers conduct electricity?
    a. humidity in the air
    b. the temperature of the air
    c. the number of free electrons

38. Most man-made noncellulosic (synthetic) fibers used in consumer goods are
    a. hygroscopic.
    b. hydrophobic.
    c. hydrophilic.

39. Despite their low absorption, many man-made noncellulosic or synthetic fibers have high
    a. electrical conductivity.
    b. heat conductivity.
    c. wicking ability.
    d. density.

40. Glass fibers have been modified to improve fabric hand by increasing fiber
    a. flexibility.
    b. tenacity.
    c. resiliency.
    d. cohesiveness.

41. Generally, mineral fibers have ______________ heat conductivity and ______________ electrical conductivity.
    a. high, low
    b. high, high
    c. low, high
    d. low, low

42. Unfinished natural cellulosic fabrics tend to exhibit __________ resiliency and __________ dimensional stability.
    a. low, low
    b. low, high
    c. high, high
    d. high, low

43. Natural cellulosic fibers are resistant to most cleaning solvents which are not
    a. alkaline.
    b. oxygen bleaches.
    c. strong acids.
    d. chlorine bleaches.

44. The natural cellulosics have a relatively high heat tolerance.
    a. true
    b. false

45. The low dimensional stability of wool is due in part to its fiber
    a. length.
    b. scales.
    c. medulla.
    d. tenacity.

46. If you launder protein fibers, you should use ______________ detergent and ______________ water temperature.
    a. built, hot
    b. built, cold
    c. unbuilt, constant
    d. unbuilt, cold
    e. unbuilt, hot

47. Wool is especially damaged by
    a. low heat.
    b. steam heat.
    c. medium heat.
    d. dry heat.

48. Rayon and acetate fibers maintain their body and color longer if
    a. dry cleaned.
    b. laundered.

49. Which of the man-made cellulosic fibers is *not* a thermoplastic?
    a. rayon
    b. acetate
    c. triacetate

50. Generally, the ______________ the heat tolerance of a man-made noncellulosic or synthetic fiber, the ______________ the likelihood of setting in wrinkles during laundering or drying.
    a. lower, less
    b. higher, greater
    c. lower, greater

51. Excessive rubbing of glass or mineral fibers should be avoided because of their low ______________ and repeated folding should be avoided because of their low ______________.
    a. tenacity, flexibility
    b. resiliency, dimensional stability
    c. abrasion resistance, flexibility

52. The care procedures for fabrics made of metallic fibers are primarily dependent on the
    a. type of metallic substance.
    b. polymer or plastic fiber coating.

53. Which fiber category has the least resiliency?
    a. man-made cellulosic
    b. protein
    c. natural cellulosic
    d. synthetic or man-made noncellulosic

54. Which fiber group has high dimensional stability?
    a. mineral
    b. protein
    c. man-made cellulosic
    d. natural cellulosic

55. Which one of the following has the highest heat tolerance?
    a. protein
    b. natural cellulosic
    c. man-made noncellulosic
    d. man-made cellulosic

56. To prevent growth of biological organisms, textile products should be stored clean in what type of environment?
    a. humid, warm
    b. humid, cool
    c. dry, warm
    d. dry, cool

57. The TFPIA requires all the following information on fabric labels except the
    a. generic name.
    b. care procedures.
    c. percent of each fiber present.
    d. manufacturer's name.

58. Permanently attached care instructions are required on textile products as a result of
    a. an act passed by Congress.
    b. a specification made by the Apparel Manufacturers Association.
    c. a Federal Trade Commission ruling.

59. Match the fiber category with the most appropriate description.
    ____ 1. natural cellulosic
    ____ 2. protein
    ____ 3. man-made cellulosic
    ____ 4. man-made noncellulosic
    ____ 5. mineral

    a. relatively expensive fibers valued for apparel and home furnishing fabrics
    b. absorbent fibers known for their comfort
    c. fibers with limited end uses
    d. known for durability and ease of care
    e. inexpensive, versatile fibers that are not exceptionally durable

Answers for Unit 2 Pretest:

1. b
2. b
3. c
4. a
5. d
6. a
7. a
8. c
9. d
10. b
11. c
12. b
13. b
14. c
15. d
16. c
17. a
18. b
19. c
20. d
21. a
22. a
23. a
24. c
25. b
26. d
27. b
28. a
29. b
30. b
31. a
32. c
33. b
34. b
35. a
36. c
37. a
38. b
39. c
40. a
41. d
42. a
43. c
44. a
45. b
46. c
47. d
48. a
49. a
50. c
51. c
52. b
53. c
54. a
55. b
56. d
57. b
58. c
59. 1-b; 2-a; 3-e; 4-d; 5-c

UNIT TWO

Part I

General Discussion of The Fiber Categories

Matrix II presents a listing of the majority of textile fibers. The fibers have been classified into general categories according to their basic sources in order to simplify the learning of fiber characteristics. The five, source-oriented fiber categories which have been established for study in this program are listed in **bold type** on Matrix II. Study Matrix II and locate the two major fiber categories which are derived from natural sources—the *natural cellulosic* and *protein fibers.*

The remaining fiber categories, composed of the *generic classes* of manufactured fibers, are grouped according to their sources, which include the *man-made cellulosic, man-made protein, man-made noncellulosic,* and *mineral fibers.* The terms man-made noncellulosic and synthetic will be used synonymously in this text. However, the phrase man-made noncellulosic is more widely used in scientific literature, while the term synthetic is more common in consumer literature.

References to mineral fibers in this text are to man-made; however there are natural mineral fibers available, like asbestos. Given the limited importance of natural mineral fibers and the health problems related to asbestos in building materials, natural mineral fibers as a category are not included for study. Further, while the man-made protein fibers complete the Matrix II outline they also are not discussed in the program due to their limited production and used.

Fiber Identification

If the consumer could tell fiber content at a glance, the problems of textile identification would be simplified. The only way one can be sure of a fiber type is to read the fabric label or tag or to resort to a laboratory test. Unless you plan to carry your studies through to textile chemistry, you will use the label method! Although consumers need to check the label as a method of positive fiber identification, the fiber categories can be classified by burning text samples of fabric. In addition to being a simple test that allows a fast, general fiber identification, the burn test· also provides an understanding of the respective flammability characteristics of the fiber categories.

1. Take out the fabric samples and read the attached instructions on how to use them throughout the program.
2. Find a suitable nonflammable container in which to burn the test fabric samples. (Bathroom sinks are ideal—and safe).
3. Be sure there is water or sand handy in the event buring gets out of hand, and make certain your test is conducted clear of other flammable materials.
4. Find a convenient source of open flame. A candle in a holder is acceptable.
5. Turn to Matrix III.
6. Study Matrix III carefully and make the following observations:
   a. The left margin of the matrix indicates the steps of your analysis. First observe the flame characteristic, then the nature of the residue and odor before you make the fiber identification.
   b. The branching allows you to follow the fiber into its appropriate category listed toward the bottom of the matrix.

MATRIX II
OVERVIEW OF TEXTILE FIBERS

I. NATURAL FIBERS

A. NATURAL CELLULOSIC

1. Seed hair
   a. Cotton
   b. Kapok
   c. Milkweed
   d. Cattail

2. Bast or Stem
   a. Flax
   b. Jute
   c. Hemp
   d. Ramie

3. Leaf
   a. Abaca
   b. Pineapple (Pina)
   c. Sisal

B. PROTEIN

1. Animal hair
   a. Wool
   b. Cashmere
   c. Mohair
   d. Camel
   e. Llama
   f. Guanaco
   g. Vicuna
   h. Alpaca
   i. Rabbit (Angora)
   j. Cowhair
   k. Horsehair
   l. Quivit (Musk Ox)

2. Animal Secretion—Silk

C. MINERAL—Asbestos

D. RUBBER

II. MANUFACTURED FIBER GENERIC CLASSES

A. MAN-MADE CELLULOSIC

1. Rayon
2. Acetate
3. Triacetate

B. MAN-MADE PROTEIN

Azlon*

C. MAN-MADE NONCELLULOSIC or SYNTHETIC

1. Nylon
2. Polyester
3. Acrylic
4. Modacrylic
5. Olefin
6. Spandex
7. Rubber
8. Vinyon
9. Saran
10. Aramid
11. Anidex*
12. Lastrile*
13. Nytril*
14. Novoloid*
15. Vinal*
16. PBI
17. Sulfar

D. MINERAL

1. Glass
2. Metallic

*Not currently produced in the United States.

c. Notice that the category for synthetic fibers is represented twice because some are self-extinguishing and some are not.
d. Notice also that cotton and linen, natural cellulosics, and rayon, a man-made cellulosic fiber, burn in the same manner.

7. Starting with sample 1, follow the instructions with the swatches. Cut smaller samples (1 1/2 by 2 inches), then burn and identify them as instructed.
8. Place sample in the flame until it ignites, then immediately remove it from flame. (Be careful not to get the sample in the melted candle wax around the wick, or the wax as well as the fabric will burn.)
9. If the fabric is not self-extinguishing and con-

MATRIX III
IDENTIFICATION OF FIBERS BY BURNING

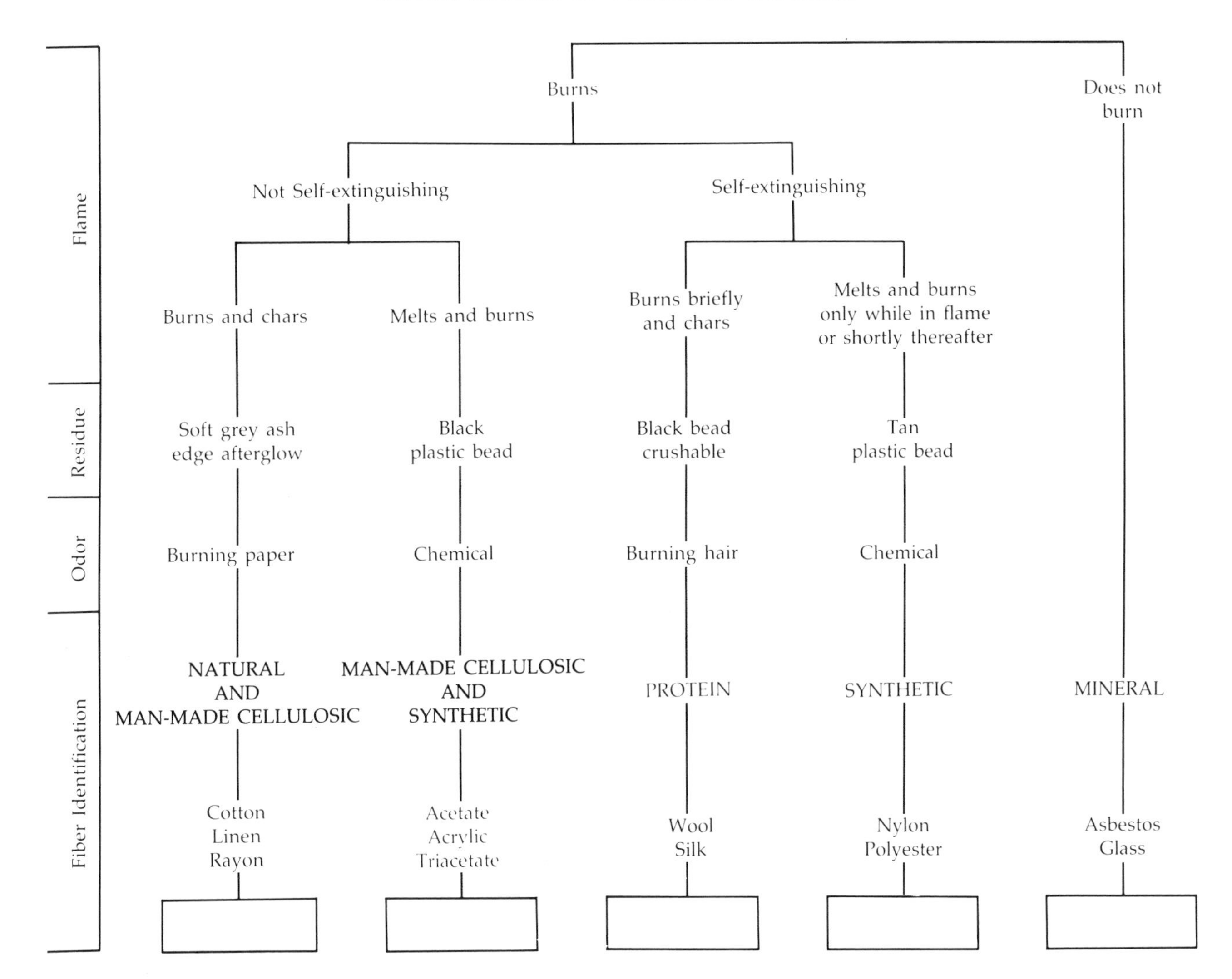

tinues to burn after removal from the flame, allow the flame to progress until it has burned half of the sample, then blow it out. If the fabric self-extinguishes when removed from the flame, return it to the flame until half of the sample is destroyed. In each case observe the flame, residue and odor as the sample burns.

Place the appropriate fabric number in the category box at the bottom of Matrix III. You will identify more than one sample in some categories.

10. Follow this procedure for burning these eight fabric samples: 1, 3, 5, 13, 21, 23, 25, 30 and yarn sample 6. As you observe the burning behavior for each, write the sample number in the appropriate box at the base of Matrix III. Then check your burn test results below.

You are not expected to be able to identify fiber name with the burn test, only the fiber category. The fiber content is given for your information.

Natural and man-made cellulosic box:
1—100% cotton, *21*—65% cotton, 35% rayon, *6*—100% rayon yarn.

Man-made cellulosic and synthetic box:
5—100% acetate, *23*—75% triacetate, 25% polyester.
(Sample 23 is not self-extinguishing and thus takes on burning characteristics of triacetate rather than polyester.)
Synthetic, self-extinguishing box:
13—100% polyester, *25*—100% polyester, *30*—100% nylon.
Mineral box:
none; no glass or asbestos is in the fabric packet.

The nature of the chemical structure of each fiber determines not only how it burns, but how its particular properties apply to the durability, comfort, care and esthetic serviceability concepts. It is important to note that dyes and finishes also may change burning characteristics, bead color and odor of fabric samples.

General Fiber Characteristics

Many properties or characteristics contribute to the behavior of fibers, and basic definitions for these terms were presented in Unit I. To be useable as a fiber, certain *primary properties* are essential, including adequate fiber length, tenacity, pliability or flexibility, spinning quality or cohesiveness and fiber uniformity. *Secondary properties,* which are desirable for certain end uses, include fiber shape, texture, luster, density, moisture regain or absorption, elastic recovery and elongation, resiliency, thermal behavior, biological-chemical reactions, and reactions to various environmental factors. The internal structure of fibers is also important in understanding their total behavior.

The properties discussed in this text are those that are most influential in determining serviceability of fibers and the ultimate textile product. Recall that for ease of discussion and to reduce complexity, fibers have been grouped into the five categories: *natural cellulosic, protein, man-made cellulosic,* which includes modified cellulose as well as regenerated cellulose; *man-made noncellulosic,* also called synthetic; and *mineral.*

You need to become familiar with the relationships between fiber properties and the four concepts of serviceability: durability, comfort, care or maintenance, and esthetic appearance.

Fiber Length

All fibers are partly described by their typical length. The general method for describing fiber length is to identify a fiber as either *staple,* which means it is short; or as *filament,* which means the fiber is very long. Staple fibers are measured in inches or centimeters, while filament fibers are measured in yards, meters, or even kilometers or miles.

From your own experience you probably already know that natural fibers such as cotton or wool are staple or short fibers, while manufactured fibers are filament or long fibers. Refer to Figure 2.1.

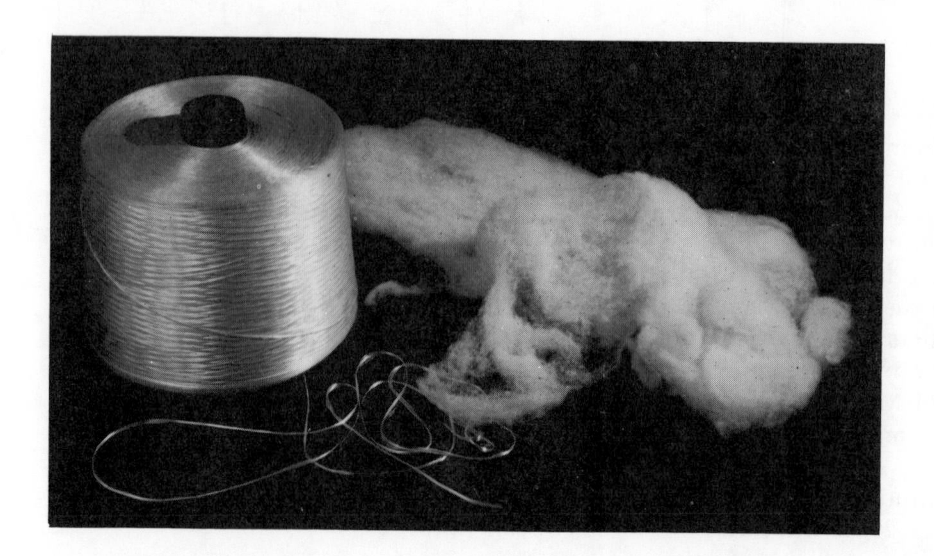

Figure 2.1. Staple fibers are seldom more than a few inches long, while filament fibers can be many yards long. Staple fibers are shipped as compressed bales to the manufacturing plant; filaments are wound on containers such as cylinders to form "cakes," as shown, in order to prevent snarling.

Natural fibers are usually short, due to the way they are produced in nature. An exception is the fiber produced by the silk worm, which spins a long continuous filament.

Internal Fiber Structure

The internal structure of a textile fiber is an important and fundamental determinant of the serviceability characteristics of textile products. A fiber is made internally of many molecular chains, polymolecules or *polymers,* which are arranged and connected in different ways, depending on the base elements and methods of production. Three basic factors relating to the internal fiber structure affect fiber properties: (1) the typical length of the molecular chains, referred to as the *degree of polymerization* (dp); (2) the *orientation of the molecular chains,* which refers to the arrangements of the polymers both in relation to the fiber axis and to each other; (3) the types of bonds or attraction forces within and between the molecular chains or polymers.

In the production of manufactured fibers it is possible to control, to some degree, each of these three factors. However, in natural fibers these are determined by various genetic and environmental factors affecting their growth and production. Manufactured fibers are often made with various modifications so that special properties are included, to provide desirable characteristics that relate to serviceability and identified end uses.

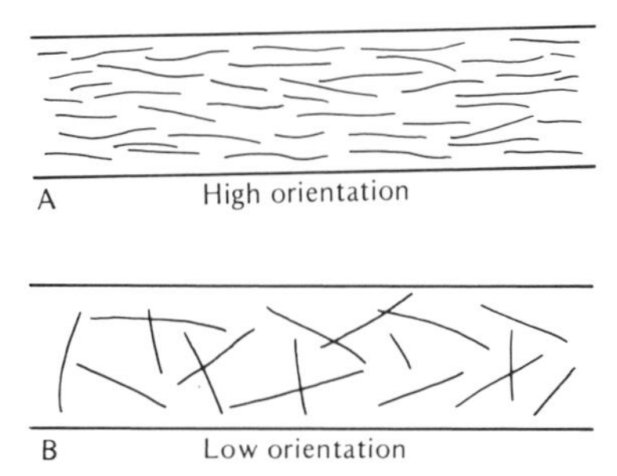

Figure 2.2. "A" represents the orderly arrangement of molecular chains characteristic of highly oriented fibers. "B" represents the random arrangement of molecular chains found in fibers with low orientation.

The first internal fiber factor that influences fiber properties is chain length. The longer the molecular chains, the more area there is for the molecules to slip over adjacent molecules before breaking. Up to a certain point, fibers with longer molecular chains (higher dp) will be stronger than fibers with shorter molecular chains.

The second internal factor affecting fiber properties is the orientation of the molecular chains. The molecular chains within a fiber may be *oriented, crystalline* or *amorphous* in nature. *Oriented* fibers are those in which the chains are parallel to the longitudinal axis and to one another (Fig 2.2A). *Amorphous* fibers are those in which the molecular chains are not parallel to the longitudinal axis nor to one another but rather are randomly arranged (Fig 2.2B). When molecular chains are closely packed in "bundles" as shown in Figure 2.3, the arrangement is described as *crystalline.* Crystalline molecules have high orientation within the crystals; but the crystals may or may not be parallel to the longitudinal axis. You should not think of any single fiber as oriented, crystalline or amorphous. These descriptions are a matter of degree and regions of different molecular arrangements may occur in the same fiber.

The degree of orientation and crystallinity in manufactured fibers can be controlled to some degree during manufacture. When synthetic fibers are extruded through the spinnerette, the fiber's molecules are in a somewhat random arrangement. However, as a part of manufacturing, these filaments are stretched and drawn out. During the stretching of the man-made fibers, the molecules slide over each other, turning in the direction of stretch which is parallel to the longitudinal axis of the fiber. The molecules become increasingly parallel to the longitudinal axis, and consequently highly oriented within the fiber. Thus, many of the manufactured noncellulosic fibers have high levels of orientation and high amounts of crystallinity.

The degree of orientation is important in determining the types of bonding which can occur between polymers. Recall that the type of bonds or attraction forces is a third internal fiber structure characteristic that can affect fiber performance. Two types of attraction forces join molecular chains. The strong forces in fibers are called *hydrogen bonds;* the weaker ones are *Van Der Waals forces.* The attraction forces work on a principle similar to magnetic attraction: the closer a magnet is to metal, the stronger the magnetic

A Low orientation of crystals.

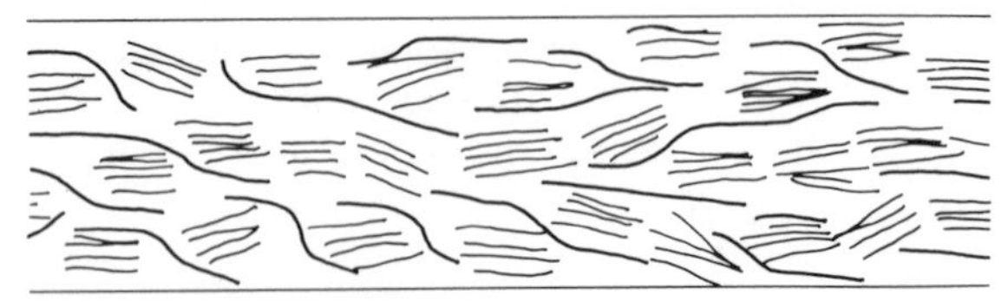

B High orientation of crystals.

Figure 2.3. When tight packing of molecular chains occurs in orderly arrangements, molecular bundles called crystals are formed. In "A" crystals are randomly arranged with respect to the fiber axis, although the molecular chains are aligned within the crystals. In "B" the crystals are oriented to the fiber axis, which yields fibers of high tenacity.

effect; likewise, the closer the molecular chains are to each other, the stronger the bonds.

A fiber with high orientation has the potential for strong bonds to form. A fiber with strong bonds will have high tenacity and low elongation potential at break. Conversely, the bonds or attractive forces in a fiber of low orientation are less strong and tend to slip more easily than the bonds in fibers which have high orientation. A fiber with low molecular orientation, and thus low tenacity, tends to have high elongation properties because the weak bonds permit the molecular chains to slip over each other, thereby allowing the fiber to elongate or stretch before breaking.

Fiber molecules with crystalline arrangements tend to have strong bonds within the crystals, which contributes to high fiber tenacity. However, fiber tenacity depends on the orientation of the molecules in relation to the fiber axis as well as the nature of the bonds in the crystal. For example, the crystal bond may have high tenacity until the bond is broken, then easy slippage to a new arrangement with high tenacity again in the new position. Thus there may be high tenacity with little slippage; and there may be easy slippage but a strong tendency to return to the original molecular arrangements. Therefore in a crystalline fiber, the fiber tenacity and molecular slippage depend on the nature and strength of the associative bonds.

Cross linking is a special chemical reaction in which strong bonding occurs across highly oriented molecules. Refer to the diagram in Figure 2.4. These special linkages contribute to increased fiber resiliency and elasticity. If cross-linking occurs in fibers that have high orientation, it can also occur in fibers that have low orientation but contain crystalline bundles; the highly oriented molecules within the crystals could be cross-linked.

The different fiber categories have varying molecular structures ranging from high to low molecular orientations, with crystalline to amorphous arrangements. In fact, amorphous, oriented and crystalline arrangements of molecules may occur in a single fiber. The advantage manufactured fibers have over the natural fibers in their molecular arrangements is that they can be designed to meet particular end uses and their molecular structures can be tailored for specific purposes.

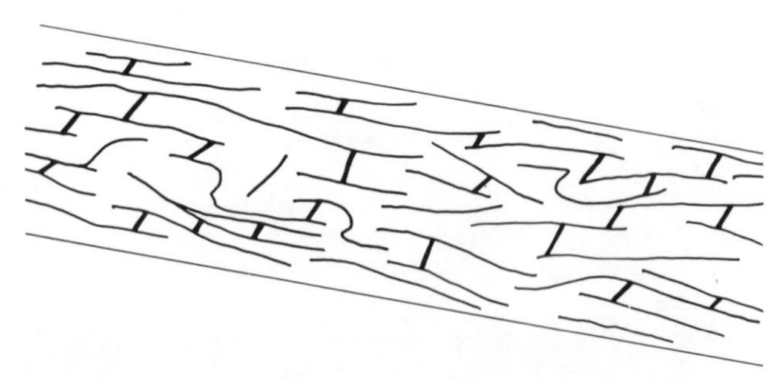

Figure 2.4. Cross-linking bonds contribute to fiber tenacity, elasticity, and resilience; the bonds act as braces to return the fiber molecules to their original positions after pulling or bending forces are applied.

QUICK QUIZ 1

1. Refer to Figure 2.5 and label the degrees of orientation and also areas of crystallinity.
2. Cross linking does not occur in areas with ______________ arrangements. Instead the weaker ______________ forces hold the molecules together.
3. Amorpous fibers with no cross linked areas have (lower, higher) resiliency and (higher, lower) elastic recovery than fibers with cross linked molecules.
4. Which has the best resiliency, cross linked fibers or those with high orientation but no cross linked molecules? ______________ Why? ______________

____

____

****

ANSWERS QUICK QUIZ 1

1. A-low orientation
   B-crystalline
   C-high orientation
2. amorphous or noncyrstalline or nonoriented; Van der Waals
3. lower; lower
4. cross linked fibers; because the cross linking bonds act like braces which help return the fiber to its original position

External Fiber Structure

The external as well as the internal physical characteristics of fibers vary considerably. For example, the smooth, round surfaces of many synthetic fibers contribute different properties to fabrics than the scaly, rough surfaces of animal hair fibers such as wool. In the preceding section, the way the internal molecular arrangements affect the serviceability characteristics of the fiber was identified briefly.

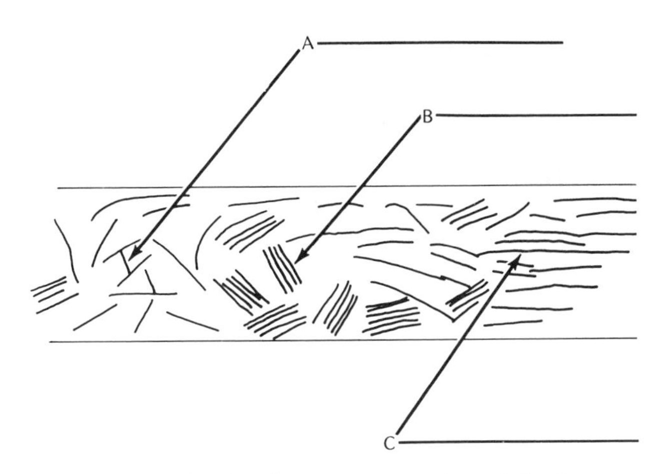

Figure 2.5. Molecular chains can pass through both crystalline and amorphous regions in a fiber.

The next selection deals with aspects of the external physical fiber structure that also affect serviceability characteristics in terms of intended end use.

Natural Fiber Categories

Figure 2.6 illustrates the microscopic physical structure of cotton. The letter "A" refers to the center canal of the cotton fiber which is called the *lumen.* The fiber length or longitudinal view of cotton appears as a flat, twisted ribbon. These twists noted at "B" are called *convolutions.* Label the lumen and convolutions in Figure 2.6.

Natural Cellulosic

The center canal or lumen running the length of the cotton fiber assists in the transmission of liquids through the fiber that provide for the food supply during growth. The twists or convolutions form an irregular surface that creates the cohesiveness necessary for spinning the relatively short cotton fibers into yarn.

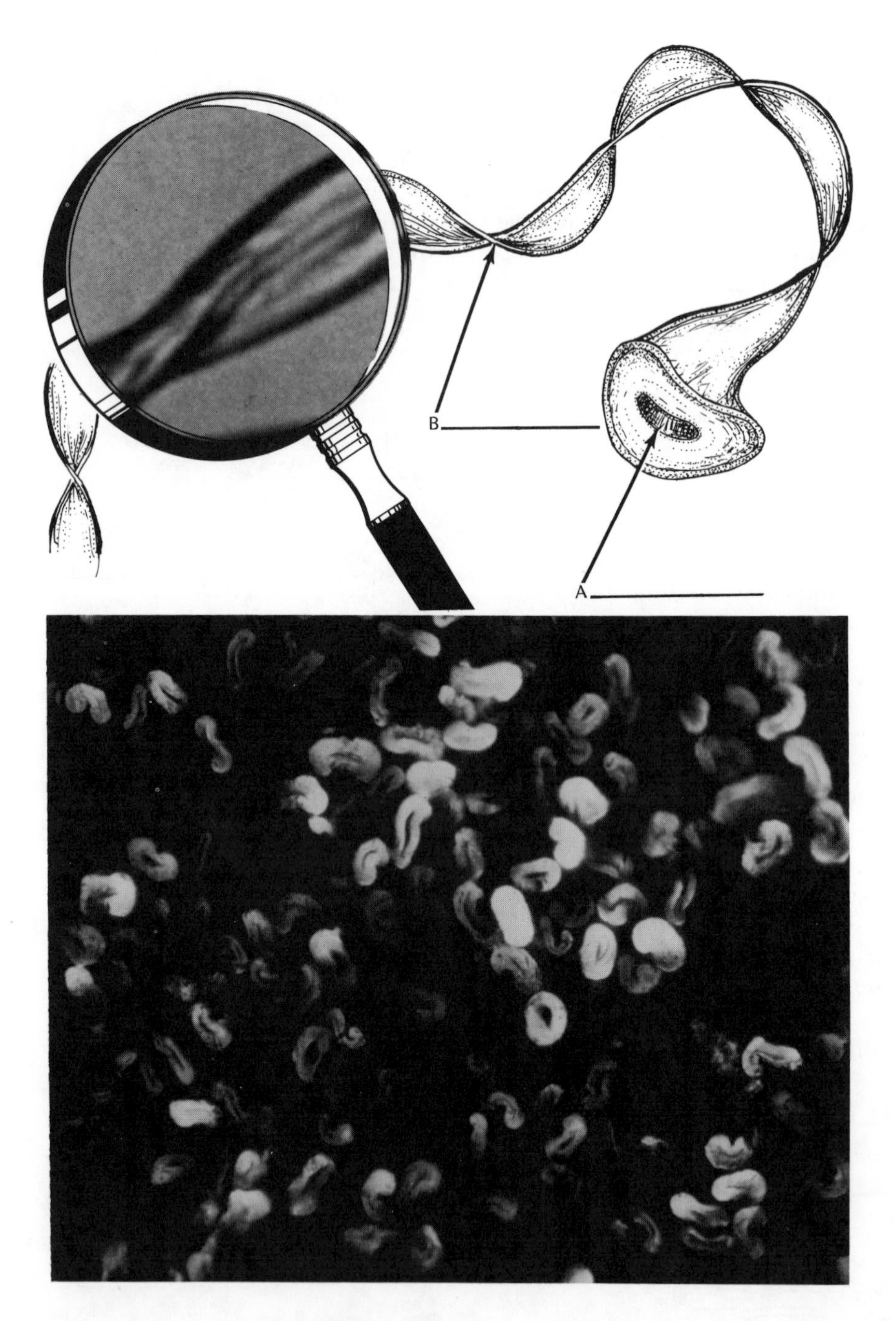

Figure 2.6. Unfinished cotton has a kidney-bean cross-section; lengthwise it resembles a twisted ribbon or spiral under the microscope. *(photomicrographs curtesy Celanese Fibers Marketing Company)*

Natural cellulosic fibers are organic compounds harvested from various plants. Three parts of plants—stems, seed pods, or leaves—provide the raw materials for natural cellulosic fibers. Fibers from stems, called *bast* fibers, include flax, ramie, hemp, and jute. Fibers from seed hairs include cotton; while those from leaves include the pineapple fiber, pina, and the banana fiber, abaca.

Figure 2.7 depicts flax, a natural cellulosic fiber derived from a plant stem. Since flax is a natural cellulosic fiber like cotton, the center canal is also called the lumen. Label the lumen in figure 2.7.

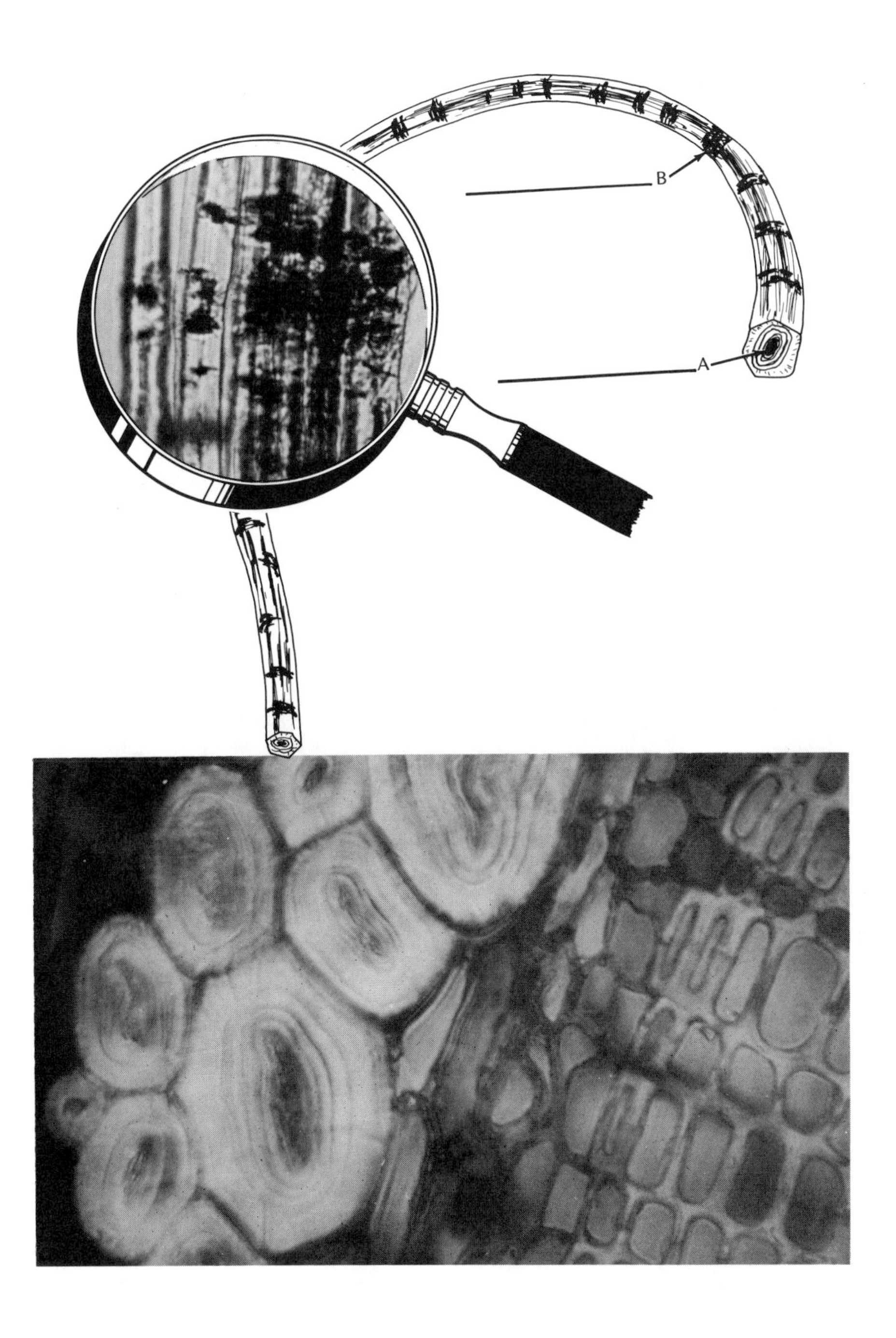

Figure 2.7. The several-sided cross section of the flax fiber looks like bamboo when viewed longitudinally under the microscope. *(photomicrographs courtesy Lambeg Industrial Research Association, Northern Ireland)*

Linen and flax are terms sometimes used interchangeably; however, to be correct, the term linen is reserved for fabrics made from flax fibers.

The stem of the flax plant has rough spots called *nodes,* which are indicated in Figure 2.7. Label the nodes at "B." These rough spots are characteristic of flax, just as joints are characteristics of a bamboo shoot. The node is partly responsible for the texture of linen fabrics.

Protein

Protein fibers are organic chemical structures derived mostly from animal hairs and, except for silk, are staple length and have the scaly external structure characteristic of animal hair; this outside layer of cells is called the *epidermis.* In Figure 2.8, locate and label the epidermis at "A." The scales vary considerably in size and texture depending on type and

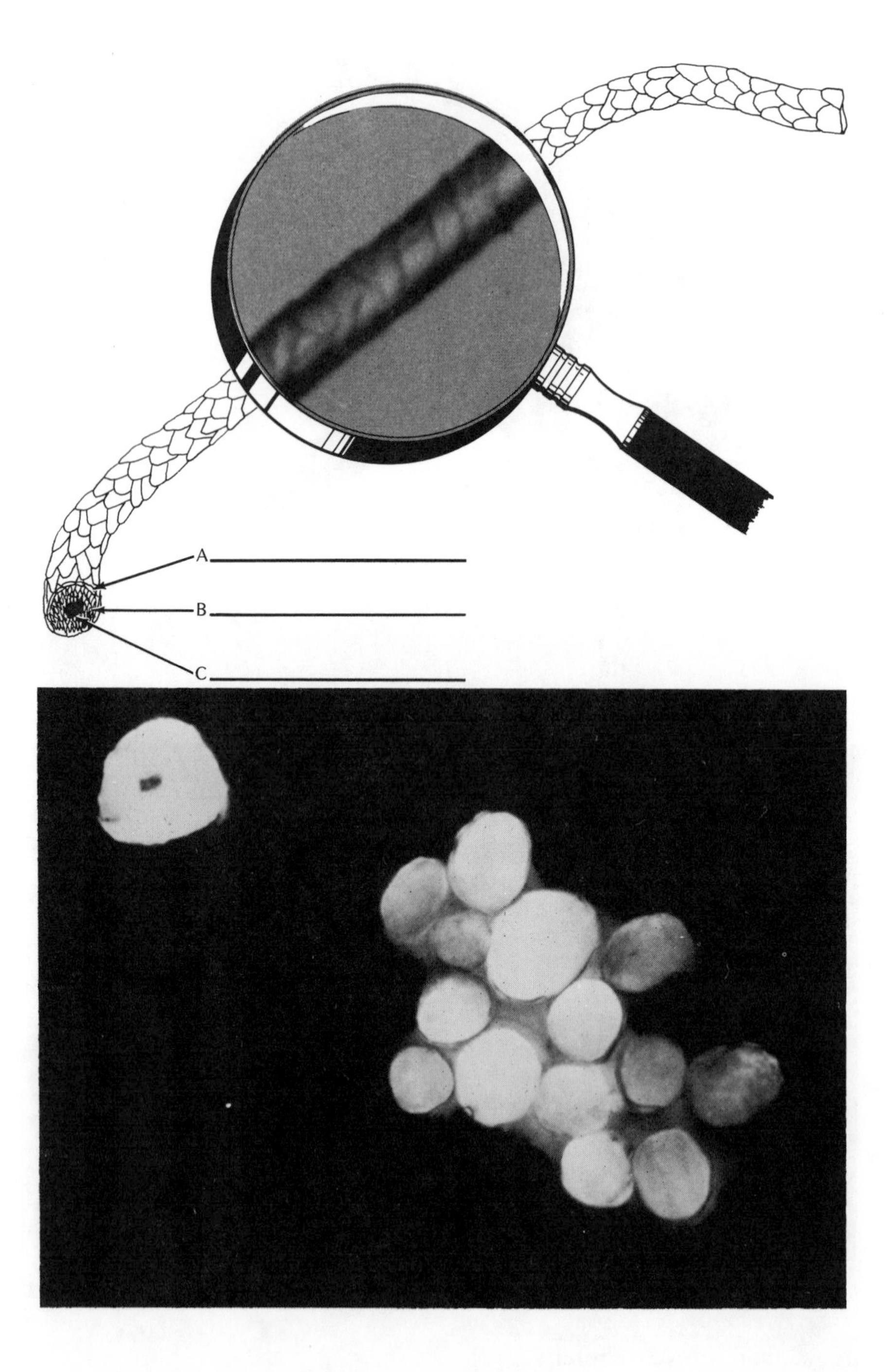

Figure 2.8. A scaly surface is the distinctive characteristic of animal hair fibers. *(photomicrographs courtesy Celanese Fibers Marketing Company)*

age of the fiber. There can be from 600 to 2,000 scales per inch along the length of a wool fiber. Numerous small scales, like fine tiles, create a smoother fiber surface than a smaller number of large scales.

The major cross sectional area of the wool fiber is the *cortex,* which is composed of small cortical cells. In Figure 2.8, label the cortex at "B." The spindle shaped cortical scales contribute to the fiber's ability to elongate and recover, to fiber elastic recovery and elongation, and to wrinkle resistance or resiliency.

The center canal of the wool fiber is called the *medulla,* rather than the *lumen* as is the central canal of natural cellulosic fibers. The size of the medulla

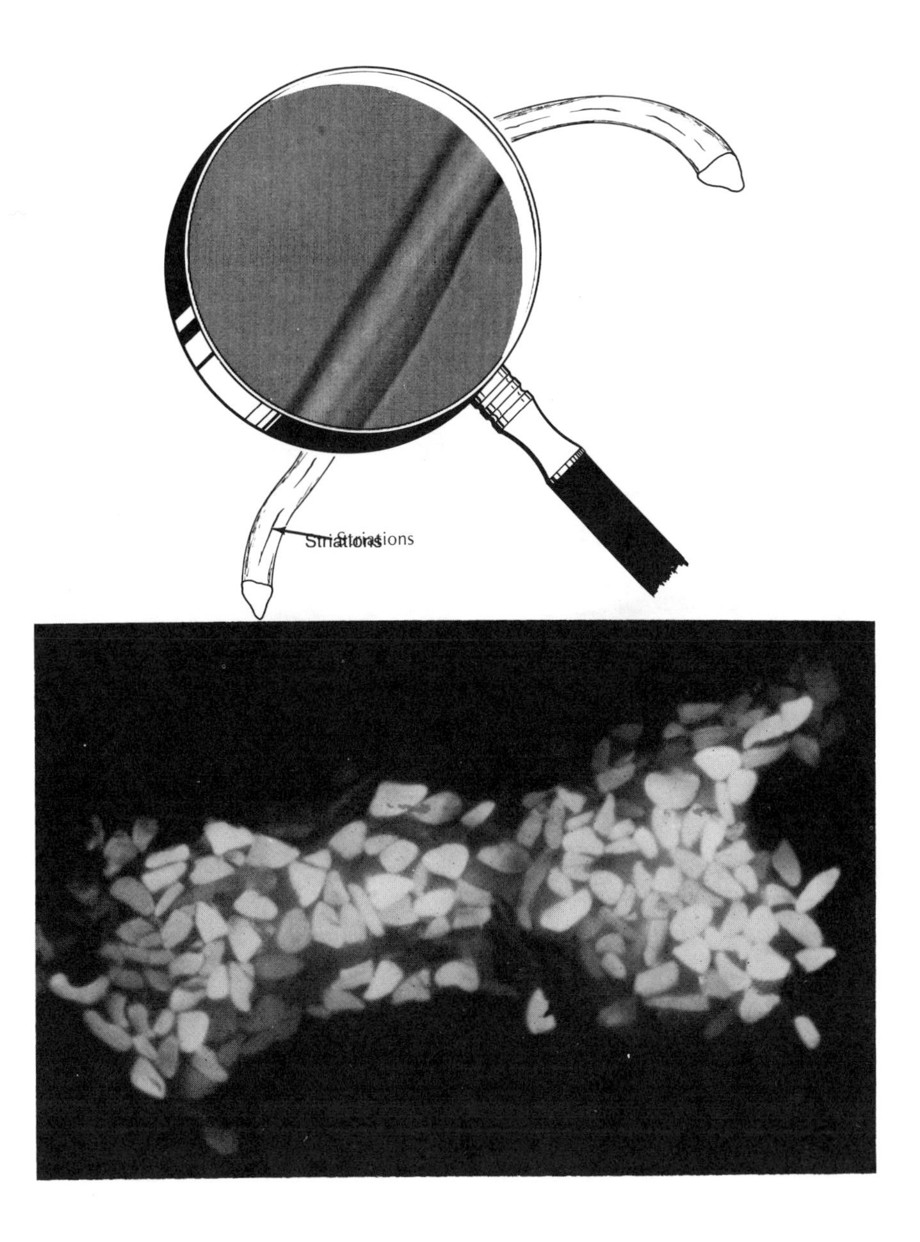

Figure 2.9. An essentially smooth surface with fine grooves and a triangular cross section is characteristic of silk. *(photomicrographs courtesy Celanese Fibers Marketing Company)*

varies with the fineness of the wool or hair fiber. In a very fine wool fiber there is a small medulla and many fine scales per inch. Label the medulla at letter "C" in Figure 2.8.

Figure 2.9 depicts the microscopic appearance of silk. Because silk is produced by a worm in a continuous filament, it does not have the scales characteristic of animal hair protein fibers.

The faint little marks or lines along the longitudinal axis of the silk filaments are due to the triangular shape of the fibers which reflects light so the lines appear. The diagram in Figure 2.9 shows that silk does not have a central canal.

Manufactured Fiber Categories

Manufactured fibers are divided into generic, or family, classifications according to each fiber's chemical composition. The necessity for a manufactured fibers classification system became apparent as more and more fibers were developed. In 1959 Congress passed the Textile Fiber Products Identification Act (TFPIA). Use Matrix II under the manufactured fibers column to review a listing of the generic terms established by the TFPIA.

Figure 2.10 illustrates the relationship of the major generic classifications, which are represented as

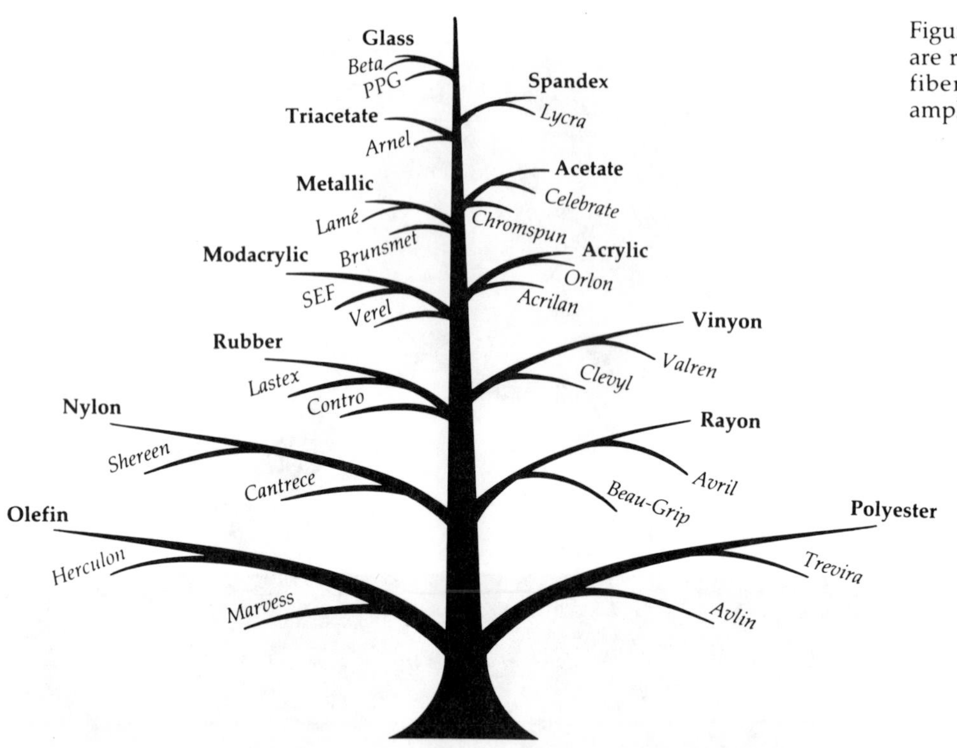

Figure 2.10. Selected generic classes are represented as branches on the fiber tree, with trade name examples given for each class.

branches on a tree. Selected trade names for the generic classes are represented as limbs on each branch. Not all generic terms have been listed on this tree.

The TFPIA requires that the *generic* class always be listed on a textile label securely attached to textile products or available to consumers when purchasing yardage. A trade name may be used with generic terms, but *trade names cannot be used in place of the generic class.* A major advantage of this legislation is

QUICK QUIZ 2

Match the following terms with their definitions.

____ 1. node
____ 2. lumen
____ 3. scales
____ 4. convolutions
____ 5. staple
____ 6. natural cellulosic
____ 7. natural protein
____ 8. amorphous arrangement
____ 9. crystalline arrangement
____ 10. cross-linking

a. center canal of natural cellulosic fibers
b. twists found along cotton fiber length
c. derived from animal sources
d. joint seen along the flax stem
e. low orientation of molecular chains to fiber axis
f. part of outer layer found on wool and other animal hair protein fibers
g. fibers measured in inches
h. molecular chains aligned in bundles (may or may not be aligned with the fiber axis)
i. derived from plant sources
j. bonding across highly oriented molecules

****

ANSWERS QUICK QUIZ 2

1-d; 2-a; 3-f; 4-b; 5-g; 6-i; 7-c; 8-e; 9-h; 10-j;

that it makes it easier for consumers to become familiar with the generic groups, instead of having to learn the hundreds of trade names available.

Rayon, acetate, and triacetate are the generic classes in the man-made cellulosic fiber category. These fibers are made from a cellulose base which is *regenerated in the manufacture of rayon,* and *chemically modified* during the production of *acetate* and *triacetate.* Any product of rayon, acetate or triacetate is required by law to have the generic name on a label securely attached; yardage must have a label available with it.

Man-made noncellulosics or synthetic fibers are different from man-made cellulosics and man-made protein (azlons) in that the synthetics are manufactured from original, organic chemical compositions. The other manufactured fiber categories are made from organic modified or regenerated cellulosic substances or protein materials. Azlon fibers, man-made from modified protein, are not currently produced in the U.S. and their category is of limited importance. There are more generic classes in the synthetic fiber category than in the categories that comprise the *man-made* cellulosics, the proteins or the mineral fibers. Mineral fibers are the only ones constructed from *inorganic* chemical substances and are generally classified as glass, asbestos or metallic. Each generic group has specific chemical similarities and compositions.

Rayon is a generic classification for man-made cellulosic fibers made from pure natural cellulose which has been regenerated or reprocessed into filament or staple length fibers. Avtex, Fibro and Courcel are trade names in the generic rayon class. You cannot possibly become familiar with every textile trade name and know its individual properties and characteristics, but you can learn the generic classes and the properties associated with each. From these basic properties it is possible to evaluate the serviceability concepts of a fiber and the fabrics designed for particular end uses.

Review the chart in Figure 2.10 to become aware of the major generic classifications. Not all generic groups are included and only a few trade names; however, these represent the generic groups commonly found in consumer goods and a few of the more frequently encountered trade names. Remember that it is the chemical composition of the fiber that determines the particular generic class or group in which a fiber will be listed.

Different manufacturers ''engineer'' special properties into their particular trade-named fiber by altering or modifying the characteristics of the base fiber. Thus, while the different trade name fibers within a generic class are chemically similar, there are variable performance claims and property differences among different types of fibers in a single generic class.

Today many manufactured fibers are designed for very specialized end uses. In apparel, home furnishing and particularly industrial end uses, fiber chemistry and structure are engineered to provide the performance properties desired. Composite textiles are textiles with dissimilar properties that are joined macroscopically to create certain desired end use properties. Composites are used in highly technical end uses such as aerospace application and biomedical textiles. Even in more familiar textile products, advanced engineering techniques are applied to improve product performance. In many instances today, end use properties are the starting point for fiber development. Fibers are then custom designed to provide the desired characteristics. As you read the following units describing fiber characteristics, you should keep in mind that these discussions refer to unaltered fibers and that properties of manufactured fibers can be changed considerably to meet consumer needs.

QUICK QUIZ 3

1. Review the manufactured fiber tree in Figure 2.10 and list six man-made noncellulosic or synthetic fiber generic classes.

   ______________ ______________ ______________

   ______________ ______________ ______________

2. What advantage do manufactured fibers have over natural fibers in meeting consumer needs?

   ____

****

ANSWERS QUICK QUIZ 3

1. any six: acrylic, modacrylic, nylon, olefin, polyester, spandex, vinyon
2. they can be engineered or altered to meet end use requirements

---

Figure 2.11. Fiber filaments are extruded from the spinneret into a controlled chemical or temperature environment, depending on the process used. *(courtesy Vernon L. Smith, Scope Associates, Inc., New York, N.Y.)*

Manufacturing Processes

Regardless of the organic chemicals used in the production of man-made cellulosic and noncellulosic or synthetic fibers, the chemical compound is extruded into long filament fibers. This is done by means of a sieve-like device called a *spinneret*. Figure 2.11 shows the spinneret which is made of a metal, often platinum. It has many tiny holes through which the fiber solution is passed to produce filaments.

There are four different processes used in forming manufactured fibers. Although the environment is different, each process involves the extrusion of a fiber solution through the openings in a spinneret. Of the four processes used in forming fibers, three are used most frequently. These three, referred to as the major processes in this text, include *wet spinning, dry spinning,* and *melt spinning*. The *emulsion spinning* method, used for highly specialized fibers such as Teflon, is not discussed here. The fibers typically found in consumer goods are seldom made by this process.

Wet spinning occurs when the fiber molecules are in solution; the solution is forced through the spinnerette into a liquid bath which has chemicals that react with the solution, forming fiber filaments. Rayons and certain acrylic fibers are made by the wet spinning process which requires additional processing after filament formation. The two processes below do not require further processing after the filaments are formed.

Dry or *solvent spinning* occurs when the fiber molecules are suspended in a chemical solvent. This fluid is forced through the spinnerette into warm air, causing the solvent to evaporate; as it evaporates, the fiber solidifies. Examples of dry spun fibers include acetates, triacetates and many acrylic and modacrylic fibers.

Melt spinning occurs when fiber molecules are melted at high temperature. The melted fluid is forced through the spinnerette into cool air, where the reduced temperature causes the fiber molecules to harden in filament form. This group includes nylons and polyesters.

The diagram in Figure 2.12 illustrates the relationships among these three major spinning processes. This spinning of manufactured fibers should not be confused with the spinning of yarn which will be discussed in Unit 3.

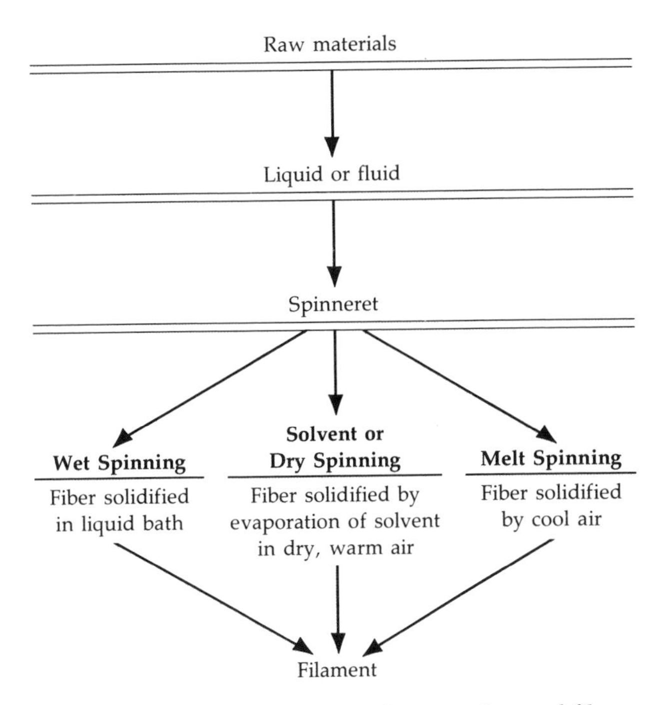

Figure 2.12. Spinning processes for manufactured fibers.

Dry spinning results in less fiber molecular breakdown than the wet spinning methods; it does not require additional processing other than drawing and stretching after extrusion from the spinnerette. The melt spinning method is considered the fastest since no solvent or other processing is required. Thus, fibers may be produced by melt spinning with somewhat reduced energy and chemical costs. However, remember that factors other than the spinning process contribute to end product cost, and fibers made by the melt spinning method are not necessarily less expensive than those made by wet or solvent spinning.

Variation in Fiber Cross-Section Shape

Figures 2.13 and 2.14 show the shape and appearance of certain manufactured fibers. Notice the varied cross-sectional shapes. Fiber cross section is determined, in part, by the shape of the openings in the spinnerette and by the spinning method used.

The shape of the spinnerette openings may be responsible for the general shape of fibers made by the melt spinning process; however, the drawing and stretching operations modify fiber cross section to some degree so that fiber cross section and spinnerette opening are not identical.

For wet spun fibers, the type of chemicals used in the solidifying bath may influence shape considerably; while in the solvent spun or dry spun fibers the chemical used as the solvent may affect fiber shape as it evaporates. Spinnerette openings are also involved in wet and dry spinning, as are the drawing and stretching operations. Obviously, final fiber cross section shape is the result of several factors involved in the manufacture of man-made fibers.

The particular cross section shapes shown with each of the generic classes in Figures 2.13 and 2.14 represent one type of fiber within each classification illustrated. The shapes of different trade-named fibers may vary within a generic class, since manufactured fibers can be modified to meet special end uses and provide better serviceability in end use. The bolder print labels under the fibers in Figures 2.13 and 2.14 give the generic classifications, and the italicized labels give examples of the trade names.

Fiber properties which may be altered by changing the fiber's cross section shape include bulk, loft, flexibility or pliability, luster, hand, and resiliency. Fiber properties of tenacity, elongation and elastic

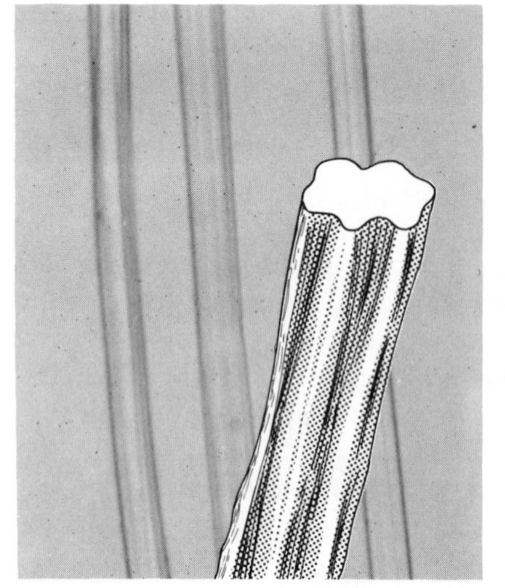

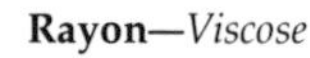

Rayon—*Viscose*

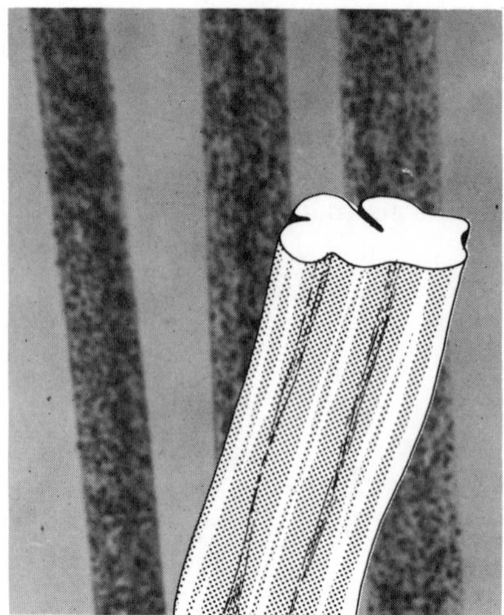

Acetate

Figure 2.13. The drawings and longitudinal photomicrographs (in the background) illustrate examples of typical shapes of the man-made cellulosic filaments represented.

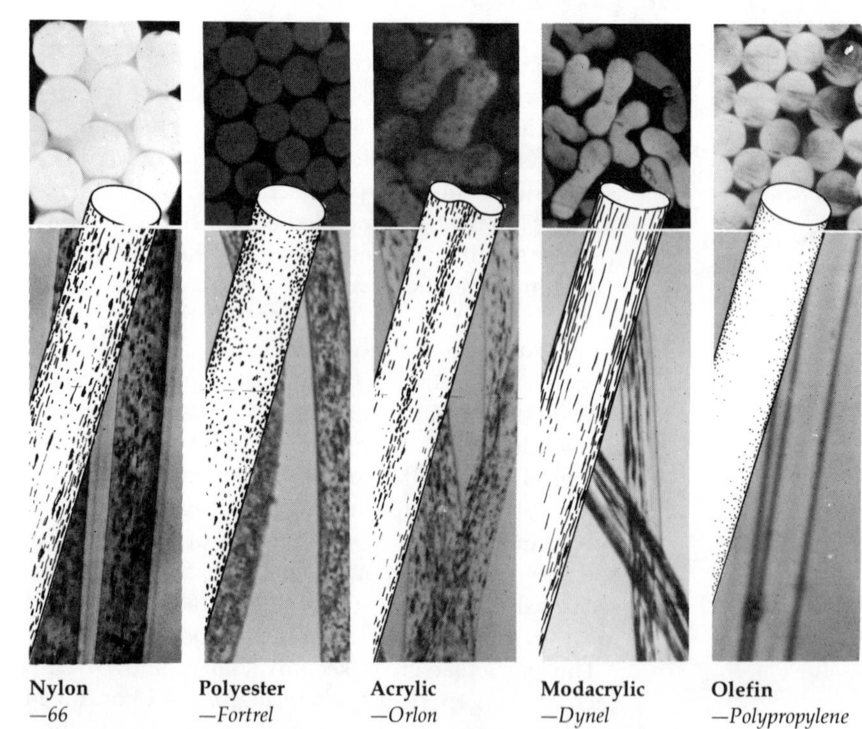

Nylon —*66*

Polyester —*Fortrel*

Acrylic —*Orlon*

Modacrylic —*Dynel*

Olefin —*Polypropylene*

Figure 2.14. There is great variety in synthetic fiber cross-sectional shapes, as seen in the photomicrographs shown in the background. *(photomicrographs courtesy Celanese Fibers Marketing Company)*

recovery, density, chemical reactivity and moisture absorbency depend more on the chemical composition, molecular arrangement, and surface characteristics of the fiber than on cross section shape.

As compared with the round nylon fiber, the cross section shapes of the acrylic and modacrylic fibers, illustrated in Figure 2.14, increase fiber bulk, loft and flexibility because irregular fiber shapes do not pack as tightly into yarns as do round ones, resulting in yarns and fabrics that are more ''springy'' and flexible.

It is important to know the chemical difference, in addition to the varying shapes, between man-made cellulosic fibers and man-made noncellulosic fibers:

Man-made cellulosic fibers are made from pure or modified cellulose and are similar chemically to natural cellulose.

Man-made noncellulosics are made from organic chemicals combined in the laboratory to yield particular properties and to form filaments from a product that is normally not fibrous in form.

QUICK QUIZ 4

Indicate by matching which of the fiber categories are made of organic and which are made of inorganic substances.

____	1. natural cellulosic	a. organic
____	2. protein	b. inorganic
____	3. man-made cellulosic	
____	4. man-made noncellulosic	
____	5. mineral	

****

ANSWERS QUICK QUIZ 4

1-a; 2-a; 3-a; 4-a; 5-b

Mineral Filaments

One of the most common mineral fibers used in textiles is glass. The glass fiber shown in figure 2.15 is made from glass marbles or cullet. The reason glass was, and still is, made from marbles by some manufacturers is that marbles can be checked for imperfections and deleted before fiber formation. Any imperfections in the marble would result in impurities in the glass filaments, producing irregular and incomplete filaments with weak points that tend to break easily. Today, with modern systems of control, glass is made by either the marble, or discontinuous process, or by a continuous process that bypasses the formation of the marble cullet.

The molten glass is not allowed to cool after passing from the spinneret until the filament has been drawn and stretched into a fine diameter. Thick glass filaments, of course, would not have the required flexibility or pliability for textile products; they would also tend to be brittle. Plainly, glass fibers are less versatile for textile products than the man-made noncellulosic fibers which can be produced in a variety of cross sectional shapes.

In subsequent parts of this unit, the serviceability concepts of durability, comfort, care or maintenance, and esthetic appearance will be discussed in relation to the fiber categories and to the basic fiber properties involved.

Figure 2.15. Molten glass is often formed into marbles to allow inspection for impurities; then the marbles are melted down again for extrusion through the spinneret.

REVIEW

1. In order to discuss the five basic categories of fibers in terms of serviceability concepts, you have to be familiar with the properties of the fiber categories. The *five* categories used in this program are listed; provide the source from which those fibers are made.
   a. Natural cellulosic: ______
   b. Protein: ______
   c. Man-made cellulosic: ______
   d. Man-made noncellulosic: ______
   e. Mineral: ______

2. Name and define *two* descriptive terms for fiber length:
   ______
   ______

3. What is an amorphous fiber? ______
   ______

4. The reaction of a crystalline fiber to various kinds of physical stress is determined by what two factors? ______
   ______
   ______

5. The attraction forces which connect molecular chains within a fiber are involved in determining the properties of the fiber. The stronger the fiber, the closer and more aligned the ______.

6. Name *two* properties that cross-linked molecules give to a fiber: ______
   ______

7. The center canal of a cellulosic fiber is called the ______ while the central canal of a protein fiber is called the ______.

8. The twists in cotton fiber are called ______.

9. Identifying marks on flax fiber, somewhat like the joints in bamboo, are called ______.

10. A distinguishing characteristic of the surface of wool fibers is their ______.

11. Why is it more important to understand the properties of each of the generic classifications than to memorize fiber trade names and properties? ______
    ______
    ______

12. What is the apparatus called through which the solution is forced during the manufacture of fibers? ______

13. Describe these *three* major spinning processes used in forming manufactured fibers.
    a. Wet: ______________________________
    ____
    b. Dry: ______________________________
    ____
    c. Melt: ______________________________
    ____

14. Why are manufactured fibers engineered with various cross section shapes?
    ____

15. The shape given to a manufactured fiber (does, does not) affect the chemical properties of the fiber.

16. The cross section shape of manufactured fibers (does, does not) affect their physical properties.

17. Indicate the fiber properties which are most influenced by altering the cross section shape by an "O," and those properties which depend more on chemical composition by an "X."

____	tenacity	____	elongation and elastic recovery
____	bulk	____	density
____	absorbency	____	loft
____	hand	____	resiliency
____	luster	____	flexibility

****

REVIEW ANSWERS:

1. a. made from natural plant cellulose, organic
   b. made from natural protein substances, organic; azlon fibers, not discussed in this text, are man-made protein fibers produced from natural protein substances, just as man-made cellulose is made from natural cellulosic substances
   c. made from pure regenerated modified cellulose obtained from nature, but reprocessed into a new organic fiber form
   d. engineered from organic compounds through manufacturing processes
   e. made from inorganic materials like glass or metal
2. staple—relatively short fibers, measured in inches or centimeters
   filament—long fibers, measured in yards, meters, kilometers or miles
3. One in which the molecules are not aligned with each other or to the longitudinal axis of the fiber; molecules are randomly arranged.
4. the nature of the bonding system holding the molecules and the crystals together; the orientation or nonorientation of the crystals to the fiber axis
5. molecular chains or polymolecules
6. any two: resiliency, tenacity, elastic recovery
7. lumen, medulla
8. convolutions
9. nodes
10. scales
11. There are limited generic classes, which have similar fiber properties within each class, while there are hundreds of trade names used.
12. spinneret

13. a—The fiber liquid is forced through the spinneret into a chemical bath which reacts with the solution to form fibers.
    b—The fluid fiber, in a solvent, is forced through the spinneret into an environment of warm air which evaporates the solvent to form the fibers.
    c—The melted fiber is forced through the spinneret into an environment of cool air where it hardens in filament form.
14. to alter fiber properties and enhance performance in selected end uses
15. does not
16. does
17. 0—bulk, absorbency, hand, luster, loft; X—tenacity, elongation and elastic recovery, density, resiliency, flexibility

UNIT TWO

Part II

Fibers and Their Durability Considerations

Matrixes IV, V, VI and VII which follow represent sections of Matrix I. They have been modified to let you fill in comments that will help you organize and learn the information in Unit 2. All matrixes are enlarged and repeated at the end of the book. In turn, each should be removed for use with the appropriate part as you work through the text.

Durability is the serviceability concept which includes the properties that influence how long a fiber will last under different use conditions. Each of the five fiber categories, classified according to the nature of its chemical composition, has varying durability properties which will be compared across the five categories. Remember to use Matrix IV as a study sheet for properties related to durability.

Fibers have varying degrees of tenacity, which is the ability to endure a strong pulling force, plus elongation and elastic recovery. Recall that elongation is the amount of extension or stretch that a fiber can undergo before it actually breaks. Some fibers have a lot of elongation, while others have very little. Fibers with high elongation will stretch a considerable amount before breaking. A fiber with high tenacity and elongation potential that tends to return to its original length immediately after being stretched, is said to have good elastic recovery or elasticity.

In addition to pulling forces, a fiber is often subjected to rubbing, as when a shirt is worn under a coat. A fiber which tolerates such forces without surface damage is abrasion resistant. If a fiber is to be successfully converted into yarns, it must have some degree of cohesiveness or spinning quality. If it does not retain its shape under normal care and wearing conditions, the fiber is said to lack the important quality of dimensional stability.

Comparison of Tenacity Among The Five Fiber Categories

Tenacity is the ability of a fiber to withstand a strong pulling force. It is the term used to denote fiber strength and is usually cited in grams per denier.

For the *natural cellulosic fibers,* the bast fibers flax and ramie have higher tenacities than cotton. These variations result from differences in molecular arrangements, molecular chain length, degree of polymerization, and differences in physical structure of the fiber. Bast fibers that come from plant stems generally have longer molecular chains than the seed hair fibers. Longer molecular chains tend to contribute to high strength, but not always. For example, jute bast fibers are not much stronger than shorter cotton seed hair fibers.

In addition, tenacity is affected by the level of molecular orientation and wet strength. Cotton fibers tend to have a somewhat amorphous or noncrystalline arrangement, while the molecules in flax fibers tend to be parallel to the fiber axis, or highly oriented, and thus stronger than cotton. The tenacity of almost all natural cellulosic fibers *increases* when wet. The exception is jute, the fiber used in burlap fabric, which loses strength when wet.

As a group, *natural cellulose fibers have a tenacity range from medium to medium high.* Remember to record this data on Matrix IV.

MATRIX IV
FIBER DURABILITY

DURABILITY PROPERTIES	FIBER CATEGORIES				
	Natural Cellulosic	Protein	Man-made Cellulosic	Man-made Noncellulosic	Mineral
OVERALL					
Strength or tenacity					
Abrasion resistance					
Cohesiveness or spinning quality					
Elongation					
Elastic recovery					
Flexibility or pliability					
Dimensional stability					

Animal hair protein fibers are generally less strong than most other fibers, but other properties compensate for the lower tenacity. On the other hand, silk in the protein category has relatively high tenacity. As a category, protein fibers are classed as generally having low to medium high tenacity. Within the range of tenacity represented by the protein family, *wool* is toward the low end and *silk* toward the high end. Compared to the natural cellulosic fiber category, silk has high tenacity comparable to the stem fiber *flax,* while wool would compare with *jute,* since both lose strength when wet. Therefore, *protein fibers as a category range in tenacity from low to medium high.*

Man-made cellulosic fibers usually have lower tenacities than natural cellulosic fibers because the molecular chains are reduced in length during manufacturing operations. Rayon is a *regenerated* cellulose fiber and is pure cellulose in a new shape; acetate and triacetate are modified cellulose fibers and are not pure cellulose. While rayon, like the animal proteins, also has decreased tenacity when wet or being laundered, *high-wet-modulus* (HWM) rayon fibers are similar to cotton in both wet and dry tenacity because the molecular chain is not broken down as much as for regular rayon. HWM rayon, currently used in apparel fabrics, goes by such trade names as Avril, Zantrel and Prima. It has been especially successful because it does not lose strength when wet as does regular rayon. Its molecular chain length is not broken down as much as in regular rayon manufacture, so it is more like that of cotton.

MATRIX V
FIBER COMFORT

COMFORT PROPERTIES	FIBER CATEGORIES				
	Natural Cellulosic	Protein	Man-made Cellulosic	Man-made Noncellulosic	Mineral
OVERALL					
Absorbency					
Wicking					
Electric conductivity					
Heat or thermal conductivity					
Heat or thermal retention					
Allergenic potential					
Density or specific gravity					

High tenacity rayons have also been developed to have a tenacity exceeding that of cotton. High tenacity rayon was originally developed for industrial uses requiring high strength, such as for tire cord. But rayons modified to improve their tenacity still cannot generally withstand heavy pulling forces. Therefore, *tenacity of fibers in the man-made cellulosic category ranges from low* for regular rayons and acetates *to medium high* for the HWM and high tenacity rayons.

Man-made noncellulosic or *synthetic fibers tend to have medium-high to high tenacities*, although there is some variation among them. Further, the manufacturing process can control tenacity to some degree so that fibers within the same generic group may differ in tenacity. Because of the wide difference in tenacity among these fibers, it is impossible to categorize all synthetics as a group. Consequently it is wise to learn the basic spread of tenacity for each of the generic groups included in this category. The man-made noncellulosic generic classifications of most importance to consumers include: *nylon, polyester, acrylic, modacrylic, olefin,* and *spandex*. The other synthetic fiber classifications are less important in apparel and home furnishings products and will not be discussed in detail.

As a group, *man-made noncellulosic fibers have high tenacity* because of the strong bonds which join long molecular chains that are highly oriented and parallel to the fiber axis. The last category—the *mineral fibers—is also associated with high tenacity*. Mineral fibers, which are inorganic, have high tenacity when

MATRIX VI
FIBER CARE

CARE PROPERTIES	FIBER CATEGORIES				
	Natural Cellulosic	Protein	Man-made Cellulosic	Man-made Noncellulosic	Mineral
General care or maintenance					
Resiliency					
Dimensional stability					
Chemical reactivity					
Detergents					
Bleaches					
Dry Cleaning					
Heat tolerance					
Flammability (recall burn test)					
Biological resistance					
Light resistance					
Age resistance					

compared with the other fiber categories, especially man-made cellulosic, the category including fibers with the lowest tenacity.

Comparison of Abrasion Resistance Among The Five Fiber Categories

Abrasion resistance is the ability of a fiber, yarn or fabric to withstand rubbing forces and friction. As with fiber tenacity, there is great variation among fibers' resistance to abrasion.

Man-made cellulosic fibers have the poorest abrasion resistance, while *man-made noncellulosic fibers,* in general, have *the highest. Mineral fibers* have comparatively *poor abrasion resistance* and *protein and natural cellulose fibers are medium to good.* Product serviceability, particularly in terms of durability and appearance, is affected by the abrasion resistance of the fibers used.

MATRIX VII
FIBER ESTHETIC APPEARANCE

ESTHETIC PROPERTIES	FIBER CATEGORIES				
	Natural Cellulosic	Protein	Man-made Cellulosic	Man-made Noncellulosic	Mineral
Color or dye affinity					
Luster					
Translucence					
Drape					
Texture					
Body or hand					
Loft or bulk					

Mineral fibers, especially glass, also tend to have low abrasion resistance. Because metallic fibers in the mineral category are usually given a plastic or polymer coating during manufacture, their abrasion resistance depends on the type of coating used.

Man-made noncellulosic or synthetic fibers have high tenacity as well as high abrasion resistance. The high crystallinity and degree of polymerization of the fiber molecules contribute to these properties. Conversely, man-made cellulosics have a more amorphous molecular orientation and therefore have lower abrasion resistance.

Animal hair fibers like wool tend to resist abrasion due to their fiber scales. The coarseness or fineness of the scales determines the degree of abrasion resistance, with coarse scales being most resistant. However, wool has less abrasion resistance than the man-made noncellulosic or synthetic fiber category.

Comparison of Cohesiveness or Spinning Quality Among the Five Fiber Categories

Cohesiveness or spinning quality is the important ability of fibers to cling together during and after spinning into yarns. A certain amount of cohesiveness *or* filament length is necessary in order to spin any fiber into yarns. Cohesiveness is particularly important for staple fibers which cannot rely on great length to contribute to spinnability.

Fiber length; the amount of crimp, convolutions or twists; scales; and the irregularity or uneveness of surface are important when determining the spinning quality or cohesiveness of fibers. Spinning quality can be built into manufactured fibers through the introduction of surface irregularity, crimp or texture, as well as by adjusting fiber length.

Because crimp and twist along the longitudinal surface of a fiber contribute to high cohesion, natural cellulosic and animal hair fibers have high cohesiveness. In the natural cellulosic and protein hair fiber categories, the irregular physical structures enable the fibers to entangle and adhere to each other. Silk, however, a relatively smooth filament length fiber, is an exception.

Filament length is so important to the spinning quality of smooth filament fibers that they are mechanically or chemically crimped or textured when cut into short staple to add the cohesiveness or spinning quality necessary for yarn making. Manufactured fibers in the man-made cellulosic and the noncellulosic or synthetic categories are often cut into staple lengths in order to make them appear more like natural fibers. Those that are thermoplastic can have crimp or texture set in by heat to provide the spinning qualities needed.

Comparison of Elongation Potential Among the Five Fiber Categories

The elongation potential of a fiber is the amount of stretch or pulling force it can withstand without breaking. Fiber elongation varies greatly from a low of two or three percent to over 600 percent. The latter represents the elastomeric fibers, while the

---

QUICK QUIZ 5

1. Match each fiber with its appropriate category:

____	1. cotton	a. natural cellulosic
____	2. nylon	b. protein
____	3. rayon	c. man-made cellulosic
____	4. acetate	d. man-made noncellulosic or synthetic
____	5. spandex	e. mineral
____	6. wool	
____	7. vinyon	
____	8. triacetate	
____	9. glass	
____	10. acrylic	

2. Give the tenacity range for each of the following categories.
   a. natural cellulosic: ____________________
   b. protein: ____________________
   c. man-made cellulosic: ____________________
   d. man-made noncellulosic: ____________________
   e. mineral: ____________________
3. Both natural cellulosic and protein categories have medium to good abrasion resistance compared to the man-made noncellulosics which are rated high. Match the following fibers to their degree of abrasion resistance.

____	1. cotton	a. low or poor
____	2. rayon	b. medium or good
____	3. nylon	c. high or excellent
____	4. glass	
____	5. acetate	
____	6. polyester	
____	7. wool	

****

ANSWERS QUICK QUIZ 5

1. 1-a; 2-d; 3-c; 4-c; 5-d; 6-b; 7-d; 8-c; 9-e; 10-d
2. a-medium to medium high
   b-low to medium high
   c-low to medium high for HWM and high tenacity rayon
   d-medium high to high
   e-high
3. 1-b; 2-a; 3-c; 4-a; 5-a; 6-c; 7-b

---

low limit is characteristic of mineral fibers and some natural cellulosic fibers.

Recall the earlier discussion of the molecular orientation of fibers. A basic determinant of the elongation potential of a fiber depends on molecular arrangement, chain length and the nature of the bonds within the fiber. High elongations are usually, though not always, found in fibers which are amorphous and have low molecular orientation. Fibers with high orientation usually have low potential for elongation, whereas fibers with low orientation tend to have high elongation characteristics. Examples of natural cellulosic fibers with high orientation include flax; and with low orientation, cotton.

High or low elongations of different fibers are relative, depending on which fibers are being compared, Although cotton has higher elongation than flax, the natural cellulosic fibers are all low compared with certain other fiber categories. Thus, as a group, *natural cellulosic fibers have relatively low elongations,* while the *elongation potential for the animal hair protein fibers is high.*

Within the protein fiber category, silk has a low elongation potential because its filament lacks the fiber crimp and the scales which, in wool, stretch and slip over each other before breaking. Silk is also composed of longer molecular chains more oriented than those in wool, and there tends to be less elongation potential with higher molecular orientation.

The *man-made cellulosic fibers have a molecular structure which allows for medium to high elongation* as compared to the natural cellulosics, which have relatively low elongation, and the animal hair protein fibers which have high elongation. These properties should be entered on the durability matrix.

You can conclude that the man-made cellulosic fibers with medium to high elongation potential and low tenacity tend to have primarily amorphous or low orientation molecular arrangements with easy molecular slippage. Recall that man-made cellulosic fibers lose tenacity when wet. Strength loss occurs because the wet fibers stretch easily as the molecules slip past each other. Therefore, the elongation potential of the man-made cellulosic fibers is higher when the fibers are wet.

The elongation potential of man-made noncellulosic or synthetic fibers varies considerably, depending on the chemical composition and molecular structure of the different generic classifications. Synthetic fibers have a nearly amorphous arrangement immediately upon extrusion from the spinneret before drawing and stretching has taken place.

After filaments are drawn out to several times their length, the fiber molecules become more highly oriented; the amount of elongation potential remaining depends upon how much the fibers were stretched during drawing and how much molecular slippage the bonds will permit. Since *synthetic fibers are engineered to have stretch before breaking,* they generally have *relatively high elongation potential.*

Another feature of certain man-made noncellulosic fibers which contributes to their high elongation potential, despite their high molecular orientation, is the presence of *molecular crimp.* Nylon is an example of a synthetic fiber which has high molecular orientation and crystallinity, resulting in high tenacity. However, due to a crimped molecular chain,

---

QUICK QUIZ 6

1. If a garment made entirely of rayon fibers were laundered and hung over a line to dry, how would the serviceability of the garment be affected? ______________________________

   ____

2. HWM (high-wet-modulus) rayon is a modification of regular rayon in which the molecular chains are longer and more like cotton. Thus, the wet elongation of HWM rayon usually (does, does not) create a laundering problem.

****

ANSWERS QUICK QUIZ 6

1. The fibers and yarns would probably stretch, and the garment fabric would lose its shape.
2. does not

---

there is still a fair amount of elongation remaining before the fiber reaches its breaking point. Think of molecular crimp as behaving similarly to fiber crimp; in both cases, there is elongation potential available before the molecular chain or fiber is extended to its breaking point.

Two generic classes of elastomeric fibers are currently produced in the United States. The elongation potential of these fibers is so high that they are not used for a relative comparison of elongation among other fibers, or all fibers would have low elongation compared to them. They are rubber and spandex, two elastomeric fibers known for their exceptional elongation potential as well as excellent elastic recovery.

Comparison of Elastic Recovery Among The Fiber Categories

Elastic recovery, or elasticity, is the ability of a fiber to return to its original length immediately after being elongated, extended or stretched. The physical and molecular structure of a fiber contribute to fiber elasticity. A molecular structure with strong hydrogen bonds and/or cross linkages prevents the molecules from slipping beyond the breaking point and contributes to high recovery from extension or stretch. In other words, strong hydrogen bonding and cross linked molecules contribute to high or superior elastic recovery.

Natural cellulosic fibers do not have strong bonds or cross linked molecules, consequently their elasticity is low. Man-made cellulosic fibers have even lower elastic recovery than natural cellulosics—regenerated cellulose, rayon, has shorter molecular chains and weaker bonds than natural cellulosic fibers. The low elasticity is more of a disadvantage in man-made cellulosics than in natural cellulosics because of their greater elongation potential. Neither group will recover very well from stretching, so the group that stretches the least will keep its shape the best.

Since natural cellulosic fibers are stronger and will stretch less than man-made cellulosics, the natural cellulosic fibers will more nearly maintain their original shape. The *low elastic recovery of both man-made cellulosic fibers and natural cellulosic fibers* contributes to the high level of wrinkling or low resiliency that is characteristic of fabrics made of these fibers if special finishes are not applied.

Protein fibers have better elongation and elastic recovery than cellulosic fibers; however, silk is not as good as wool within this category. Remember that cross linking is a special type of linkage between parallel molecular chains that increases the ability of a fiber to recover from pulling forces and from folding or creasing. Thus, wool, which has a high proportion of cross linking, has superior elasticity and resiliency. Wool fibers have a somewhat spiral arrangement of the molecular chains which further contributes to the ability of wool to extend, to recover from stretching, and to spring back into shape after creasing or wrinkling.

Man-made noncellulosic fibers have varying levels of elongation potential and elastic recovery. However, due to the relatively high orientation of molecules, *the majority of synthetic fibers have a high level of elastic recovery for the amount of elongation characteristic of these fibers.*

Glass fibers, in the mineral group, have a low elongation potential, but for that elongation they have *excellent elastic recovery.* Metallic fibers have low elongation and varying levels of elastic recovery, depending somewhat on the type of covering used on the fibers being made into yarns and fabrics.

Most people prefer clothing that stretches and gives with movement; therefore, high or medium elongation with good to high elastic recovery are desirable properties for wearing apparel. The two fiber categories providing the most elastic recovery and giving the best recovery from extension stretching are man-made noncellulosic and protein.

Garments worn for sports need exceptionally high elongation and elastic recovery. In order to provide this *power* and *action stretch,* manufacturers now use spandex fibers much more than the older elastomeric fiber, rubber. Since spandex has such high elongation and elasticity properties, a small amount of it can be combined with other fibers in a fabric to make a product with high amounts of power or action stretch.

For upholstery fabrics to have "give" under heavy use and yet maintain their shape, they need high elongation and high elastic recovery. Thus, the two fiber categories most suitable for upholstery fabrics are man-made noncellulosic and protein.

A fiber which has medium to high elongation, but low recovery, will soon stretch out of shape. Among the man-made cellulosic fibers, rayon especially has

---

QUICK QUIZ 7

1. Indicate the relative degree of elongation for each fiber category listed.

___	1. natural cellulosic	a. low
___	2. man-made cellulosic	b. medium
___	3. animal hair protein	c. high
___	4. man-made noncellulosic	
___	5. mineral	

2. Since elastic recovery is hard to evaluate unless elongation is also considered, correlate the two properties for the fiber categories.
   a. Natural cellulosic ____________________
   b. Protein ____________________
   c. Man-made cellulosic ____________________
   d. Man-made noncellulosic ____________________
   e. Mineral ____________________

****

ANSWERS QUICK QUIZ 7

1-a; 2-b; 3-c; 4-c; 5-a

2. a. low elongation and low elastic recovery
   b. high elongation and high elastic recovery, except silk, which has only moderate elongation and medium high elastic recovery.
   c. medium to high elongation and low elastic recovery; when wet, high elongation with low elastic recovery, except high-wet modulous rayon, which has been modified to have properties similar to cotton.
   d. high elongation for some, medium to high elongation for others, with generally high elastic recovery for all
   e. low elongation with elastic recovery, especially for glass

---

this tendency to lose shape because of its poor dimensional stability.

Comparison of Flexibility Among The Fiber Categories

A fiber which is flexible or pliable will be able to withstand continual bending forces. The *natural cellulosic, man-made cellulosic and mineral fibers have low flexibility,* which means if they are subjected to continual bending forces, the fibers will weaken and eventually break. Flax and ramie, both bast fibers, are especially stiff, brittle fibers. Blending flax or ramie with other, more flexible fibers reduces possible damage that may result from creasing or bending. Both *protein* and *man-made noncellulosic* or *synthetic fibers have good flexibility.*

The factors which contribute to high elastic recovery in fibers are also responsible for high flexibility. The nature of the molecular bonds is primarily responsible. The ability of the fiber molecules to slide over each other and return to their original relationship is dependent on the strength and linkage of the bonds. In wool, cross linkages contribute to high elastic recovery; in synthetics, strong hydrogen bonding returns the fiber to its original position.

Mineral fibers have low but adquate flexibility for some end uses. To produce more flexible mineral fibers, a glass filament of very small diameter, called *Beta* glass, was developed. Beta opened up new uses for glass fibers because of its increased abrasion resistance and flexibility.

Comparison of Dimensional Stability Among the Fiber Categories

A fabric that possesses good dimensional stability is able to retain its original shape, over time, under normal use and care. *Man-made cellulosic fibers,* except HWM rayon, lose strength and tend to elongate when wet; thus, these fibers *have low dimensional stability.*

Natural cellulosic fibers increase in strength when wet, so the fibers tend to be *dimensionally stable,* although they frequently shrink when laundered. Fiber stability and fabric shrinkage may appear to be inconsistent. However, while natural cellulosic fibers are relatively stable and do not stretch or shrink, fabrics made from these can shrink as a result of the processes used in yarn and fabric construction. Consequently, *natural cellulosic fabrics* which have not been treated for shrinkage, i.e. preshrunk, *are not dimensionally stable.*

In the protein fiber category, *animal hair fibers have low dimensional stability* and *silk has good stability.* The scales of the animal hair fibers contribute to the low stability because they become increasingly overlapped in the presence of heat, agitation and moisture. This process is called felting and causes the shortening and shrinking of the fiber length. Silk, on the other hand, is smooth and does not tend to shrink into a shorter form.

Synthetic fibers have excellent dimensional stability because they are thermoplastic and can be "heat set" into shape. However, incorrect finishing during manufacturing or the application of excessive heat following heat setting can damage thermoplastic fibers and cause them to melt or to shrink and change shape. With proper heat setting procedures and with normal use and care, the majority of the noncellulosic fibers are dimensionally stable.

In addition, man-made cellulosic fibers such as acetate and especially triacetate that have been modified can be heat set to some degree. Thus, these fibers also can be made relatively dimensionally stable.

Mineral fibers are not altered in shape by the normal temperatures and stress encounted in everyday use. Therefore, they have *high dimensional stability.*

REVIEW

1. Summarize the *natural cellulosic fibers* in terms of durability properties.
   a. tenacity ____________
   b. abrasion resistance ____________
   c. cohesiveness or spinning qualities ____________
   d. elongation potential ____________
   e. elastic recovery ____________
   f. flexibility or pliability ____________
   g. dimensional stability ____________

2. In the presence of water, most natural cellulosic fibers increase in ____________.
   The exception in the category is ____________.

3. Summarize the *protein fibers* in terms of durability properties.
   a. tenacity ____________
   b. abrasion resistance ____________
   c. cohesiveness or spinning quality ____________
   d. elongation potential ____________
   e. elastic recovery ____________
   f. flexibility or pliability ____________
   g. dimensional stability ____________

4. Silk and animal hair fibers such as wool have somewhat different properties, though they are both from protein sources. Give three physical differences between the two that account for many of their different properties. ____________
   ____
   ____

5. Match the degrees of durability found in *man-made cellulosic fibers.*

____	1. tenacity	a. high
____	2. abrasion resistance	b. medium
____	3. elongation potential	c. low
____	4. elastic recovery	
____	5. flexibility	
____	6. dimensional stability	

6. By altering the manufacturing process of rayon, the strength of the fiber can be greatly increased. What is the resulting fiber called that was originally used in tire cord?
   ____

7. With most man-made cellulosic fibers, wetting markedly (increases, decreases) durability, especially the properties of ________________________ and ________________________.

8. What type of rayon has been modified to provide increased wet strength and resulting properties similar to cotton? ______________________________

9. Summarize the *man-made noncellulosic fibers* in terms of durability properties.
   a. tenacity ______________________________
   b. abrasion resistance ______________________________
   c. cohesiveness or spinning qualities ______________________________
   d. elongation potential ______________________________
   e. elastic recovery ______________________________
   f. flexibility or pliability ______________________________
   g. dimensional stability ______________________________

10. Since synthetic fibers rate higher in durability than other fibers, you may ask why they are not used in all textile products. From personal experience, account for some of the limitations of synthetic fibers, particularly in wearing apparel. ______________________________
    ____
    ____
    ____

11. Match the durability measures of *mineral fibers* for the following properties.

____	1. tenacity	a. high
____	2. abrasion resistance	b. medium
____	3. elongation potential	c. low
____	4. elastic recovery	
____	5. flexibility	
____	6. dimensional stability	

12. What is done to metallic fibers during manufacturing to alter the properties evaluated above, and to make them more serviceable for apparel? ______________________________
    ____

13. A high measure of ______________________________ is necessary for good serviceability *only* when the ______________________________ demands it.

14. The serviceability of a particular fiber must be evaluated not only in terms of durability, but also in terms of what three other serviceability concepts?

_______________ _______________ _______________

****

REVIEW ANSWERS:

1. a. medium to medium high—flax tends to be medium high and cotton is medium
   b. medium
   c. high
   d. low—cotton elongates more than flax
   e. low
   f. low—especially flax
   g. high for fibers, but low for fabrics which have not been preshrunk
2. tenacity; jute, (which decreases in strength)
3. a-low to medium high—wool is low, silk is medium high
   b-medium to good—wool is higher than silk
   c-high
   d-medium high to high—wool has greater elongation potential than silk
   e-high, wool higher than silk
   f-high
   g-low to medium high—wool is low; silk is medium high
4. wool is a staple length fiber; silk is a filament
   wool has scales and fiber crimp; silk does not
   wool has cross linked molecular chains and a spiral molecular structure; silk does not
   wool has a lower molecular orientation than silk; silk molecules are more parallel and aligned to the fiber axis
5. 1-c, especially when wet, except for HWM rayon; 2-c; 3-b, except high for rayon when wet; 4-c; 5-b; 6-c, except for HWM rayon
6. high tenacity rayon. (Now replaced by steel, nylon and polyester cords, this fiber is far less common today.)
7. decreases; tenacity and dimensional stability
8. high wet modulus (HWM) rayon
9. a. medium high to high
   b. high
   c. high; especially when filament surfaces are modified through texturing, etc.
   d. medium to high
   e. high
   f. medium high to high
   g. high, when appropriately heat set
10. Synthetic fibers are not comfortable to wear in certain situations because of their low absorbency—they may feel hot and clammy in hot humid weather, and cold in cold weather. They may have a less lofty hand than natural fibers. They cause allergenic reactions in some individuals.
11. 1-a; 2-c; 3-c; 4-a; 5-c; 6-a
12. Fibers are coated with plastic or impregnated into plastic.
13. durability; end-use
14. comfort, care, and esthetic appearance

UNIT TWO

Part III

Comfort Properties of The Five Fiber Categories

Factors involved in fabric comfort include:
- air permeability or circulation—warmth or coolness
- heat conductivity or retention
- moisture absorbency and wicking
- static charge formation
- texture, roughness or smoothness of surface
- fiber and fabric weight

The comfort properties that influence the warmth or coolness of a fabric are related to how well its fibers are able to maintain a comfortable body temperature by allowing air to circulate or flow freely through the fiber or fabric and by how well the fiber absorbs, wicks or takes up moisture. The comfortable feel of a fiber next to the body is determined in part by the fiber's texture and chemical composition. Fibers that abrade the skin and cause an adverse reation due to either texture or chemical composition are said to have high allergenic potential.

The property of *absorbency* relates to the ability of a fiber to take up moisture and transmit it to the air for evaporation. Wicking, the ability of a fiber to transmit moisture along its length or around its circumference, can substitute for fiber absorbency and aid in body cooling by moving moisture away from the body. However, wicking may also transmit moisture from the outside fabric surface inward toward the body, which can create discomfort.

Absorbency also influences the conducting of electric charges; fibers with good moisture absorbency do not usually build up static charges. An exception is wool. While it has good absorbency, wool builds up static charges when humidity and temperature are low. When this charge of electrons builds up to a certain potential and is suddenly released, a static shock results. Thus, *electric conductivity* affects comfort.

Thermal or heat conductivity refers to the ability of a fiber to transmit heat energy along its length. Its opposite, thermal retention, refers to a fiber's or fabric's ability to hold heat. The physical structure of a fiber helps determine whether heat is conducted away or retained. With the increasing focus on energy conservation, demand for fabrics that retain heat will increase. For example, as homes and offices are maintained at lower temperatures, warmer indoor clothing will be needed. Some special fiber modifications and finishes that increase insulation potential have already been developed for use in such products as draperies.

The weight of a fabric, determined by fiber density, yarn type, and compactness or looseness of the fabric structure and any finishes applied to it, also affects thermal conductivity. Fiber with low densities can be used to create fabrics which are lightweight yet warm, by structuring the fabric so that insulating air is trapped in the fibers and yarns.

Comfort Properties of the Natural Cellulosic Fibers

The *absorbency of natural cellulosics is high* because of the somewhat amorphous molecular structure and the central canal or lumen through which moisture travels or is transmitted. Although flax and ramie have high molecular orientations compared to the more amorphous arrangement of cotton, both are still very absorbent fibers.

QUICK QUIZ 8

1. A fiber with high thermal conductivity is generally comfortable in __________ weather; a fiber with high thermal retention is desirable in __________ weather.
2. A fabric with a rough surface texture tends to __________ the skin.
3. A determining factor in the allergenic potential of a fabric is the fiber's __________ and __________ composition.

****

ANSWERS QUICK QUIZ 8

1. warm; cold
2. irritate or abrade
3. texture, chemical

While natural cellulosic fibers absorb moisture readily, they do not generally transmit moisture along the fiber length except for flax. Therefore, *natural cellulosics, with the exception of flax, have low wicking ability.*

Natural cellulosic fibers are one of the heavier groups, with a *density of 1.5.* (You need not memorize fiber densities, but do note them on Matrix V to provide a basis for comparing high and low fiber densities.) Products made from cotton fibers, for example, have a heavier weight than comparably structured fabrics made from lower density fibers.

Comfort Properties of the Protein Fibers

Protein fibers have excellent moisture absorbency; recall from Unit I the discussion about the types of absorbency. Animal hair fibers such as wool absorb moisture in water vapor form without feeling wet to the touch, unless the fabric becomes saturated. This special type of absorbency of protein fibers is known as hygroscopicity.

Since wool can absorb a lot of water without feeling damp or clammy, it is an especially desirable fiber for wear in cool, humid climates. The absorbed water vapor then assists in serving as an insulator. The physical structure of wool is partly responsible for its ability to absorb so much moisture. View again the illustration of the cross section of wool in Figure 2-8. Although the epidermis actually repels liquid due to a layer of wool grease called lanolin, the cortex absorbs large amounts of moisture from the air in vapor or gaseous form. In addition to the cortex, which makes up the biggest portion of the fiber, the amorphous, spiral molecular configurations of animal hair protein fibers and their chemical composition also contribute high moisture absorbency.

While silk does not have scales or a cortex like the animal hair proteins, or an amorphous molecular orientation, it still absorbs moisture readily. This is because silk comes from a protein source, like wool, and therefore has a similar chemical composition which allows moisture vapor to be absorbed internally. As a group, protein fibers are known for their absorbency and can be referred to as water loving or hydrophilic.

Just as moisture is not conducted along the length of protein fibers, neither is electricity. Consequently, *wool and silk are poor conductors of electricity,* and a wearer can receive shocks from static electricity, especially in dry weather.

Based on the above discussion, wool and silk seem to have desirable properties contributing to the serviceability concept of comfort, especially in cool, humid weather. Yet, wool in particular is not considered comfortable by everyone because the harsh surface texture and/or chemical composition of some woolens can abrade sensitive skin, causing itching or rashes. Also, static charges that develop in dry cold environments are objectionable.

The protein fibers have a medium density of 1.3 grams per cc, which is less than the density of natural cellulosic fibers. In two fabrics identical in both yarn and fabric structure, the one made of cotton fibers would weigh more.

Comfort Properties of the Man-Made Cellulosic Fibers

The comfort of man-made cellulosic rayon fibers is very similar to that of natural cellulosic fibers because both contain cellulose in similar form—the regenerated cellulose in rayon is changed physically, but not chemically.

Acetate and triacetate, however, are made from modified and chemically changed cellulose and are thermoplastic and less absorbent than the natural cellulosics and rayon. The cellulose triacetate is the least absorbent of the three man-made cellulosics, so it would be less comfortable to the wearer.

Recall that rayon loses tenacity when wet and stretches out of shape. Rayon molecules can absorb so much moisture that you can see an obvious difference in the length of rayon draperies on a damp day. This "elevator" ability of rayon to stretch and shrink depending on moisture absorption is not apparent in other fabrics. Rayon is one of the most absorbent man-made fibers.

Electrical conductivity is associated with absorbency, since moisture is necessary for electrostatic charges to become adequately grounded. Among the man-made cellulosic fibers, rayon *conducts electricity* the best; *triacetate,* because it has the least absorbency of the three man-made cellulosics, *tends to build up the most static charges.*

Absorbency, related to the molecular structure of a fiber, also affects thermal retention or conductivity. Fibers with an amorphous molecular arrange-

---

QUICK QUIZ 9

1. How would you rate the general comfort potential of natural cellulosic fibers? (high, medium, or low) List three properties that support this rating. ____________________
____
2. Although a cotton fabric would weigh more than a comparable wool fabric, why would the wool have more thermal retention and consequently be warmer than the cotton fabric?
____
____
3. Therefore, fabric weight, which depends in part on fiber ____________ is not the determining factor in how warm or cool a fabric will be for the wearer.
4. The relationship of fiber density to fabric warmth depends on the amount of thermal retention the fibers possess. Fibers that have low density and high thermal retention are said to provide warmth without ____________.
5. Of the three generic classes in the man-made cellulosic category, which fibers are most comfortable? ____________
6. Based on their absorbency abilities, acetate and triacetate are not as comfortable as fibers in the ____________________ category; however, they tend to be more comfortable and absorbent than many of the synthetic fibers, perhaps excluding acrylics.
7. Like most natural cellulosic fibers, man-made cellulosics absorb moisture in a manner similar to (cotton, flax).

****

ANSWERS QUICK QUIZ 9

1. high; any three: high absorbency, good electrical conductivity, high thermal conductivity and low allergenic potential
2. wool fibers trap air and absorb water vapor in the fiber structure
3. density
4. weight
5. rayon
6. natural cellulosic
7. cotton

---

ment usually have high absorbency and the ability to transmit or conduct heat away from the body. Since rayon has high absorbency like natural cellulosic fibers, it is also likely to have *high thermal conductivity.*

Rayon has the same *high density of 1.5* as natural cellulosics; however, the additional chemical modification of acetate and triacetate creates fibers with *medium density of 1.3.* For two fabrics of equal thickness and identical structure, then, a rayon fabric would have more weight than an acetate fabric and would be warmer.

Since the chemicals added to cellulose for manufacturing man-made cellulosic fibers do not create a chemical sensitivity for most people, *man-made cellulose fibers have a low allergenic potential* similar to that of the natural cellulosic fibers.

Comfort Properties of the Man-Made Noncellulosic (Synthetic) Fibers

Generally, *synthetic fibers are not considered comfortable to wear* in hot weather because they have high orientation and high crystallinity and, thus, low absorbency. They do not have many amorphous areas with low crystallinity that allow for moisture absorption in the spaces between unoriented molecules. In recent years, however, textile researchers have improved upon the absorbency of some synthetic fibers, most notably polyester. Some modified polyesters absorb moisture much more readily to allow for increased comfort in apparel.

To cut down static electricity, especially during cool, dry weather, a fabric softener can be added to the rinse cycle or put in the dryer. Other static controls include hanging the garment in a steam filled bathroom, running a damp sponge over the fabric surface, or spraying it with an antistatic finishing compound. These methods give only temporary relief but are useful.

Since the *absorbency of synthetic fibers is low,* you would expect that their *thermal conductivity is low* because neither moisture nor heat is transmitted away from the body. Any wicking that might occur would increase thermal conductivity to some degree.

Low heat conductivity combined with low absorbency make for uncomfortable garments in hot, humid weather. Olefin fibers have very low absorbency, but the wicking property inherent in olefin allows for the use of these fibers in apparel. Because olefin fibers are light weight and wick readily, they are often used in activewear such as leotards and tights. Olefin (sometimes referred to as polypropylene) activewear allows moisture to be drawn away from the body and to quickly evaporate. You will learn later how fabric structure can partially alleviate some of the discomfort of other synthetic fibers that have not been altered or do not wick.

The chemical composition of synthetic fibers varies, so some individuals may be allergic to certain fiber polymers. However, discomfort while wearing synthetic fibers is more often due to low absorbency and low thermal conductivity than to chemical allergenic reactions.

Synthetic fibers are often manufactured in relatively smooth filaments, thus typically creating little discomfort from skin abrasion or irritation, as could be expected from rough-textured animal hair fibers. Nylon, an exception to the abrasion factor, is such a tough fiber and so resistant to abrasion, that it tends to abrade other fibers and surfaces, including the skin. For this reason, some individuals prefer not to wear nylon lingerie.

Because the chemical composition of synthetic fibers differs in each generic classification, their densities vary from polyester's medium density of 1.38 to the very low densities characteristic of olefin, nylon and acrylic. But, none has a density as high as the natural cellulosics; therefore, a cotton fabric of equal structure, thickness and bulk would weigh more than one of a synthetic fiber. Cotton density is 1.54. As a whole, synthetic fibers are known for low densities.

Nylon and polyester, two of the older generic classes of synthetic fibers, are representative of the difference in densities within the synthetic category. Regular nylon has a density of about 1.14, which is considered to be low. Qiana nylon, which was an exception to the nylon group, has a density of 1.03, which is even lower than regular nylon. Based on density and absorbency properties *only,* a nylon fabric would be more comfortable than unmodified polyester which is uncomfortable on hot, humid days due to greater fabric weight and less absorption of perspiration. While synthetic fibers can be quite warm to wear, especially in certain types of weather, they are not associated with the high thermal retention properties of animal hair fibers like wool. There are three reasons for this:

1) The scales, cortex and medulla trap still air, which is an excellent insulator.
2) Wool is hygroscopic and water vapor is absorbed internally; this creates heat energy.
3) The molecular spirals and amorphous oriented areas hold air, and the external scales prevent air transfer through the fiber.

QUICK QUIZ 10

1. Although the low absorption of synthetics tends to limit their comfort in wearing apparel, a fabric with very low absorbency has some advantages. Cite at least two advantages:
   ____
   ____
2. Moisture travels along the outside length of certain hydrophobic fibers; therefore, synthetic fibers typically have high __________ ability.
3. Cut a small sample (1 1/2″ × 2″) from fabric 29 and drop a spot of food coloring or ink on the surface. Observe the wicking properties exhibited by the polyester fibers. How can wicking contribute to the comfort of a synthetic fabric?
   ____
   ____
4. Based on the preceding discussion, fabrics made of synthetic fibers are likely to be most satisfactory to wear on (cool, warm) (dry, humid) days. Why? __________
   ____
5. Despite the disadvantages from a comfort viewpoint, synthetics continue to be popular fibers in apparel and home furnishings because of their ease of care and durability. List three properties relative to durability that contribute to their use.
   ____
   ____

****

ANSWERS QUICK QUIZ 10

1. liquid stains are not readily absorbed, fabrics are easy to launder and dry quickly, and they travel well
2. wicking
3. moisture can be wicked away from the body and passed into the air for evaporation
4. cool, humid; in cool weather, moisture absorption is not as crucial as in warm weather, and humidity in the atmosphere reduces the development of static electricity
5. any three: high elasticity and elongation; tenacity; abrasion resistance; dimensional stability; flexibility or pliability

Synthetic fibers can be textured, shaped or crimped to trap air and make them good insulators. Acrylic fibers, a synthetic classification specially engineered to be wool-like; are usually cut in staple length and have high bulk potential. Another characteristic of acrylics contributing to their insulative properties is their dog bone or irregular cross section shape which allows more air to be trapped as the fibers are spun into yarns. Refer to Figure 2.14.

The synthetics are sometimes referred to as having ''bulk density,'' since the fibers are often bulky but very lightweight for the volume they occupy. The reason that the lightweight, bulky fibers are so warm is that they trap still air which is an excellent insulator.

Individuals are frequently willing to accept low absorbency and less comfort in synthetics in order to have easy-care properties lacking in more absorbent cellulosic fibers.

Comfort Properties of the Mineral Fibers

The absorbency and wicking properties of mineral fibers are low, making for uncomfortable garments. However, since glass is not used in apparel, its zero absorbency is not a disadvantage. In fact, this nonabsorbency is actually an advantage in such end uses as draperies, because glass draperies are very quick drying and can be rehung immediately after laundering.

Heat conductivity and electrical conductivity of mineral fibers, especially glass, are low because electrons cannot move freely along the fiber surfaces. With low heat conductivity, the mineral fibers make good insulators. The *allergenic potential of glass is especially* high because small particles of glass within the fiber can break and irritate the skin, causing intense itching and burning. Anyone who has worked with "angel hair" made of spun glass, knows the sensations that glass filaments can create as a result of handling.

Although a polymer coating around a metallic fiber core increases the comfort potential of metallic yarns, the fabric can still have a harsh texture. While metallic fabrics are not noted for pliability and comfort, they do lack the allergenic potential of glass fibers.

The density of mineral fibers is high; in fact, it is the highest of the fiber categories. Glass and aluminum, the metal most used in metallic yarns produced today, have similar densities ranging from 2.5 to 2.7. Fabrics made of these fibers, therefore, are comparatively heavy and not very comfortable to wear. Glass fiber draperies may require special rods for large windows in order to bear their excess weight.

REVIEW

1. Rate the comfort properties for fibers in the *natural cellulosic* category, and comment on their overall comfort.
   a. absorbency ______
   b. wicking ______
   c. electrical conductivity ______
   d. thermal conductivity ______
   e. allergenic potential ______
   f. texture ______
   g. density ______
   h. overall comfort ______

2. Rate the comfort properties of animal hair and silk *protein* fibers.

	animal hair	*silk*
a. absorbency	______	______
b. wicking	______	______
c. electrical conductivity	______	______
d. thermal conductivity	______	______
e. allergenic potential	______	______
f. texture	______	______
g. density	______	______
h. overall comfort	______	______

3. Wool has both desirable and undesirable comfort features. How do wool fibers contribute to comfort in wearing apparel? ______

4. Compare the comfort properties of *man-made cellulosic* fibers. Rate regenerated cellulose (rayon) and the modified cellulosics (acetate and tricetate) separately.

	rayon	*acetate-triacetate*
a. absorbency	______	______
b. wicking	______	______
c. elecrical conductivity	______	______
d. thermal conductivity	______	______
e. allergenic potential	______	______
f. texture	______	______
g. density	______	______
h. overall comfort	______	______

5. The man-made cellulosic fiber, rayon, has comfort properties similar to fibers in the ______________________________ category.

6. Rate the comfort properties of *man-made noncellulosic or synthetic* fibers.
   a. absorbency ______________________________
   b. wicking ______________________________
   c. electrical conductivity ______________________________
   d. thermal conductivity ______________________________
   e. allergenic potential ______________________________
   f. texture ______________________________
   g. density ______________________________
   h. overall comfort ______________________________

7. Synthetic fibers are not considered exceptionally comfortable, yet they are widely used due to properties that contribute to the serviceability concepts of ____________________ and ______________.

****

REVIEW ANSWERS:

1. a. high; the fibers are hydrophilic
   b. low; except for flax which is high
   c. high
   d. high
   e. low
   f. somewhat smooth
   g. comparatively high, 1.5
   h. high, especially in warm weather

	animal hair	*silk*
2. a.	high, hygroscopic properties	high
b.	low	low
c.	low	low
d.	low (high thermal retention)	low
e.	high	low
f.	harsh or rough	smooth
g.	medium	medium
h.	high, except for allergenic potential	high

3. They are good insulators and hygroscopic, contributing to comfort in cold weather.

	rayons	*acetate-triacetate*
4. a.	high	medium to low
b.	low	low
c.	high	medium to low
d.	high	medium
e.	low	low
f.	smooth	smooth
g.	high, 1.5	medium, 1.3
h.	high, good for warm weather wear	medium

5. natural cellulosic

6. a- low; most synthetics are hydrophobic
   b- high
   c- low
   d- low
   e- low, except for nylon which is abrasive to sensitive skin
   f- smooth, unless the filament has been textured or cut into staple lengths and structured into rough-textured yarns
   g- low to medium
   h- generally rather low, particularly in warm weather
7. durability; ease of care

Since mineral fibers are not used where comfort is a major factor, their comfort evaluation has been omitted.

UNIT TWO

Part IV

Comparison of Care Considerations for The Five Fiber Categories

Care has been defined as any activity which maintains the original appearance of a garment or other textile products. To maintain a new look, textile products should receive proper care during three periods: during wearing or using; during laundering or dry cleaning, including accompanying stain removal, bleaching, or brightening; and during storage. Of the three processes, care during laundering and dry cleaning appropriately receives the most attention, since that is when the most damage can occur to alter the durability and appearance of textile products. A common term used for care is maintenance.

Fiber properties involved in care or maintenance include dimensional stability; wrinkle recovery or resilience; elongation and elastic recovery; chemical reactivity; thermal or heat reactions; biological, light, and age resistence; plus reactions to other environmental conditions. All of these properties may be modified to some degree by the application of various finishing compounds. Some properties may also be affected and altered by the procedures used in making the yarns, as well as the methods used in fabric construction.

Resiliency and crease recovery are important in the maintenance of a fresh-looking, attractive and easy-care fabric. Dimensional stability is important in determining if the fabric will stretch or shrink out of shape during use and care. Fabric structure and finish can noticeably affect dimensional stability.

The chemical reactivity of the fiber helps determine what types of detergents and other laundry additives or cleaning solvents should be used, and how the fiber reacts to these materials. Thermal reactions are involved in the choice of water temperatures used in laundering, the temperatures of cleaning solvents, and related drying or ironing temperatures. Added compounds and dyestuffs are important considerations in care—they affect choice of cleaning agents and temperatures required. Thermal reactions are responsible for the flammability of fibers.

Biological properties determine a fiber's reactions to insects and microorganisms, such as moths, silverfish, fungi or mildew and bacteria. The reactions of fibers to light, smog and other environmental conditions are important in identifying care procedures recommended for storage.

The following discussion is directed at the properties and behavior of fibers in relation to care. It is extremely important to recognize that the other components of a final textile product can noticeably modify or alter fiber behavior during care. Remove Matrix VI from the back of the book and fill it in as you study care properties across the fiber categories.

Care Procedures for the Natural Cellulosic Fibers

The *resiliency of cellulosic fibers is generally low;* however, it can be increased by the application of suitable finishes. Unless finished, low resiliency is a major disadvantage during wear and use of natural cellulosic products, since they wrinkle easily and do not retain their original appearance without ironing.

The *dimensional stability of natural cellulosic fabrics also tends to be low unless a shrink-resistant finish* such as Sanforized® has been applied to the fabric. While dimensional stability of fabrics made of natural cellulosic fibers is satisfactory during wearing and use, it tends to be a problem during laundering.

The type of finish applied to a fabric, combined with the chemical reactivity of the fibers present, determines the care procedures necessary for cleaning and maintaining textile products. For example, because *natural cellulosic fibers react favorably with alkalies,* they should also respond favorably to laundering with built or heavy duty detergents.

On the other hand, *natural cellulosics react unfavorably with acids* and are even sensitive to certain acidic foods that might be spilled on them. Consequently, orange juice or tomato stains should be removed immediately from cellulosic fabrics to avoid weakening and permanent stain and damage.

Natural cellulosic fibers are quite resistant to most cleaning solvents as long as they are not acidic in nature. Although dry cleaning solvents are acceptable, many people prefer to home launder fabrics made from natural cellulose. Cotton can be maintained very satisfactorily by laundering, which is considerably less expensive than dry cleaning.

Natural cellulosic fibers can be successfully bleached by chlorine or oxygen (perborate) bleaches, providing the bleach is correctly diluted prior to laundering. Since *natural cellulosics are not sensitive to bleaches,* chlorine would be the best choice for badly soiled cottons, because it has more bleaching and whitening power than oxygen or perborate bleach. However, some *finishes* used on cotton and other natural cellulosic fabrics may react negatively with chlorine bleaches, so it is important to read all care labels attached to the product. Some warn the consumer not to use chlorine bleach.

Natural cellulosics are not sensitive to heat normally encountered in laundering; therefore, hot water can be safely used for optimum cleaning power without fear of damage to the product. You can tell by the thermostat setting on an iron for cotton and linen that the natural cellulosics have a relatively high heat tolerance to pressing. However, an excessively hot iron can scorch a cellulosic fabric. Again, finishes can change fiber reactions and alter safe temperatures that may be used in laundering or ironing. Because natural cellulose burns rapidly, with little or no ash, it is necessary to adhere to sensible precautions when these fibers are around flame.

Constant *exposure to sunlight will eventually cause yellowing and weakening of natural cellulosic fibers.* Therefore, to maintain their appearance, fabrics of these fibers should be stored in a sun-free, low light, and dry and cool environment.

Another storage consideration for natural cellulosic products is to prevent damage from organisms, such as mildew and silverfish, which attack cellulose. The one exception is ramie, which is resistant to microorganisms, insects and rotting. See Figure 2.16 for a comparison of mildew and silverfish, two pests which thrive on dampness, warmth and darkness.

There is an inconsistency in this situation relative to light and storage. Since sunlight damages natural cellulose over time, cellulosics should be stored in low light or dark conditions. Yet darkness encourages the growth of microorganisms, mildew, and insects like silverfish.

Therefore the best solution for storing cellulose fabrics is to make certain they are kept in a dry, cool area to inhibit the growth of biological organisms, with protection from direct sun. A low level of light is acceptable. Natural cellulosics also need to be stored clean and free from added starch because soil, stains, perspiration and starch all tend to attract biological organisms.

Care Procedures for Protein Fibers

You can tell from the wool garments you own that *wool wrinkles less and is more resilient than cotton* fabrics not given an easy care finish. *Silk has moderately good resiliency* though not as high as wool because its physical structure lacks wool's fiber crimp and cross linked molecules that contribute to high resiliency.

The *dimensional stability of wool fabrics is low* because wool fibers shrink through felting caused by fiber scales overlapping, matting and tangling. Wool can also be seriously damaged by strong alkaline detergents. In fact, it can be completely dissolved in a boiling alkaline solution; thus if laundered, it is necessary to use unbuilt detergents and rinse well. Water temperatures used in the care of wool fabrics must be carefully controlled. Hot water should not be used, and wash and rinse tempera-

Figure 2.16. These biological pests attack cellulosic (starch-based) fibers.

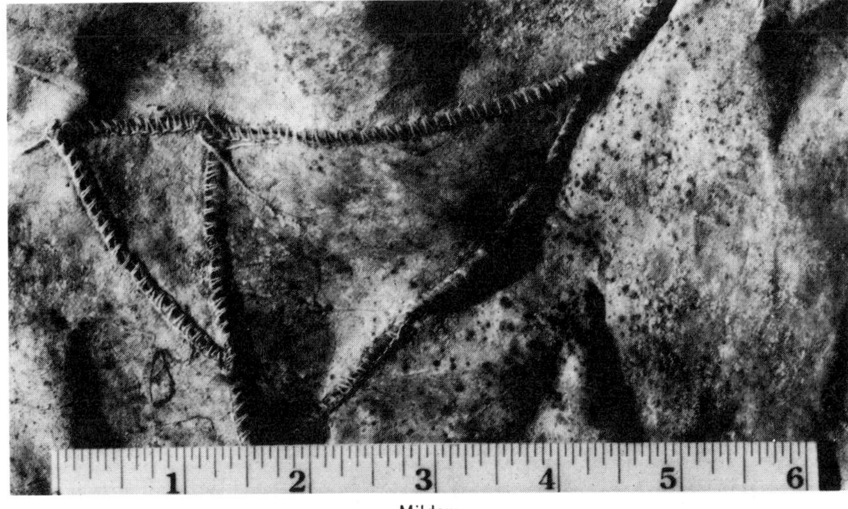

Mildew

Silverfish

tures should be consistent, since changes can cause fiber felting. Warm or cool water may be used if the same temperature is maintained for all laundering steps. In addition, agitation should be kept to a minimum.

The thermostat setting on irons indicates lower temperatures for wool and silk than for cotton or linen. Thus, one can conclude that *protein fibers have more heat sensitivity than natural cellulosics,* but they are still nonthermoplastic and scorch rather than melt when exposed to excessive temperatures. Silk scorches even more easily than wool, turning yellow if pressed with too hot an iron.

Wool becomes harsh, brittle and shiny in the presence of dry heat; therefore, when pressing wool products a steam iron or a damp press cloth should be used, but used in moderation to avoid felting.

Protein fibers are less flammable than cellulosics, as observed in the burn test conducted earlier in this unit. Protein fibers do not create as high a fire hazard as natural cellulosic fibers because they tend to be self extinguishing and do not support combustion as readily as cellulose fibers; for review refer to the Burn Matrix presented earlier. Thus, protein fibers are more sensitive to dry heat than the natural cellulosics, but are not as flammable.

In addition to maintenance procedures required during cleaning, pressing and use, storage of fabrics made of protein fibers demands special care to avoid possible damage from insects and microorganisms, as shown in Figure 2.17. Moths will eat wool and other animal hair fibers because of keratin, a protein present which contains sulfur. Carpet beetles also destroy both hair fibers and silk. Precautions that

QUICK QUIZ 11

1. Natural cellulosic fabrics have (good, poor) launderability, but require ironing following laundering due to low fiber ______________ unless proper finishes have been applied.
2. Storage of natural cellulosics presents no real difficulties either, since closet, chest or trunk conditions are usually satisfactory if the environment is ______________ and ______________.
3. Since natural cellulosics are not sensitive to alkalies or to extremely high temperatures, fabrics of these fibers can be safely laundered in (built, unbuilt) detergents using hot water to restore the appearance of the garments. When are hot water and strong detergents especially necessary? ______________
4. The preferred method for care of all protein fibers is (laundering, dry cleaning).
5. If silk products are home laundered, warm or cool water, not hot, and ______________ detergent are recommended.
6. Strong bleaches are damaging to all protein fibers. If bleaching is required for whitening, ______________ bleach should be used.
7. Laundering protein fibers in hot water stiffens and weakens them over time. What other problem can excessively hot water or temperature changes cause in wool products? ______________

****

ANSWERS QUICK QUIZ 11

1. good; resiliency
2. dry; cool
3. built; when fabrics are badly stained or have greasy stains
4. dry cleaning
5. unbuilt
6. oxygen (perborate)
7. shrinkage from fiber felting

help prevent these damages to protein fibers include: use of insect poisons such as moth crystals; storage of fabrics free of soil, perspiration and food residues; storage in cool dry environments; and presence of fabric finishes with some type of insect repellent.

Microorganisms attack protein products, especially stains left in the fabric. Mildew will form and damage wool stored in a humid environment, but it is more likely to attach to natural cellulosic fibers than to protein fibers. *If wool is properly stored* with adequate provisions to inhibit moth and mildew damage, *age has no destructive effect.* However protein fibers have low light resistance, and they are weakened and yellowed by continued exposure to sunlight and ultra-violet rays, so aging is a factor in certain end uses, like draperies.

On the other hand, silk requires even more careful handling and storage to last through the years. Oxygen from the atmosphere causes a gradual decomposition of silk and certain types of silk may crack at fold lines. To avoid damage, silk should, ideally, be stored flat in sealed containers away from air.

Plainly, care procedures for protein fibers tend to be more expensive and demanding than those for the natural cellulosics. Both wool and silk products require dry cleaning or very careful hand laundering and careful storage to prevent damage from aging and moth larvae or carpet beetles.

Care Procedures for Man-Made Cellulosic Fibers

The discussion of care procedures for man-made cellulosic fibers is designed to provide specific information on care of regenerated man-made cellulose fiber—rayon—and modified cellulose fibers—

Figure 2.17 Protein fibers are damaged by carpet beetles and noth larvae.

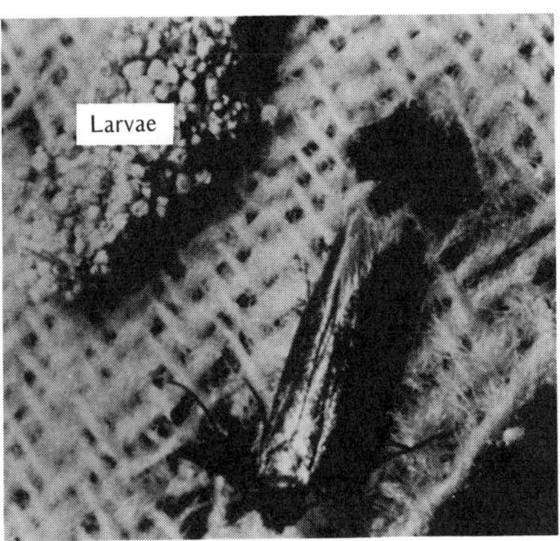

Moth and Larvae

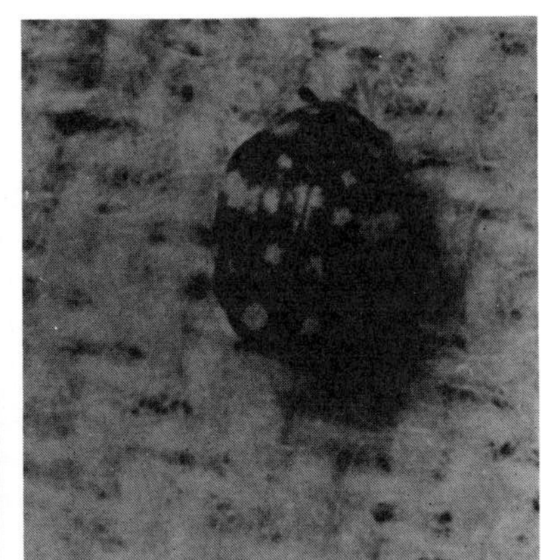
Carpet beetle

acetate and triacetate. Due to differences in their chemical structures, there are differences in the care procedures required.

Since rayon and cotton are chemically pure cellulose, the fiber properties which affect their care procedures are similar, with the notable exception of differences in their wet strength and resulting differences in dimensional stability. Due to the physical differences in length of the cellulose molecules in the two fibers, cotton increases in strength when wet and is not damaged by laundering; rayon (except for the HWM type) decreases in strength when wet. Thus, rayon tends to stretch out of shape and lose strength considerably more than cotton. *Therefore,* rayon fabrics exhibit low dimensional stability.

High wet modulus (HWM) rayon, modified during manufacture to increase molecular chain length and make it more like natural cellulosic fibers, can be cared for by the same procedures as the natural cellulosics.

When rayons and acetates need whitening for the sake of appearance, either chlorine or oxygen (perborate) bleaches may be used if correctly diluted. However, strong concentrations of bleach will weaken and damage the fibers. Actually, man-made cellulosics, especially acetate and triacetate, keep their whiteness quite well with a minimum of bleaching, so it would probably be wise to choose a milder oxygen or perborate bleach rather than a chlorine one to assure long life.

Acetone, the basic ingredient in fingernail polish remover, can completely dissolve acetate. Triacetate is also affected, but not as rapidly. Another reason to be aware of fiber sensitivity to acetone is that it is often used as a "spotter" or cleaning agent in many home stain removers. In fact, acetone is especially effective on ink stains. However, labels on stain removers must be read very carefully to determine if they are safe to use on fabrics made of acetate and triacetate fibers.

Rayon has good resistance to all dry cleaning solvents, and acetate and triacetate have good resistance to petroleum solvents; thus, as many manufacturers recommend, they can be successfully dry cleaned rather than laundered to keep their body and appearance for a longer period of time.

Another reason many people prefer dry cleaning to laundering of rayon and acetate is that the fibers are noted for their low resiliency, so washing tends to set wrinkles which require a lot of ironing to remove unless the fabric has a wrinkle resistant, easy care finish. However, acetate does not accept easy care finishes well, so it especially benefits from dry cleaning, since acetate becomes very wrinkled during laundering.

Triacetate and acetate are both made from modified cellulose, but the chemical modifications are different enough to result in acetate and triacetate being separate generic classifications. Consequently, different fiber properties are present, especially related to fabric care.

Because it has lower moisture absorption and is more resilient than rayon or acetate, triacetate launders considerably easier than acetate and has many easy-care properties of the noncellulosic synthetic fibers. Triacetate also has higher heat resistance than acetate; it is thermoplastic and can be "heat set" to hold permanent pleats and creases. It also has more resistance to sunlight than either rayon or acetate.

Acetate is damaged by sunlight exposure and atmospheric gases, yet shiny acetate fabrics are more resistant to sunlight than delustered or dull ones, since they tend to reflect light away from fabric and do not absorb as much light.

Air tends to age acetate products over time, and fume or gas fading can result. Tests have shown that fume fading is caused by nitrogen oxide, which comes from gas heaters, industrial furnaces and chemical plants. Consequently, acetate draperies in houses, theaters, or auditoriums may show color loss even if not exposed to sunlight, although fading inhibitors developed to curb acetate fume fading may reduce the danger. However, *acetate* still becomes weak during general use and is said to have *poor aging resistance.* It should be stored away from light and open air.

In addition to fading problems, another factor that influences storage is a fiber's resistance to biological organisms such as mildew and silverfish. These attack both cotton and rayon, since both are made of cellulose.

By contrast, acetate and triacetate are thermoplastic fibers with properties somewhat like those of the

---

QUICK QUIZ 12

1. The modification of rayon that increases its wet strength also makes it more resistant to sunlight. What is this modification called? ______________
2. Why are triacetate fabrics usually dimensionally stable? ______________
3. Now, make certain that Matrix VI has been completed for the care properties of the man-made cellulosic fibers. Cite the general cleaning procedures that should be used for rayon, acetate, and triacetate fibers.
   a. Rayon: ______________
   b. Acetate: ______________
   c. Triacetate: ______________
4. Explain how man-made cellulosic fibers differ in flammability and heat sensitivity.
   a. Rayon: ______________
   b. Acetate: ______________
   c. Triacetate: ______________

****

ANSWERS QUICK QUIZ 12

1. high wet modulus rayon
2. they can be heat set to maintain shape and appearance
3. a. dry clean; or launder with mild detergent, avoiding long soaking periods and agitation when wet
   b. dry cleaning preferred, as wrinkles from laundering are difficult to press out; if laundered, use a mild detergent and handle so as to avoid excessive wrinkling
   c. launder normally, no special problems except that high temperatures should be avoided
4. a. burns readily and scorches at medium to medium-high temperatures
   b. melts at medium-low temperatures and burns readily
   c. melts at slightly higher temperatures than acetate; burns readily

synthetics or man-made noncellulosic fibers. Therefore, *acetate and triacetate tend to have good resistance to damage from biological pests.*

Care Procedures for Man-Made Noncellulosic (Synthetic) Fibers

Synthetic fibers generally produce easy-care products, due in part to their high resiliency; they *are wrinkle resistant* and require little or no ironing. Another property which contributes to their ease of care is thermoplasticity, which allows the fabric to be heat set into desired shapes. Not all man-made noncellulosics are truly thermoplastic, but most synthetic fibers used in consumer goods are heat sensitive so that they can be heat set.

A fabric that is heat set tends to retain its shape during laundering as long as the original heat setting temperature and time are not exceeded. Thermoplastic fabrics also tend to maintain shape and appearance during wearing; therefore, they require minimum care because of their properties of resiliency and dimensional stability.

The low absorbency of synthetic fibers, which makes them uncomfortable to wear in warm or hot weather, is actually an advantage in caring for them. This low absorbency contributes to ease of fabric care in three ways. They are quick drying; they do not absorb water-borne stains; and they do not lose strength or shape when wet.

Although most man-made noncellulosic fibers do not absorb water-borne stains, many have an affinity for oily stains, especially cooking and body oils. Consequently, fabrics made of synthetic fibers should be cleaned frequently to prevent absorption and buildup of oils which tend to get set and are difficult to remove. The "ring around the collar" stain, often hard to remove from polyester and cotton blend fabrics, is a typical oily stain.

The proper way to clean a synthetic fiber depends on its chemical reactivity. Thus, the choice of detergents and bleaches might vary with the different generic classes of fibers, although all *synthetic fibers react favorably to mild alkalies* and can be satisfactorily laundered—in fact, laundering is preferred for most of these products.

Since built detergents are necessary for removing stubborn oily stains, especially if they are allowed to become set, it is important to know which generic classifications cannot withstand built detergents. Most synthetic fibers used in apparel and home furnishings can be laundered as necessary with built detergents and all are resistant to unbuilt detergents. To be sure, though, you should check for special instructions on the care labels which are required by law.

Synthetic fibers are not damaged as noticeably by strong detergent or bleach as by excessive heat, as can be seen in Figure 2.18. The highest heat levels most textile products are exposed to under normal

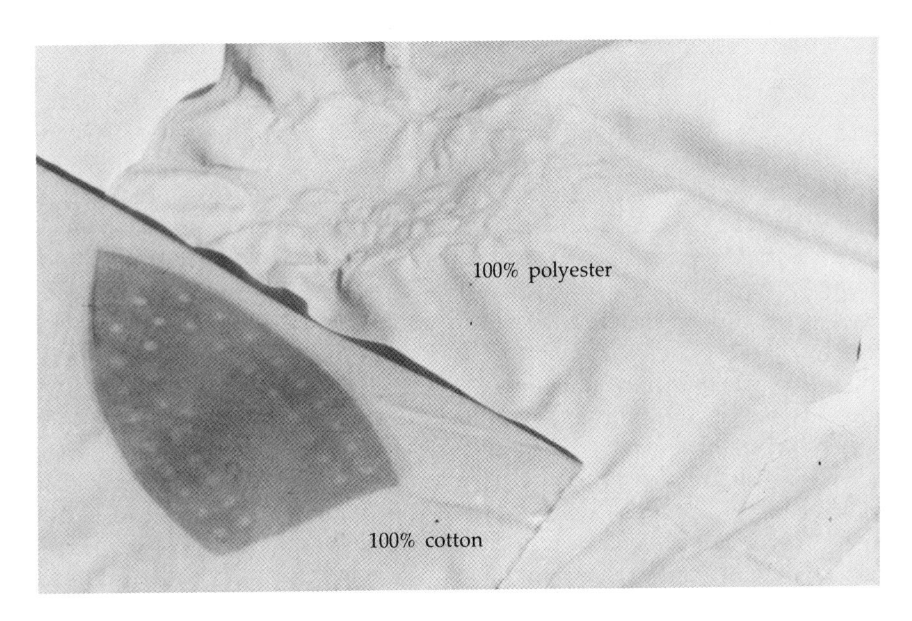

Figure 2.18. Synthetic fibers melt with excessive heat, while natural fibers scorch.

home care conditions occur in ironing and tumble drying. On many appliances a thermostat guide for the different fibers is given; therefore, the recommended ironing and drying temperatures need not be memorized. Rather, you should remember that very low heat tolerance levels generally are *associated with olefin, vinyon and modacrylic* generic classes. However, these fibers will probably not create ironing problems because they most often are used in products that do not require ironing, like carpets, rugs, wigs, hairpieces and special vinyl films.

As a generic class, *nylon has the highest heat tolerance* of the synthetics commonly used in apparel items, but still is pressed at a lower heat than natural cellulosic fibers because it is a heat sensitive synthetic. Polyester and acrylic fibers also have relatively good heat resistance and all three can be laundered satisfactorily in medium to hot water with a variety of detergents, and dryer dried.

Also, *synthetic fibers are resistant to bleach if it is correctly used and diluted.* Nylon, polyester, spandex and vinyon are somewhat more resistant to bleaching than olefin, acrylic and rubber fibers. To ensure that no damage will be done to fibers which are somewhat less resistant to bleach, oxygen bleaches could be used because these are milder and safer for all fibers.

Synthetic fibers are generally resistant to acids, especially the ones encountered in normal use and care of textile products. However, strong, concentrated acids at high temperatures are detrimental to all fibers. Excellent resistance to acids is associated with the olefin, modacrylic and vinyon fiber groups. These generic classes are especially useful in industrial applications where strong chemicals such as concentrated acids or strong alkalies may be encountered.

Synthetic fibers are relatively resistant to damage from most laundering chemicals, but some, notably *olefin, rubber, acrylic and modacrylic, may be harmed by dry cleaning solvents.* Since synthetic fibers launder beautifully and can be maintained with a minimum of ironing, there is no real advantage to dry cleaning them. In fact, that only adds expense and cleaning solvents may even damage some fibers.

Pests generally do not attack fabrics made of synthetic fibers because the organic chemicals in them have no food value. However, if no other food is available, insects may attack nylon or polyester fabrics, especially if soil, food or perspiration stains are left in them. Microorganisms may also attack certain finishes on synthetic fabrics such as starch or sizing; however, there is no need to take special precautions in care and storage other than normal procedures, which include storing textile products clean, in a dry, cool place.

Just as synthetic fibers have different sensitivities to heat, they have varying degrees of light sensitivity. Most fabrics made of *synthetic fibers can generally be associated with low heat tolerances, but good resistance to sunlight.* However, there is no relationship between fiber heat sensitivity and light sensitivity. Nylon fibers have relatively good resistance to heat but lose up to half their strength when exposed to sunlight. Recall how acetate fibers can have their resistance to sunlight increased; nylon likewise is more resistant if manufactured with a shiny, reflectant or smooth surface. (Special finishes can also increase nylon's light resistance.)

Polyester fibers have good resistance to sunlight when behind glass, although they may be weakened by direct sun, and are preferred over nylon for drapery or curtain fabric. Acrylic and olefin fibers also have excellent resistance to sunlight. Nonabsorbent olefin fibers make excellent indoor-outdoor carpet; however, when first developed, they aged with sun and weathering. The improved olefins provide just one example of how fiber chemical additions can alter properties, a subject you will learn more about later.

Spandex and rubber have been competitors in the textile field because they both add stretch or elasticity to textile products. However, spandex has replaced rubber in many applications since it is superior in its holding power; takes dyes better, and has greater resistance than rubber to sunlight, abrasion, oils, chemicals and heat. Thus, fabrics with spandex are not likely to be damaged through routine use and care procedures. The elastomeric fibers are rarely used alone; typically, they are combined in small amounts with other fibers in fabrics to give desired stretch.

Good basic practices for sorting articles prior to laundering are essential: keep badly soiled items from those only slightly or lightly soiled; keep white articles separate from dark or brightly colored ones; and keep delicate articles separate from those that need heavy-duty cleaning.

These procedures should be followed before laundering any textile product, but are particularly

---

QUICK QUIZ 13

1. See that Matrix VI is completed for the synthetic fiber category. Two synthetics which require special use and care to maintain their properties are nylon and rubber. Describe the limitations of each.
   a. Rubber: ____________________
   b. Nylon: ____________________
2. Briefly outline recommendations for care of synthetic fabrics in terms of:
   a. Detergents: ____________________
   ____
   b. Temperatures: ____________________
   ____
   c. Bleaches: ____________________
   ____

****

ANSWERS QUICK QUIZ 13

1. a. sensitive to sunlight, heat and laundering chemicals, including built detergents and chlorine bleach
   b. loses strength in sunlight; picks up color and soil during laundering
2. a. mild or unbuilt preferred, but heavily soiled products can withstand heavy-duty detergents if thoroughly rinsed
   b. warm to hot water; moderate dryer temperature; and if ironing is needed, low to medium setting
   c. oxygen or chlorine bleaches can be used without damage, though oxygen is safe for all fabrics

---

important for some synthetics, especially nylon. It is a scavenger and picks up color and soil from other articles in the laundry. Although synthetic fibers are hydrophobic and therefore difficult to dye, tiny spaces in the oriented molecular chains can hold color molecules picked up in washer loads. Nylon poses a serious problem in this respect as it will pick up loose color from other items in the laundry. Thus, it is recommended that white nylon fabrics be laundered alone.

As a group, synthetic fabrics are considered easy care because they can be machine laundered and dryer dried and used with minimum or no ironing. To cut down on the amount of pressing required after laundering, avoid overloading the washing machine because overloading tends to set in wrinkles. The dryer poses the most serious problem, as the heat can set in wrinkles that are almost permanent if items are not removed and hung up as soon as the dryer stops. Cool-down cycles on washer and dryer help reduce these problems but cannot replace prompt removal.

In summary, products made of synthetic fibers can and do give good service with minimum care if the following correct sorting, laundering, drying and pressing procedures are followed:

1. Separate white from colored fabrics; separate badly soiled items from lightly soiled ones; and separate heavy from delicate fabrics.
2. Launder often enough to prevent buildup of hard-to-remove body oils as well as other soil.
3. Choose detergent according to the fiber's chemical reactivity—an unbuilt detergent is always safe; a built detergent may be safely used when an article is badly soiled, providing the detergent is thoroughly removed during rinsing.
4. Oxygen or perborate bleaches are recommended and are safe for all synthetic fabrics; but if more whitening is needed, chlorine bleaches usually are safe if properly diluted (unless a label specifically states that chlorine bleach can harm the product).
5. Laundering at moderate temperatures is recommended; for drying medium temperatures are

also usually preferred, but follow the recommended temperature setting given in the dryer instruction book; a cool-down cycle is desirable to reduce wrinkling; if ironing is required, use a moderate to low heat setting.
6. Avoid overloading the washing machine and particularly the dryer. Remove articles from both as soon as cycle is complete.
7. Generally, stain removal solvents do not harm synthetic fibers; however, some solvents used in commercial cleaning can damage fibers, notably acrylics, olefins, and vinyons. Thus, laundering is preferred for synthetics whenever possible.

Care Procedures for Mineral Fibers

Actually, care procedures for mineral fibers are the simplest of any fiber category, primarily because little care is required, given their typical end uses. In home furnishings the major end use for glass fibers is in draperies, which need only periodic or seasonal cleaning. Metallic fibers are manufactured for esthetic purposes for use in special apparel items and in home furnishings such as upholstery, draperies and bedspreads. These are also cleaned only seasonally, and apparel with metallic yarns is typically worn only for special occasions.

Major care considerations for glass fabrics relate to precautions during use. Their low abrasion resistance and flexibility are the properties which determine appropriate uses, as well as suitable methods of care.

Excessive rubbing and repeated folding of glass fabrics cause fiber abrasion and breakage, and folded edges are potential weak spots. Draperies should be short enough so that the hems do not rub against windowsill or floor. Further, draperies should be hung away from walls, ceilings and sills to prevent contact and resulting abrasion.

When glass fabrics need laundering, which should be only infrequently, some special procedures must be followed. *Glass can be damaged and discolored by strong alkaline solutions*; thus unbuilt or mild detergents are recommended. *Glass fibers are resistant to acids, oxygen bleaches, unbuilt detergents and most other chemicals.* Chemicals that can damage glass seem to be those concentrated ones used in some care procedures. It has already been stated that glass can be discolored and yellowed by concentrated built detergents; it can also be damaged by certain dry cleaning solvents and insufficiently diluted chlorine bleach solutions. Therefore, dry cleaning is not recommended. If whitening is required, oxygen bleaches are recommended.

Another important consideration in laundering glass fiber fabrics is to avoid agitation or abrasion which breaks and damages the fibers. This is why fabrics of glass fibers are usually laundered by hand rather than in a washing machine whose agitation tends to break glass fibers and leave a residue of glass particles in the machine.

Due to skin discomfort which may result from handling or laundering glass fabrics, the Federal Trade Commission established a ruling in 1967 that they must be labeled to warn customers of possible skin irritations.

Water temperatures are no problem with glass: it can take brief exposures to temperatures as high as 1500° F. Although glass fabrics have high heat resistance, they should not be dryer-dried because tumbling in a dryer breaks down the fibers, leaving particles of glass residue that could be picked up in later loads.

In addition, ironing is not recommended since pressure from an iron also tends to break glass fibers. Glass fabrics are best laundered, gently shaken to remove excess water and hung up immediately. If necessary they can be wiped dry with a towel. However, drying is not generally a problem because glass has zero absorbency, and wrinkling is no problem because it is highly resilient.

Since glass is nonflammable, it is an excellent choice for various industrial uses in fire protection and in electrical and insulating equipment. An interior designer might also select glass fabrics for household use in order to provide nonflammable and insulative draperies.

Just as the durability and comfort of metallic yarns are mostly dependent on the plastic or polymer material used with the metallic material, care is also dependent on the other material used with the metal. Therefore, in order to determine care procedures to be used for a metallic fabric, labels should be checked to determine identified care procedures and the content or composition of the other fibers involved and the plastic or polymer coating used.

Polyester or modified acetate films are used as the outer coating on metallic yarns or as the base for metallic granules. Polyester coatings launder more satisfactorily and have higher age and light resistance than modified acetate coatings, although both combinations have high biological resistance.

While polyester films combined with metallic yarns are generally superior to acetate coatings in durability properties and ease of care, modified acetate-coated yarns are less expensive and probably would be serviceable because metallic fabrics are primarily used for decorative or esthetic purposes.

Comparison of Care Considerations Among The Fiber Categories

The laundry procedures that help ensure washable fabrics will perform according to expectations include: use medium water and drying temperatures, since high temperatures tend to set in wrinkles; avoid overloading washer or dryer, as wrinkles are set by crowding fabrics; and remove laundry immediately from both washer and dryer and hang up to prevent setting in wrinkles.

The ability of fibers and fabrics to retain their shape and wrinkle-free appearance is affected by both dimensional stability and resiliency. The two properties are usually related; however, wool is an exception because it has low dimensional stability and high resiliency.

During normal consumer use and care, fabrics made of natural cellulosic fibers are not likely to shrink excessively if they are adequately finished or preshrunk. However, fabrics made of animal hair fibers are likely to shrink whenever they are laundered with a built detergent and agitated in hot water or when there are noticeable temperature changes in laundering processes. This shrinkage is a result of fiber felting.

Using home laundering methods, shrinkage can be prevented when caring for protein fibers by using warm or cool water, maintaining the same temperature at all steps in the laundry process. Using unbuilt neutral detergent and a minimal amount of agitation will prevent felting shrinkage.

Of the man-made cellulosic fibers, some rayon products can be a problem during laundering, since the fibers lose strength when wet; consequently long soaking periods should be avoided.

Acetate is also somewhat of a laundering problem, but not for the same reason as rayon. Acetate fibers wrinkle excessively when wet; therefore, many people prefer to care for both rayon and acetate fabrics by dry cleaning.

In addition to a fabric's dimensional stability, alkaline reactivity must also be considered when deciding on laundering procedures. Built detergents are always acceptable for natural cellulose. Unbuilt detergents are essential for protein fibers and are preferred for mineral fibers. For man-made noncellulosic or synthetic fibers, built detergents can be safely used when extra cleaning power is required; but for gentle laundering, unbuilt detergent is usually recommended.

Protein fibers are damaged excessively by built detergents and normal laundering conditions; therefore, dry cleaning is the preferred method for care.

Although strong concentrated acid will damage any fiber, two categories are sensitive to even mild acidic substances: natural cellulosic and man-made

---

QUICK QUIZ 14

Check which of the following care procedures should be followed to maximize the life of fabrics made from glass fibers.

____ built detergents
____ unbuilt detergents
____ iron with cool iron
____ iron with hot iron
____ do not iron
____ dryer set at medium heat
____ do not dryer dry
____ avoid agitation
____ never wring or spin dry
____ machine wash on regular cycle

****

ANSWERS QUICK QUIZ 14

unbuilt detergents, do not iron, do not dryer dry, avoid agitation, never wring or spin dry

---

cellulosic fibers. As noted earlier, this means that stains such as acid fruit juices should be removed immediately with cool water and detergent.

All fiber categories react satisfactorily to oxygen or perborate bleaches, but certain fibers, such as the protein, are damaged by chlorine bleaches. Further, some synthetic fibers and glass can be damaged by chlorine bleaches if used in a strong solution. However, if properly diluted in water, chlorine bleach can be used satisfactorily when its whitening power is necessary.

Basically, dry cleaning solvents are safe for most fibers except rubber, olefin and glass. However, due to the expense of dry cleaning, laundering is preferred for products that will react favorably and maintain a new look when laundered. Laundering is preferred for natural cellulosic and man-made noncellulosic, mineral and glass. Dry cleaning is preferred for protein; and either method can be safely used for man-made cellulosic.

After reviewing the properties that affect fiber care, you will note that protein is plainly the category considered the most sensitive to normal care procedures. However, any textile product can serve well if a wise selection of fiber content is made for a given end use, and if the product receives proper care.

REVIEW

1. Match the fiber categories to their degree of resiliency:

____	a. natural cellulosic	1. low
____	b. animal hair protein	2 medium
____	c. animal secretion protein	3. high
____	d. man-made regenerated cellulosic	
____	e. man-made modified cellulosic	
____	f. man-made noncellulosic or synthetic	
____	g. mineral	

2. Check the fiber categories that require only a minimum of pressing care:
   ____ natural cellulosic
   ____ man-made noncellulosic
   ____ protein
   ____ man-made cellulosic
   ____ mineral

3. Which fiber category requires the least amount of special attention during laundering, and little or no pressing after wearing or laundering? ____________________

4. List two fiber categories that yield fabrics of high dimensional stability and three that produce low dimensional stability.
   high: ____________________
   low: ____________________

5. Indicate procedures that should be followed in laundering fabrics for these fiber categories.
   a. natural cellulosic: ____________________
   b. man-made noncellulosic: ____________________
   c. mineral: ____________________

6. Why is it important to be familiar with the heat sensitivities of the fiber categories in order to care for textile products properly? ____________________
   ____
   ____

7. In what type of environment should fabrics be stored to inhibit the growth of biological organisms? ____________________

8. Describe the sunlight resistance of the following fiber categories:
   a. natural cellulosic ____________________
   b. man-made cellulosic ____________________
   c. most man-made noncellulosic ____________________

9. For each of the five fiber categories, indicate the potentially destructive elements that can affect serviceability.

____	a. natural cellulosic	1. moths and carpet beetles
____	b. man-made cellulosic	2. silverfish
____	c. protein	3. mildew and mold
____	d. man-made noncellulosic	4. sunlight
____	e. mineral	5. atmospheric gases or fumes
		6. none, excellent resistance

****

REVIEW ANSWERS

1. a-1; b-3; c-2; d-1; e-2; f-3; g-3
2. man-made noncellulosic and mineral
3. man-made noncellulosic or synthetic
4. high: man-made noncellulosic and mineral
   low: natural cellulosic, protein and man-made cellulosic
5. a. most resistant category to normal laundry procedures
   b. mild alkalies and oxygen bleaches preferred, but built detergents and properly diluted chlorine bleaches can be used; heat sensitivity of many synthetics requires medium temperatures
   c. mild alkalies and oxygen bleaches recommended; a minimum of agitation is essential to prevent fiber breakage, particularly for glass
6. in order to select the correct temperatures for both wash and rinse water, for drying, and for ironing when needed
7. dry, cool, light environments, but away from the direct sun. (Dark or low light areas may be used, providing there is adequate air circulation and no dampness.)
8. a. eventual yellowing and weakening from the sun, except for flax which has generally good resistance.
   b. general weakening over time—more major for acetate; less for rayon
   c. generally high resistance, except for nylon and rubber
9. a-2, 3, 4; b-2, 3, 4, 5; c-1, 4; d-4; e-6

Textile Labeling and Care Implications

Textile labeling is regulated in two ways: 1) through laws passed by Congress, and 2) through rulings of the Federal Trade Commission (FTC), known as Trade Regulation Rules, which have the force of law. The Textile Fiber Products Identification Act (TFPIA) which officially established the generic fiber classifications became effective in 1962. The TFPIA stipulated that the generic name must be included on fabric labels, along with the percent of each fiber present and the manufacturer's name.

Although the TFPIA is the most comprehensive textile law, it is not the only one. The Wool Products Labeling Act (WPL) of 1939, the Fur Products Act of 1952, and the Flammable Fabrics Act amended in 1967 (discussed in Unit 5 with flame-retardant finishes), all provide valuable consumer protection. None of these laws required that care procedures be provided nor that care labels be permanently attached.

When the TFPIA went into effect in 1962, it incorporated the entire field of textile fiber content labeling, including articles of wearing apparel, draperies, floor coverings, furnishings, bedding and other household textile products. There were some specific exceptions, such as products already covered by the Wool Products Labeling Act.

The WPL is still used as a guide to what must go on the labels of wool and other animal hair products. Required information includes type of wool, percentage of each fiber type present and the manufacturer's name.

Wool type refers to whether the wool has ever been used in a fabric before—the three types specified by the WPL are *wool, reprocessed wool and reused wool.* The quality of wool within the types is not specified by law. Modifications to this legislation have changed the terminology and replaced the terms reprocessed and reused with *recycled.* The term, wool, which would remain for use on new fibers, refers to new or virgin wool that has never been used in a fabric before. Reprocessed wool is wool that has been made into fabric and then reprocessed into fiber again without being used by the ultimate consumer. When the wool has actually been used by the consumer and is then broken down into a fibrous state to be used again in fabrics, the label indicates reused wool. It is these latter two terms that have been replaced with the term *recycled.*

In the mid 1960s, to promote products of 100 percent high quality virgin wool, the Wool Bureau, Inc., developed the Woolmark logo shown as "A" in Figure 2.19. The Woolmark is not required on labels but was developed as a marketing or promotional device. The Wool Bureau also developed the Woolblend Mark, "B" in Figure 2.19, designed to promote products made of wool plus some other fiber type.

Although the labeling laws and FTC rules are far from perfect or inclusive, they do provide valuable information for the consumer to use in selecting textile products and in caring for them property.

Figure 2.19. The woolmark and woolblend mark are registered certification marks of the Wool Bureau, Inc. (*courtesy of the Wool Bureau, Inc.*)

Care Labels

In 1972, an FTC Trade Regulation Rule was put into effect that requires permanently attached care labels on finished articles of wearing apparel, including hosiery, but excluding shoes, headgear and handgear.

This care labeling ruling also requires that yard goods produced for wearing apparel be sold with a care label or tag which can be permanently attached by the home sewer to the finished article. Some home furnishing items, including draperies, rugs and carpeting, are now required to be labeled with care instructions.

The Labeling regulation was amended in 1984 to more accurately define care procedures. An important change is that directions should be given concerning laundering, bleaching, and drying. If any directions are missing it is assumed that the fabric will withstand the most rigorous care—hot water, strong detergent, bleach, and tumble dry on high temperature setting. Specific details can be obtained from the Federal Trade Commission.

QUICK QUIZ 15

1. Why was the TFPIA passed by Congress?
   ____
   ____
2. What other textile areas are covered by textile legislation?
   ____
3. What did the 1972 FTC Trade Regulation Rule require for consumers?
   ____

****

ANSWERS QUICK QUIZ 15

1. to establish fiber generic classifications and require products to be so labeled
2. wool products, furs and flammable fabrics
3. permanently attached care labels on selected apparel

UNIT TWO

Part V

Esthetic Appearance Of Fibers

Esthetic properties of fibers relative to their effect on fabric appearance are discussed in this section. The characteristics covered include *color, luster, transulence, texture, drape body or hand* and *loft*. Matrix VII should be used in developing your study guide for comparing esthetic fiber properties.

Among the most apparent characteristics of a fabric are color and design applications. Physical and chemical structures of fibers determine the type of dyestuff that can be used and to some degree the durability of the dye, as well as *dye affinity* of a fiber. Good dye affinity is found in fibers which are absorbent and have a porous molecular structure that allows the dye to penetrate into the fiber. Fibers may have a central canal that enables the dye bearing solution to pass through the fiber structure; and fibers may react chemically with the dye; dyestuffs may actually dissolve within the fiber.

The *natural cellulosics, proteins, and man-made cellulosic, rayon, all have porous molecular structures that allow dyes to be absorbed.* Also, they may react chemically with the dye molecules.

Synthetic and modified cellulosic acetate and triacetate fibers have molecular structures which do not readily admit the dye solution. However, these fibers may react chemically with the dye, or be modified to admit the dye molecule; or they dissolve the dye into the fiber.

Dyestuffs may be added to the chemical solution before it is forced through the spinneret. This is called *solution dyeing or mass pigmentation*. It can also be added to the fiber, to yarns, or to completed fabrics.

Another visual quality of a fabric is the manner in which light is reflected from its surface, a quality called *luster*. The luster of a fiber is determined, in part, by its texture. A smooth fiber tends to reflect light directly rather than diffusing it; thus smooth fibers appear to have a higher luster than rough ones.

Except for silk, mineral and synthetic fibers tend to be shinier than natural fibers because of their smoother surfaces. In fact, special chemical treatments are sometimes used to reduce the luster of man-made fibers. *Titanium dioxide* is a white pigment that can be added to the fiber spinning solution to alter the pattern of light reflected from the fiber surface, which reduces luster or shine.

If a fabric is held up to the light, it will generally have a certain amount of *translucence*, or the ability to let light through. The degree of translucence is more related to yarn and fabric construction than to the individual fiber; however, fiber color and luster can influence light transmission or translucence.

The manner in which a fabric hangs is referred to as fabric *drape*. Drape is a fabric quality that can be seen, and to some degree felt; thus, it has both visual and tactile qualities.

A general, subjective textile term which is used to describe how a fiber feels or handles is known as fabric *hand or body*. Generally, cotton fabrics are associated with a very desirable hand or body that is more comfortable and pleasing than the hand of many synthetic fabrics. Body or hand relates to the lightness or heaviness of a textile product, as well as to its firmness or looseness.

As you become used to handling fabrics, it is possible to develop subjective judgments in describing hand or body. Listed below are some adjectives used to describe representative fibers in the five categories:

natural cellulosic:
cotton—soft, fine, smooth
flax—crisp, stiff, firm
protein:
wool—rough, scratchy, fuzzy, heavy
silk—smooth, silky, slippery, light
man-made cellulosic:
rayon—cottonlike or silklike, limp
acetate—sometimes crisp, smooth
synthetic:
hand varies with structure—some synthetics are smooth and silklike, others are rough and woollike; still others are crisp, firm, or stiff
mineral:
glass—harsh, smooth, slippery, irritating

Fill in selected descriptive terms on Matrix VII. Other terms may be used for describing the hand or body of certain fibers based on previous experience.

A more detailed discussion of fabric body is included in some of the following units, since this property can be modified to a great degree by yarn and fabric structures, by finishes, and by some types of design applications. Almost any amount of weight or firmness can be given to a product during yarn and fabric structuring or finishing. The *body* of a fabric is difficult to perceive until the fabric is handled; but the eye sees *drape,* which gives some clues as to what weight and firmness can be expected.

Another esthetic quality in which the visual and tactile senses are almost equally important to perception is the texture of a fabric. *Texture,* perhaps the most all encompassing esthetic property of textiles, since it is perceived visually as well as by feel or handling, helps determine how the fabric surface reflects light. Therefore, the roughness or smoothness of a fiber is one determinant of fabric luster.

It is essential to recognize that most tactile and visual aspects of textiles are influenced to a tremendous degree by the type of yarn structure and the method of fabric construction. Thus, fiber properties, while important, can be modified to provide the consumer with almost any type of fabric from almost any type of fiber. Complete the Matrix VII for the general textural qualities of fibers in each category and check your responses against these texture descriptions: natural cellulosic—relatively smooth; protein—wool is rough, silk is smooth; man-made cellulosic—smooth; man-made noncellulosic or synthetic—smooth or, if textured, somewhat rough and irregular; mineral—smooth appearance but irritating or harsh feel.

If a fabric is crushed and springs back to its original shape, it has high compressional resiliency or high *loft*. Compressional resiliency or loft differs from resiliency, which refers to the ability of a fiber or fabric to recover from wrinkling. Loft, which refers to the ability of a thick, bulky fiber or fabric to recover from compression or flattening, is primarily an esthetic consideration, since a fabric is less attrac-

---

QUICK QUIZ 16

Review the fiber densities recorded on the Comfort Matrix. Considering fabrics of the same identical fabric and yarn structure, indicate which fiber categories would contribute to heavy, medium and light weight fabrics.

a. heavy: ______________________________
b. medium: ______________________________
c. light: ______________________________

****

ANSWERS QUICK QUIZ 16

a. natural cellulosic, man-made regenerated cellulose (rayon), and mineral
b. protein, man-made modified cellulosics (acetate and triacetate), and some synthetics (polyester)
c. most synthetics

---

tive when its fibers have been packed down or flattened.

The natural fiber crimp, medulla, and scales of animal hair fibers create air spaces that trap air and prevent fibers from packing. The protein cross-linked molecular forces and spiral configurations also help return the fibers to their original position after bending or wrinkling. Therefore, wool is characterized as having high crease recovery as well as high loft.

Silk has somewhat different properties from other protein fibers because of its different fiber structure. Since silk fibers are smooth and have no scales or fiber crimp, they have less loft and resiliency than the animal hair fibers. In fact, silk has medium resiliency due to its basic protein structure, but it has low loft due to smooth filaments that tend to pack together when compressed.

Mineral fibers have a somewhat rigid molecular structure which contributes to good wrinkle or crease recovery. However, since they cannot tolerate crushing forces, they have low loft.

REVIEW

1. The appearance of a fabric provides insight as to how it will feel; for example, the luster of a fiber can give a visual clue as to how the fabric texture will handle. Generally, the more lustrous the fibers in a fabric, the (smoother, rougher) the texture will be.

2. Drape and body are fabric characteristics that depend more on yarn and fabric construction than on the particular qualities of the ______________.

3. What quality describes a fabric's ability to transmit light? ______________________

4. The absorbency, molecular structure, and chemical composition of the fiber determine, in large measure, how well a fabric will take color, or its ______________________.

****

REVIEW ANSWERS:

1. smoother
2. fiber
3. translucence
4. dye affinity

Posttest

1. The best method for identifying a fiber is the
   a. burn test.
   b. microscope.
   c. fabric hand.
   d. fiber content labeling.

2. The superior resiliency of wool is not dependent on which of the following?
   a. fiber crimp
   b. fiber scales
   c. fiber cross-linking
   d. cortical cells
   e. fiber length

3. Protein fibers have which of the following characteristics?
   a. hygroscopicity
   b. hydrophobicity
   c. dry heat tolerance
   d. alkaline tolerance

4. What is the most outstanding property of cotton fibers?
   a. absorbency
   b. resiliency
   c. elasticity
   d. tenacity

5. All of the following properties apply to cellulosic fibers except
   a. resistance to dilute bleaches
   b. high density
   c. low resistance to wrinkling and creasing.
   d. good resistance to acids

6. Fibers with amorphous molecular arrangement and no special bonding tend to have (high, low) tenacity and (high, low) elongation potential.
   a. high, high
   b. high, low
   c. low, high
   d. low, low

7. Convolutions in cotton fibers contribute to which durability property?
   a. resiliency
   b. cohesiveness
   c. density
   d. flexibility

8. Wool and silk are similar in what way?
   a. fiber length
   b. surface appearance
   c. center canal
   d. fiber category

9. Fibers are divided into generic classifications according to chemical composition.
   a. true
   b. false

10. What is a disadvantage of the wet spinning process?
    a. Fibers are weakened by the coagulating acid.
    b. Fiber raw materials are dissolved.
    c. Additional processing is required.
    d. Fiber shape cannot be extruded uniformly.

11. Glass fibers are stretched when they leave the spinneret to increase fiber
    a. elasticity.
    b. flexibility.
    c. resiliency.
    d. density.

12. Which fiber category is known for losing strength when wet?
    a. man-made cellulosic
    b. natural cellulosic
    c. man-made noncellulosic
    d. mineral

13. Which fibers typically increase in tenacity when wet?
    a. man-made cellulosic
    b. protein
    c. natural cellulosic
    d. synthetic
    e. mineral

14. In comparison with other fiber categories, which one is associated with the highest tenacity?
    a. natural cellulosic
    b. protein
    c. man-made cellulosic
    d. man-made noncellulosic (synthetic)

15. High tenacity of crystalline fibers is primarily dependent upon the
    a. slippage between the crystals.
    b. nature of the crystal bonds.
    c. low orientation of the crystals to the fiber axis.
    d. number and size of the crystals.

16. In relation to fiber length, the ______________ the fiber, the ______________ need for cohesion.
    a. longer, more
    b. shorter, less
    c. shorter, more

17. Which generic class of synthetic fibers is known for its elastomeric properties?
    a. acrylic
    b. spandex
    c. olefin
    d. polyester
    e. nylon

18. Generally, the ______________ a fiber is elongated, the ______________ the elastic recovery.
 a. more, greater
 b. less, greater
 c. less, smaller

19. Which fiber group has the highest flexibility and resiliency?
 a. protein
 b. natural cellulosic
 c. man-made cellulosic
 d. man-made noncellulosic (synthetic)

20. The dimensional stability of synthetic fibers is most likely to be adversely affected by excessive
 a. heat.
 b. moisture.
 c. pulling.
 d. wear.

21. Since natural cellulosic fibers and rayon have high absorbency, you also would expect them to have
 a. poor electrical conductivity.
 b. good electrical conductivity.
 c. high allergenic potential.
 d. high thermal retention.

22. The high thermal conductivity of natural cellulosics means that they
 a. are comfortable in cold, dry weather.
 b. have high thermal retention.
 c. have low thermal retention.
 d. have low electrical conductivity.

23. What development in the manufacture of rayon has increased its use in wearing apparel?
 a. differential shrinkage of component parts
 b. fibers created with multilobal shapes
 c. high-wet-modulus variations
 d. fibers created with bilobal shapes
 e. cross-linked fibers

24. The ______________ the fiber density, the ______________ warmth without weight possible.
 a. higher, less
 b. lower, more
 c. higher, more

25. The absorbency of rayon is better than that of acetate because
 a. rayon has superior wicking properties.
 b. the cellulose in rayon is not changed chemically.
 c. acetate has a rougher surface.
 d. acetate is a nonthermoplastic.

26. Silk has lower thermal retention than wool because silk
 a. is always made into thin, smooth fabrics.
 b. lacks a scaly surface.
 c. has a lower degree of absorption.
 d. is less dense.

27. The low absorption of the synthetics is due to (low, high) orientation and (low, high) crystallinity.
 a. low, high
 b. high, high
 c. high, low
 d. low, low

28. The inability of man-made noncellulosics to absorb moisture readily and conduct electricity make them best suited for wear on ______________ days.
 a. cool, dry
 b. warm, damp
 c. warm, dry
 d. cool, damp

29. What enables metallic fibers to be used successfully in apparel?
 a. their long length
 b. a plastic coating
 c. an adhesive
 d. a backing fabric

30. Pilling is a fabric problem caused by
    a. strong fiber ends balling up on fabric surface.
    b. highly twisted ply yarns in fabrics.
    c. resin finishes on cotton.
    d. highly elongated fibers in yarns.

31. Which properties are applicable to man-made noncellulosic fibers?
    a. high abrasion resistance and low tenacity
    b. good electrical conductivity and low absorbency
    c. low moisture absorption and high dimensional stability
    d. high resiliency and high absorbency

32. A low absorbency fiber is considered easy-care because it
    a. is quick drying.
    b. does not conduct electricity.
    c. easily absorbs water-borne stains.

33. Movement of moisture along a fiber is referred to as
    a. moisture regain.
    b. hydrophobicity.
    c. absorbency.
    d. wicking.

34. Resiliency and dimensional stability of protein fibers present a problem in fabric
    a. laundering.
    b. ironing.
    c. wrinkling during wear.

35. What is the best cleaning agent for heavily soiled, white cotton clothes in hard water?
    a. unbuilt soap
    b. built soap
    c. unbuilt detergent
    d. built detergent
    e. heavy-duty soap

36. The sensitivity of protein fibers to (hot, cold) water and (acidic, alkaline) chemicals makes dry cleaning a preferred method of care.
    a. cold, acidic
    b. hot, acidic
    c. hot, alkaline
    d. cold, alkaline

37. Which of the following factors is least likely to cause gradual decomposition of silk?
    a. oxygen
    b. sunlight
    c. dry cleaning
    d. carpet beetles

38. Spandex has largely replaced rubber in apparel because spandex fibers have greater ______________________ than rubber.
    a. density
    b. recovery
    c. holding power
    d. elongation
    e. elasticity

39. Which of the following properties makes olefin a good choice for activewear apparel?
    a. thermal retention
    b. inexpensive to produce
    c. wicking
    d. absorbency

40. Man-made noncellulosics have an affinity for
    a. acidic solutions.
    b. water-borne stains.
    c. oily stains.
    d. alkaline solutions.

41. White nylon fabrics should be laundered alone because they
    a. require special detergents.
    b. pick up color and soil.
    c. cannot stand hot water.
    d. need to be bleached.

42. To prevent fabric wrinkling in products made of synthetic fibers it is important to
    a. dry clean them.
    b. let clothes remain in the washer or dryer after laundering.
    c. remove clothes immediately from washer and dryer.

43. Man-made noncellulosic fibers, such as nylon or polyester, are often blended with wool to increase product
    a. loft.
    b. flexibility.
    c. wrinkle resistance.
    d. dimensional stability.

44. Which fiber category requires the least pressing?
    a. man-made cellulosic
    b. natural cellulosic
    c. man-made noncellulosic
    d. protein

45. Fabrics with low dimensional stability are most likely to be damaged by
    a. ironing.
    b. laundering.
    c. dry cleaning.
    d. spotting.

46. Which fiber categories are generally not bothered by biological organisms?
    a. protein and synthetic
    b. natural cellulosic and protein
    c. mineral and synthetic
    d. man-made cellulosic and mineral

47. Which environmental conditions are recommended for storage of natural cellulosic and protein products to inhibit the growth of biological organisms?
    a. humid, cool, dark
    b. dry, cool, light
    c. humid, warm, light
    d. dry, warm, dark

48. The Wool Products Labeling Act requires that the manufacturer's name be on the label of a wool product, in addition to the
    a. percent and type of wool.
    b. type of wool and finish applied.
    c. percent of wool and care procedures.
    d. type of animal and percent of wool.

49. What information is not required on the hangtag by the TFPIA?
    a. trade name and manufacturer's address
    b. fiber generic name and percent by weight
    c. manufacturer's name

50. A 1972 Federal Trade Commission Ruling requires that apparel be supplied with
    a. the fiber generic name.
    b. permanently attached care instructions.
    c. the manufacturer's name and address.
    d. the trade name.

Answers for Unit 2 Posttest:

1. d	26. b
2. e	27. b
3. a	28. d
4. a	29. b
5. d	30. a
6. c	31. c
7. b	32. a
8. d	33. d
9. a	34. a
10. c	35. d
11. b	36. c
12. a	37. c
13. c	38. c
14. d	39. c
15. b	40. c
16. c	41. b
17. b	42. c
18. b	43. d
19. d	44. c
20. a	45. b
21. b	46. c
22. c	47. b
23. c	48. a
24. b	49. a
25. b	50. b

UNIT THREE

Yarns

In the preceding unit, characteristics of the fiber categories and their implications for serviceability were discussed. A review of the matrixes will provide a summary of the durability, comfort, care and esthetic considerations as a basis for understanding how fiber properties relate to yarns and their serviceability.

This unit is divided into three parts: 1) *spinning of fibers into simple yarns;* 2) *characteristics and types of novelty or complex yarns;* and 3) *serviceability considerations of yarns.* Comments will be made frequently that the serviceability of yarns is intricately tied to the types of fibers used, the way in which the yarns are structured into fabrics, and any finishes that are added. Therefore, yarns are just one component of a final textile product and only partially explain why different fabrics perform differently.

Objectives for Unit 3

1. You will be able to identify and describe the manufacturing processes and characteristics which differentiate the following yarn types:
   A. Simple versus complex or novelty yarns
   B. Singles, plies and cord yarns
   C. Types of complex or novelty yarns, such as slub, bouclé, flock or flake, ratiné, core, loop
2. You will be able to identify different types of yarns used in making fabrics.
3. You will be able to evaluate fabric serviceability in terms of fabric properties that are influenced by yarn structure.

Pretest

1. Generally, staple fibers produce ____________ yarns.
   a. short
   b. dull
   c. shiny
   d. long

2. The carding and combing processes are used with ____________ length fibers.
   a. staple-
   b. filament-

3. Which list below gives the order of steps for forming a singles yarn?
   a. roving, sliver, mat, singles
   b. sliver, roving, mat, singles
   c. sliver, mat, roving, singles
   d. mat, sliver, roving, singles

4. Singles, ply and cord yarns are ____________ yarn types.
   a. simple
   b. novelty

5. Which yarn structure would best offset the undesirable properties of certain fibers?
   a. combination or ply yarn
   b. blended singles yarn
   c. core yarn
   d. combination cord yarn

6. You would expect a __________ yarn to be stronger than a __________ yarn.
   a. ply, singles
   b. singles, ply

7. The more complex and large the effect yarn, the __________ durable the resulting fabric.
   a. more
   b. less

8. Core yarns frequently contain __________ fibers.
   a. polyester
   b. metallic
   c. elastomeric

9. How are the yarns in a "combination" fabric structured?
   a. Two or more fibers are spun together into a singles yarn.
   b. Two or more plies of different fiber content are twisted together into a larger yarn.
   c. Two or more yarns, each of a single fiber component, are used in a fabric.

10. A slub yarn is produced by
    a. eliminating the drawing process.
    b. varying the amount of yarn twist.
    c. varying the sliver size.
    d. eliminating the roving process.

11. A tweed or flake yarn is similar to a singles yarn except that the tweed has
    a. longer fibers.
    b. several plies.
    c. flecks of color.
    d. more yarn twist.

12. Crepe effect yarns made of synthetic fibers are produced by
    a. tightly twisting the yarns.
    b. texturing the yarns.
    c. adding effect yarns.
    d. crepe weaving.

13. Yarns which consist of an inner yarn completely covered by an outer cover are referred to as __________ yarns.
    a. seed
    b. core
    c. tweed
    d. crepe

14. A yarn that is crinkled and heat-set is referred to as a __________ yarn.
    a. slub
    b. bouclé
    c. tweed (flake)
    d. textured

15. The lower the yarn twist, the __________ the yarn tenacity.
    a. lower
    b. higher

16. Which yarn would be most affected by laundering agitation?
    a. ply
    b. single
    c. ratiné
    d. core

17. The ________ the yarn twist, the ________ the fabric absorbency.
    a. higher, higher
    b. lower, higher
    c. lower, lower

18. Yarn texturing tends to
    a. increase fiber pilling.
    b. decrease fiber stretch.
    c. decrease fiber luster.
    d. decrease fabric heat retention.

Answers for Unit 3 Pretest:

1. b	10. b
2. a	11. c
3. d	12. b
4. a	13. b
5. b	14. d
6. a	15. a
7. b	16. c
8. c	17. b
9. c	18. c

UNIT THREE

Part I

Fibers into Yarns: The Spinning Process

Fibers are classified by their length into two basic types—short fibers are called *staple fibers,* while long continuous fibers are *filaments*. Natural fibers, except for silk, are of the staple variety; silk is a filament fiber. Manufactured fibers are made in filament form, but they can be cut into staple length whenever appropriate for the end use.

Figure 3.1 illustrates yarns structured from the two types of fibers. The smooth shiny yarn is made from filament fibers; the dull, somewhat uneven yarn is from staple fibers.

The conversion of short, staple fibers into yarns requires considerably more processing than needed for making yarns from filament fibers. This is because staple fibers are short and without numerous straightening and twisting processes they would fall apart. Filament fibers are spun into yarns easily because their length holds them together securely. Filament and staple fibers are shown in Figure 3.2. As you can see, the staple fibers are randomly arranged and not parallel to each other, while the filament fibers are already aligned and therefore do not require straightening processes. Whether staple or filament, the continuous strand resulting when fibers are twisted together is called yarn, and the operation of making fibers into yarn is called *spinning*. The following section describes the steps involved in producing yarns of staple fibers.

Traditional Spinning of Staple Fibers

The first step in making yarns of staple fibers is called *opening and blending*. During this stage fibers are loosened and blown so that the fibers are separated (Figure 3.3). In the modern textile mill, much of this stage is automated, with fibers transported through a system of chutes and pipes. This allows for higher processing speeds as well as a cleaner work environment.

In the next stage, the fibers are formed into a *picker lap* which can be described as a mat of fibers several inches thick. The average length of these fibers is never more than a few inches. The lap is then fed into a *carding* machine, which removes further impurities and begins to align the fibers into a parallel arrangement. Figure 3.4 shows the carded fiber web coming from the carding machine and being formed into the *carded sliver.* If the staple fibers are not sufficiently aligned and parallel after carding for a given end use, a second straightening process called *combing* is used. Figure 3.5 shows the combing process where additional dirt and short fibers are removed and the remaining fibers are made more parallel.

A fabric made from yarns that have been combed is smoother than fabric made from yarns that have only been carded. Figure 3.6 illustrates the difference in the relative amount of alignment represented by carded and combed yarns. The difference between *muslin* and *percale* sheets is a good example of the differences in products of the two yarn types. Percale sheets tend to have a smoother, softer hand and a higher cost, while muslin sheets are rougher, yet less expensive. The equivalent of carded and combed yarns in the wool industry is *woolen* and *worsted* yarns. Woolen yarns, which produce fabrics with a rough texture, are comparable to carded yarns. Likewise, worsted yarns are similar to combed yarns that have the short fibers removed, yielding a smoother fabric.

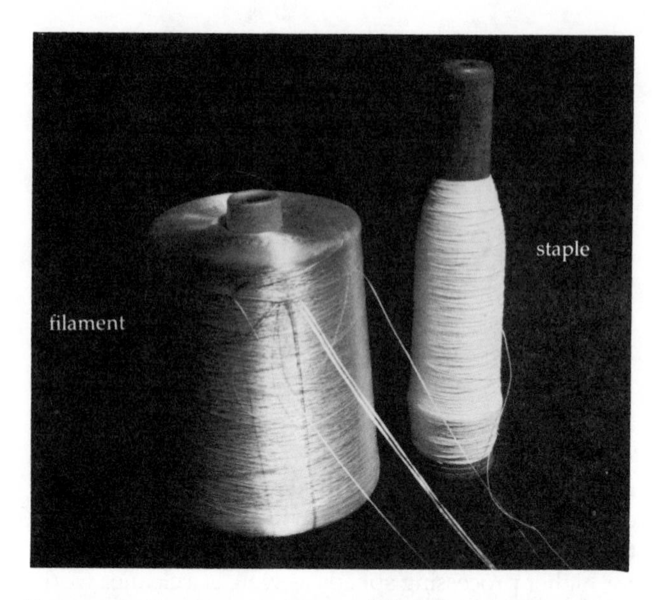

Figure 3.1. Staple yarns have a duller, less shiny appearance than untextured filament yarns.

After carding and combing, the slivers are drawn. The *drawing* operation may occur once, twice or even three times, depending on the type of fibers and the quality indicated for the final yarn. In the drawing process several slivers are combined into a new strand, and then pulled out so the new strand is no larger in diameter than one of the slivers used. The drawing process makes the fibers more parallel, aligned and compact in arrangement. For most high quality yarns, fibers are subjected to several drawing processes.

Drawing occurs after carding, even if combing is to take place. For example, to produce smooth combed yarns, fibers are carded and drawn; then they are combed and put through another process called *finisher drawing*. From this latter drawing process, the sliver goes to the *roving* operation. Combed fibers, therefore, go through carding, drawing, combing, and finisher drawing processes, prior to roving and spinning. Yarn sample 4 is a staple, single yarn which has gone through the additional straightening process called combing.

When fibers such as polyester and cotton are to be blended into spun yarns of staple fibers, the blending usually occurs during the first drawing operation. After the first drawing, which partially blends the fibers, the slivers go through a second drawing operation to ensure a more uniform blend of fibers. The final blended sliver goes to either the combing unit or directly to the finishing drawing on the roving frame. Figure 3.7 shows the feeding of slivers into the drawing frame.

The finisher drawn sliver is fed into the roving frame where it is extended into a fine, slimmer form. A small amount of twist may be inserted during the final drawing operation. The purpose of the roving frame is to extend the slivers into a thin unit, futher align the fibers, and provide sufficient twist to hold the fibers together during the spinning process. The strand which leaves the roving frame,

Figure 3.2. Unlike filament fibers, staple fibers must be straightened and cleaned before they can be spun into yarn.

Figure 3.3. An opening carousel uses automation to open and separate large bales of cotton—a procedure that used to be performed by hand. *(courtesy Burlington Industries, Inc.)*

called the *roving,* is considerably smaller in diameter than that which enters the roving frame. See Figure 3.8. The roving strand is fed onto roving bobbins which are taken to the spinning frame, where the single yarn is formed. In the spinning operation the roving is pulled out to the extent required to produce the size of yarn desired, with the specified amount of twist.

Figure 3.9 shows the spinning frame with the roving package that is hung in the frame. The yarn formed at this stage is called a *single yarn* whose durability depends on fiber length, tenacity, and cohesiveness or spinning quality. Up to a certain point, the greater the yarn twist, the greater the yarn strength. Apart from increased strength, however, a tightly twisted single yarn will be somewhat less flexible than one that is loosely twisted.

The auto-coning machine in Figure 3.10 is one additional step after yarn spinning that helps make yarn ready for fabric construction.

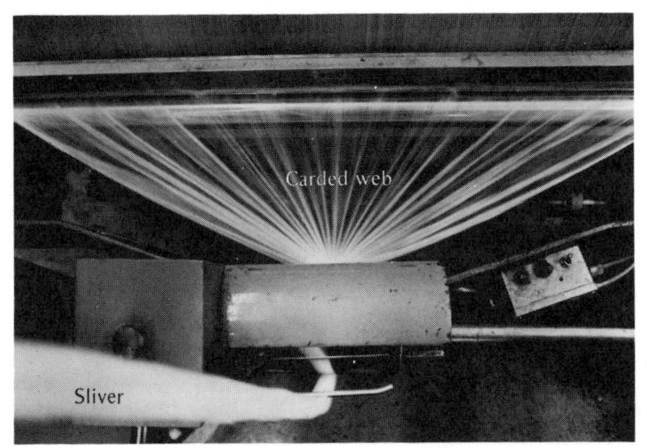

Figure 3.4. The carded fiber web is drawn from the machine and gathered into the ropelike sliver. *(courtesy American Textile Manufacturers Institute, Inc.)*

Figure 3.5. In combing, a finer version of the carding process, carded laps are further straightened to remove additional short fibers, thus leaving the longest and strongest staple fibers in the combed slivers. *(courtesy National Cotton Council)*

Improvements in Yarn Spinning

The process of forming a yarn just described is called ring spinning. In an effort to improve efficiency and decrease costs in yarn spinning, alternative methods have been employed in some yarn mills. These include open-end and friction spinning. The most popular of these, open-end spinning, takes fibers directly from a sliver to form the final yarn. The yarns formed in this manner, called open-end yarns, differ from ring spun yarns in that they may be somewhat rougher in appearance, slightly coarser and lower in strength. In recent years, automation and higher speeds have improved ring spinning also, which is often considered the preferred process for higher quality, fine count yarns. For many end uses open-

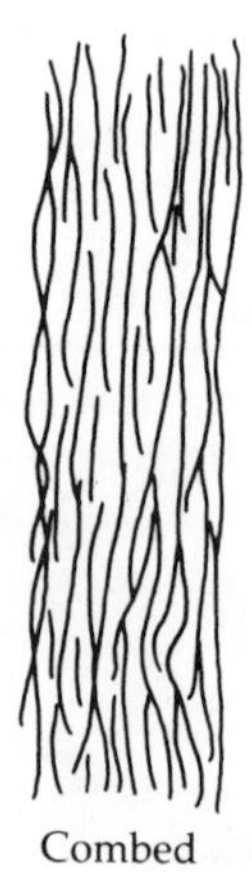

Figure 3.6. Remaining fibers are longer and more aligned in combed yarns than in carded yarns.

QUICK QUIZ 1

Briefly describe the initial processing steps in yarn formation that have been discussed so far:

1. Opening and blending ____
2. Lap or mat formation ____
3. Carding ____
4. Combing ____

****

ANSWERS QUICK QUIZ 1

1. Fibers are loosened and fluffed.
2. Fibers are formed into a sheet or mat of fibers.
3. Initial straightening process where fibers are partially aligned and arranged in a somewhat parallel relationship, then pulled into rope form, called a sliver.
4. Optional straightening process where fibers are made increasingly parallel; short fibers are combed out and the fibers are pulled into a combed sliver used for making fine quality yarns. (See Figure 3.6)

---

end yarns are very successful, especially in end uses where coarser yarns are acceptable. As a consumer, you will not be able to distinguish yarns spun on the traditional ring system from those spun on the open-end system. Instead, you must rely on the manufacturer to use the appropriate yarn for the end use.

Spinning of Filament Fibers

For manufactured filament fibers, the number of filaments in the yarn is determined before the fiber liquid is forced through the spinneret. Thus, the spinneret has the number of holes equal to the number of filaments to be used in forming the yarn as shown in Figure 3.12.

After the filaments are extruded through the spinneret, they are pulled together and given whatever amount of drawing and stretching is required for the particular fiber type. Then the multifilament strands are given the amount of twist designated for the particular yarn, based on its serviceability and end use requirements.

Figure 3.13 illustrates a comparison between filament and staple yarns. Any fiber produced in filament form may be cut into short lengths and then processed into staple yarns using the procedures described for spinning staple fibers. For this type of operation, hundreds, even thousands, of filaments are extruded simultaneously through a large spinneret head; this mass of filaments is called *tow.*

Traditionally, these filaments are then cut into the "staple" length desired for the particular type of yarn-making process to be used. If the cut length is similar to the length of cotton fibers, the tow is processed using the cotton system. If the filaments are cut to the length of wool fibers, they are made into yarns on machinery designed for wool fibers; either coarse woolen-type yarns or smooth, worsted yarns are produced.

A newer and even simpler way of turning filaments into short staple tow is to run the tow through a machine in which rollers randomly break the filaments. The breakage occurs at various points so throughout this procedure *parallel fiber arrangement is maintained* and no straightening processes

Figure 3.7. Several slivers are collected from the cans of sliver and combined into a single strand, which is drawn out and reduced to about the same size as one of the original slivers. Blends of two or more fiber types are sometimes made at this point. *(courtesy American Textile Manufacturers Institute, Inc.)*

are necessary. The tow can then be fed to regular drawing and spinning equipment where the actual yarn is formed.

When this cost cutting and now popular *tow-to-yarn* system is used, several steps in the conventional yarn-making process are eliminated. There is no formation of mat or lap, nor are the straightening processes of carding, first drawing and combing required.

The procedures used to construct yarns from filament fibers differ considerably from those used for staple fibers. In the case of silk, the two filaments produced by each cocoon are reeled from the cocoons. The size of the yarn is determined by the

Figure 3.8. Drawn slivers go to the roving frame where they are further drawn out and given a carefully measured twist. The resulting roving is wound onto a roving bobbin, shown at the bottom of the picture. *(courtesy American Textile Manufacturers Institute, Inc.)*

Roving Bobbin

Spinning Bobbin

Spinning Frame

Figure 3.9. A worker is loading a roving bobbin on the spinning frame, which handles the final operation in making yarn. The rovings are further twisted to required specifications and then wound on the row of small yarn bobbins at the bottom of the frame. If the roving should break, some machines automatically detect the problem and repair the break without stopping the operation. *(courtesy American Textile Manufacturers Institute, Inc.)*

number of filaments combined, which depends on the number of cocoons used. Then the filaments are given the desired amount of twist, and the yarn is completed. Silk may also be used in staple length. In this case, the spinning would occur in the same manner as other staple yarns. A spun silk yarn has the bulkier, fuzzier texture of a staple yarn.

Manufactured filament fibers may be used in either *monofilament* or *multifilament* form. While a monofilament may be used as a yarn, structuring it strong enough reduces flexibility considerably due to the large yarn diameter required. Thus, most filament yarns are multifilament structures, made by twisting together several small diameter filaments to provide increased yarn flexibility and strength. The diameter of filament fibers is determined by the size of the spinnerette opening, the amount of drawing and stretching given the fiber following extrusion and the

Figure 3.10. This machine automatically rewinds yarns from the small bobbins, which come from the spinning operation, onto larger cones. *courtesy Burlington Industries, Inc.)*

method of spinning. Monofilament yarns are suited for tire cord, fishing line, nylon hosiery and similar items. For these end use, monofilaments give good service under heavy weight or abrasion. Multifilament yarns are better choices than monofilaments for apparel and home furnishing fabrics because they contribute adequate strength and flexibility for these purposes.

Figure 3.11. Large textile manufacturers are making major capital expenditures to continue modernization projects such as this *open-end spinning* installation which eliminates intermediate processing steps in the production of yarn. *(courtesy Burlington Industries, Inc.)*

QUICK QUIZ 2

1. Describe these final stages in yarn processing:
   a. Drawing ______________________________
   b. Finisher drawing ______________________________
   c. Roving ______________________________
   d. Spinning ______________________________
2. The durability of the single yarn formed on the spinning frame depends on what three fiber properties? ______________________________
3. Yarn serviceability is also greatly affected by the amount of __________ added during spinning.
4. Open-end spinning skips the __________ stage of yarn formation and produces yarns that are somewhat __________ and __________ .
5. The strength of a multifilament yarn depends on the ______________________________
6. If a monofilament yarn is to be made, how many spinneret holes will there be? ______
7. Yarns produced from filament fibers require (more, less) processing than yarns made from staple fibers.

****

ANSWERS QUICK QUIZ 2

1. a. Further straightening of slivers in preparation for either combing or finisher drawing.
   b. A final drawing following carding or combing, which pulls the slivers into a smooth form ready for the roving process.
   c. Drawn slivers are fed into the roving frame where they are further pulled out and extended; a small amount of twist is added.
   d. Roving bobbins are placed on the spinning frame, where the roving is further extended and given the specified amount of twist.
2. length, tenacity and spinning quality
3. yarn twist
4. roving, coarser, weaker
5. type of fiber, size and number of the filaments, and amount of twist
6. one
7. less

Simple Yarn Types

Simple yarns are made by the processes described in the preceding discussion. They are characterized by being even in width, uniform in strength and stretch, and having the same number of turns or twists per inch. They tend to produce relatively smooth, flat fabrics. Simple yarns may be the *single, ply* or *cord* types described in this section.

Regardless of the system used in making yarns—cotton or woolen staple carded processes, combed or worsted operations; filament or tow spinning and the newer open-end spinning or tow-to-yarn operations—the type of yarn described to this point is a *single* yarn. The term ''singles'' may be used by some authors.

A single or singles yarn can be combined with other single yarns to form *ply yarns.* See Figure 3.14. If two

Figure 3.12. Filaments are extruded from the spinneret, drawn out the desired amount, and spun into multifilament yarn. *courtesy Industrievereinigung Chemiefaser E. V., Informationsstelle, Frankfurt/Main, Germany)*

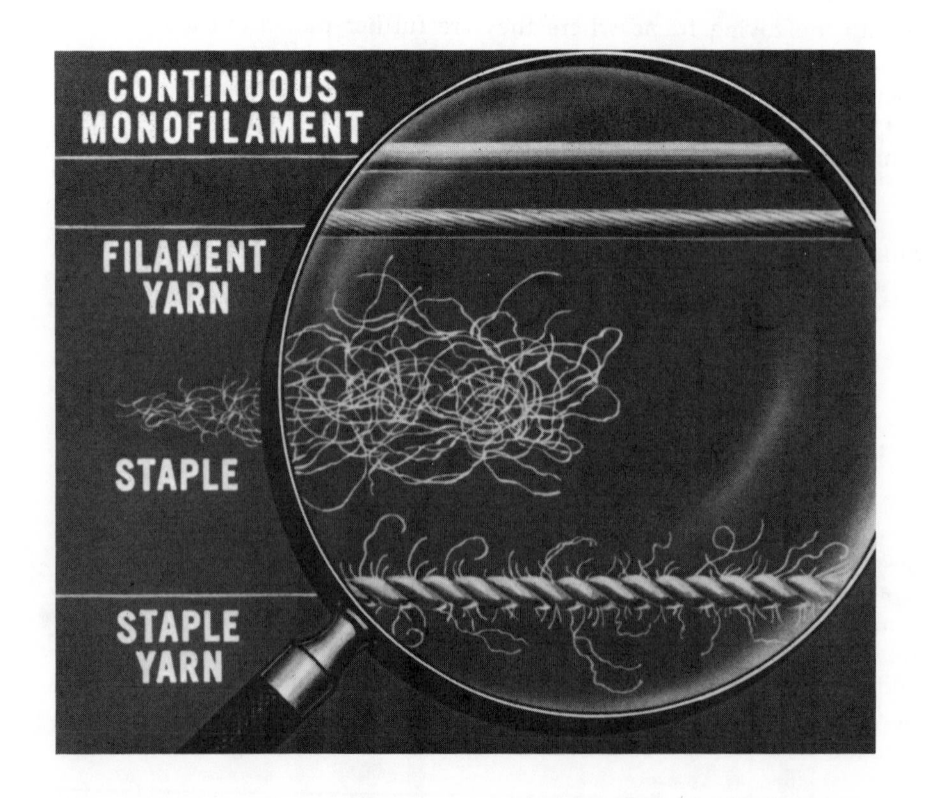

Figure 3.13. Staple yarns are made from staple fibers, which must be aligned before being spun into yarn; filament fibers require no straightening for yarn processing, and can be spun into yarn immediately upon extrusion from the spinneret. *(courtesy Man-Made Fiber Producers Association, Inc.)*

singles are used, the ply is called a two-ply yarn. Three singles make a three-ply yarn.

When three or more single yarns are combined into a ply, the term *multiple ply* may be used. Thus, a ply yarn composed of five singles could be called either a five-ply or multiple ply yarn.

When two or more ply yarns are combined, the resulting yarn is called a cord, also shown in Figure 3.14. The three basic procedures involved in the formation of a cord yarn are: staple or filament fibers are formed into a single yarn; two or more single yarns are formed into ply yarns; and two or more ply yarns are formed into a cord yarn.

Yarn Twist and Size

The amount of twist at each stage is an important factor in yarn serviceability. Generally speaking, the higher the twist the stronger the yarn. This is true up to a certain point, however, when very high levels of twist cause the yarn to become brittle and weak. Abrasion resistance and soil resistance are also improved by a greater amount of twist. Higher twist tends to decrease absorbency as there is less surface area and space within the yarn for moisture to be absorbed. For end uses in which durability is a consideration and absorbency is not as critical, high twist yarns are a better choice than low twist yarns. Upholstery fabrics are an example of an end use in which strength, abrasion resistance and soil resistance are desirable, while absorbency and flexibility are not as important. For this end uses, therefore, fabric constructed of high twist yarns should be chosen.

Yarn twist may be identifed as "S" or "Z." These describe the direction in which the yarn is twisted. An "S" twist yarn is twisted so that the turns are in the direction of the center portion of an S; a "Z" twist yarn is twisted in the direction that conforms to the center part of the Z. The direction of twist does not affect performance properties to any great extent.

A relatively complex system for measuring yarn size is currently used by the textile industry. Yarn size is measured differently depending on fiber type and length; that is, measuring techniques depend on whether a yarn is cotton, wool or man-made and on whether a yarn is staple or filament. This yarn measuring system can be summarized by dividing it into direct methods and indirect methods.

The direct approach to measuring yarn size evaluates the weight of a specified length of yarn. The most common measure used in the direct method is *denier.* This technique is most often used for filament yarns. In the direct measure system, the higher the number, the larger diameter (or coarser) the yarn. A 200 denier yarn, therefore, is coarser than a 50 denier yarn. Because many man-made noncellulosic yarns are used in filament length, denier is a term used to refer primarily to the size of man-made noncellulosic yarns.

The indirect method relies on the measurement of the length of a fixed weight of yarn. Simple yarns are most often measured on the indirect system. "Count" is frequently the terminology used to identify yarns measured indirectly. In this case, the higher the number, the smaller the yarn diameter or the finer the yarn. An 11 count yarn is coarser than a 35 count yarn. Cotton is often measured using this system.

You may occasionally encounter the terms denier and count in technical publications or promotional

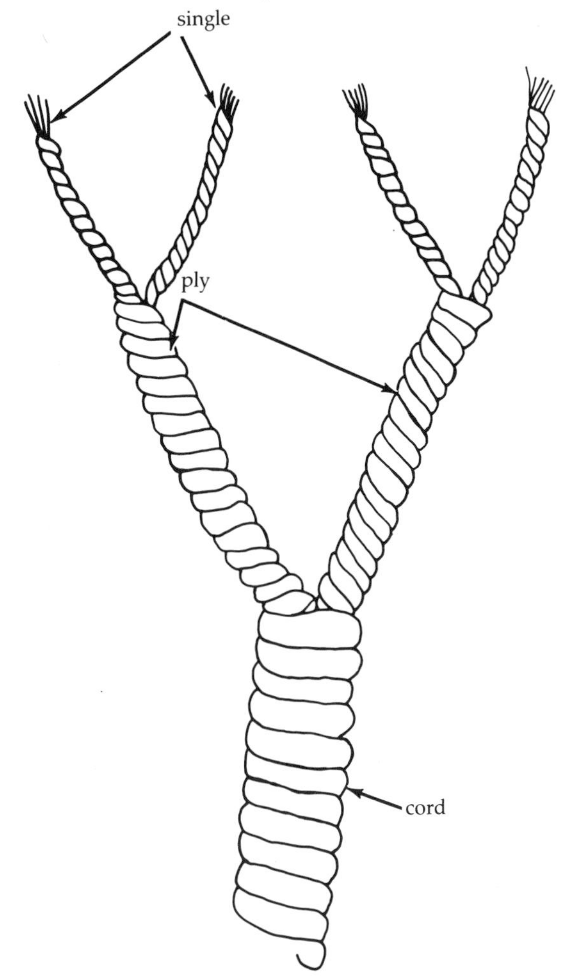

Figure 3.14. Simple yarns include single, ply or cord types; they may be made of staple fibers or filaments.

Figure 3.15. Crepe yarns are highly twisted simple yars used for decorative purposes.

literature distributed by textile companies. In the literature, these yarn numbers are sometimes followed by the abbreviation "Ne" which stands for number. Many proposals have been made to simplify the complicated system for measuring yarn size. No specific changes have been made as yet, however.

One type of simple yarn made with a very high amount of twist is called a *crepe* yarn. While some authorities classify crepe yarns as a novelty type, they are by definition simple yarns because the amount of twist is uniform throughout the structure. Additionally, yarn size and fiber distribution are uniform throughout, as can be observed in Figure 3.15.

Crepe yarns have so much twist that they tend to twist back on themselves and develop a crinkled fabric surface. As discussed earlier, excessive twist may contribute to reduced strength. However, for crepe fabrics produced primarily for esthetic purposes, the high twist yarns are satisfactory. In addition, crepe yarns can be combined with yarns of less twist in order to maintain adequate fabric performance.

Blended and Combination Yarns

When a yarn is composed of only one fiber type, such as cotton, it is called a single fiber component yarn. When fibers are blended together prior to the spinning operation, the result is a blend of two or more different fiber types, called a *blend yarn.* It is characterized by uniformity of fiber arrangement throughout the yarn. Blend yarns are single yarns composed of different fiber types—for example, a single yarn of wool and nylon.

Through blending, a yarn can be made to combine the advantages of different fibers, as well as to reduce their disadvantages, such as in the popular cotton/polyester blends. Polyester fibers contribute easy care and wrinkle resistance to the fabric, while the cotton contributes absorbency and comfort.

Figure 3.16 is a diagram of how two types of yarn with more than one fiber can be formed. The blend is characterized by uniform dispersion of the different fiber types throughout each single yarn, while the combination yarn has two single fiber component yarns of different fiber types combined in making a two-ply yarn.

Blend yarns are generally preferred to combination yarns since blends offer the best properties of each fiber present in each individual single yarn. Combination fabrics may be preferred for special appearance rather than performance; fiber properties are not as integrated within combination yarns as in blend yarns.

When a blended, single yarn is composed of natural cellulosic cotton, which gains strength when wet, and a man-made cellulosic which loses strength when wet, like rayon, the resulting product is improved over one composed of only rayon yarn. Also, the low heat tolerance of certain synthetic fibers can be offset by blending them with fibers not as heat sensitive.

Extensive research has gone into determining how much of each fiber in a blend is necessary to provide certain performance characteristics. There are no absolute percentages which are guaranteed to improve fabric qualities the optimum amount.

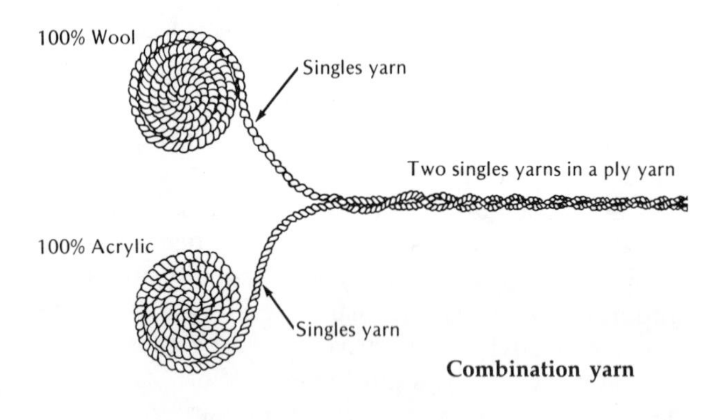

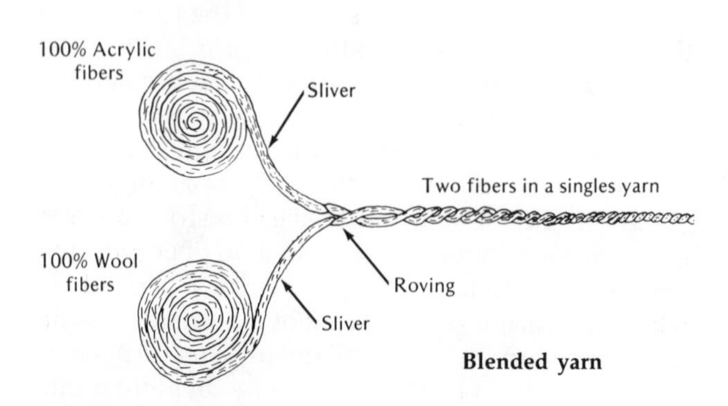

Figure 3.16. Blended and combination yarns produce different fabric characteristics.

Unfortunately, the real value a fiber offers to a blend does not come out in the same proportion as its respective percentage in the blend, due to the variables introduced as a result of varying fiber properties, such as density.

The popular blend of 50 percent polyester and 50 percent cotton does not mean that each fiber will contribute exactly half of its characteristics to the blend. The blending percentage refers to the percent by weight of the fiber present; thus, since cotton has a higher fiber density and greater weight than polyester, there will not be as much cotton fiber as polyester present.

Certain optimum blend levels are fairly well established as a result of research. In polyester/cotton blends, polyester content generally ranges from 35 to 80 percent, depending on the desired serviceability properties required for the end use. Generally, with increasing proportions of polyester and decreasing proportions of cotton, the fabric will have increased ease of care properties but decreased comfort potential.

Generalizations about optimum percentages are impossible since there are so many factors to consider. Yarn and fabric structures modify the effect of blending levels, as well as the fiber properties. A blend of lower-cost nylon and more expensive wool provides another illustration that the exact percentage does not determine the amount of contribution of each fiber. A small amount of nylon—15 percent is typical—increases the strength of washable wool, which is weakened because of the finishing treatment. Much greater amounts of nylon are needed, however, to increase the tenacity of fabrics with rayon. Despite the fact that specific percentages do not contribute precise amounts of fiber qualities to the fabric, blend levels have been perfected sufficiently to yield a superior product by combining the assets of two or more fibers.

To form a blended yarn, the various fibers involved are combined at the drawing stage in order that each single contains the different fibers. When a ply yarn is made from two or more singles where each is a different single fiber component yarn, the resulting ply yarn is a combination rather than a blend. Refer to Figure 3.16.

A ply yarn is a 100 percent nylon yarn if all the singles which make up the ply contain nylon fibers. It is a blend yarn if each of the singles which make up the ply contains two or more fiber types.

In summary, a blend yarn results when two or more fiber types are combined in each single; a combination yarn is made when two or more singles, each composed of different single fiber components, are joined into a ply or cord construction.

The manner in which yarns are woven into fabrics will also affect fabric performance. Sometimes different fiber contents are combined by using yarns of one fiber content in the lengthwise direction of the fabric, and yarns of a different fiber content in the crosswise direction. See Figure 3.17.

QUICK QUIZ 3

1. High twist yarns generally have (more, less) strength and (more, less) absorbency than low twist yarns.
2. Denier is a (direct, indirect) method of measuring yarn size.
3. List three advantages for blending fibers in making yarns:
   1. ____________________
   2. ____________________
   3. ____________________

****

ANSWERS QUICK QUIZ 3

1. more; less
2. direct
3. Any three:
   Blending maximizes the advantages of more than one fiber.
   It neutralizes the disadvantages of a given fiber.
   It contributes to more economical manufacturing of some yarns.
   It improves spinnability of short, noncohesive fibers.

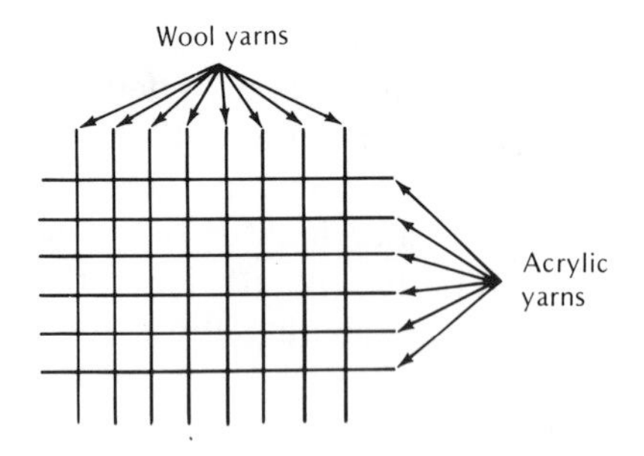

Figure 3.17. A combination fabric has lengthwise and crosswise yarns made of different fiber types.

When an end use demands that one direction of the fabric withstand great stress and strain, fibers with high tenacity may be used in the yarns for that direction. For example, a single fiber component yarn of a strong fiber such as nylon can be used in the lengthwise direction, with yarns of less tenacity in the crosswise direction, like cotton, to allow for increased absorbency or comfort.

Upholstery fabric is often made with the lengthwise yarns composed of high tenacity fibers and the crosswise yarns composed of another fiber type for esthetic appearance, for comfort or to cut costs. Although these combination fabrics may be advantageous for a particular end use, the most thorough means of integrating two or more fiber properties in a fabric is achieved through blend yarns.

QUICK QUIZ 4

Check the blends which would be desirable combinations. In making these decisions, evaluate if the fibers combine desirable properties and compensate for fiber limitations.

___ a. rayon and polyester
___ b. polyester and nylon
___ c. cotton and polyester
___ d. wool and polyester
___ e. rayon and glass
___ f. acrylic and wool

****

ANSWERS QUICK QUIZ 4

a, c, d, f are desirable blends

NOTE: The polyester and nylon blend (b) is not particularly advantageous since both fibers have similar properties; neither is absorbent. The rayon and glass blend (c) would be undesirable because the glass fibers would abrade the rayon.

UNIT THREE

Part II

Novelty, Complex and Textured Yarns

Novelty or complex yarns are characterized by irregular surfaces, uneven diameters, uneven twist, or other irregularities that result in fabrics with special appearance characteristics. Observe the variety pictured in Figure 3.18. Complex yarns may be either single, ply, or cord constructions. The majority of complex yarns used in modern fabrics are ply or a modified cord construction. The modified cord occurs when a yarn is composed of at least one ply yarn and one or more single yarns.

The types of complex yarns discussed in this text include *slub, flake or flock* (sometimes called *tweed*), *bouclé, loop, ratiné, nub, seed* and *chenille* yarns. *Core* and *metallic* yarns are also included in this section since they are frequently used for special or novelty effects in fabrics; however, they may appear to be simple yarns.

Single, Complex Yarns

Slub

A slub yarn is a single, complex yarn created by varying the amount of twist added during the spinning operation. The resulting yarn has thick-and-thin areas where the turns per inch are varied, as sketched in Figure 3.19. Number 3 of the yarn samples is a slub with characteristic thick-and-thin areas.

Flock or Flake

A tweed effect is created with a flock or flake yarn, which is usually a single yarn in which small tufts of fiber of a color that differs from the base yarn color is incorporated during the spinning operation. These small tufts of fiber are blended with the base yarn fibers either during the final drawing process, or at an intermediate step between drawing and roving formation. For those flake or flock yarns made of two-ply construction, the tufts of extra fiber are incorporated when the ply is formed. The extra fibers are held in place by twisting together two singles into the ply yarn.

Flock or flake yarns are used in the manufacture of tweed fabrics, which is why they are sometimes called tweed yarns. They are generally made from staple-length fibers, and the extra fibers added to form the flake or tweed-effect are also staple-length fibers. The flecks of contrasting color are included in the final spinning stages involved in the formation of either a single yarn, or in combination of two singles to form a two-ply yarn. Tweed yarns are generally fuzzy because the fibers are not highly straightened or tightly twisted. Therefore, it can be reasoned that flock or flake yarns are made by a system that includes only the carding straightening process. Wool fibers, or fibers cut to the typical length of wool, are usually used. The characteristics of wool fibers make them a wise choice for use in tweed fabrics; the fiber scales and crimp contribute to good cohesiveness, which helps hold the short tufts in place. Yet, the small color flecks or tufts can be pulled out of loosely twisted areas.

Complex Ply Yarns

Complex yarns, frequently of ply construction, are produced with a variety of novelty effects. Types

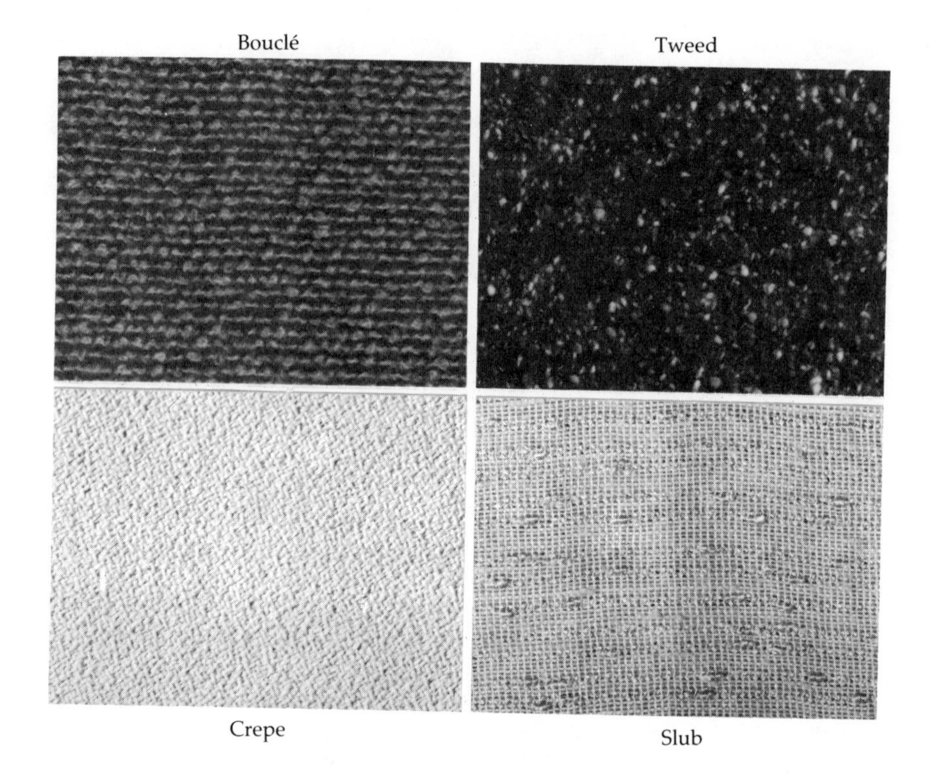

Figure 3.18. Different textures and designs result from the use of yarns—notice the fabrics created from bouclé, tweed, crepe and slub yarns.

Figure 3.19. Slub yarns create nubby fabrics.

of complex ply yarns include bouclé, loop, ratiné or gimp, nub, knot, seed, corkscrew, and chenille. Figure 3.20 shows a diagram of a bouclé, a typical complex ply yarn. The *three ply structure* includes a *base or core* single, an *effect* single, and a *binder* or tie

---

QUICK QUIZ 5

1. Other methods can be used for achieving tweed effects besides flake yarns, where small tufts of colored fibers are added. Examine the yarn samples and indicate which yarn has a multicolored tweed effect. ______________ Untwist the yarn; how many plies does it have? ______________
2. How was the multi-colored effect achieved? ______________________________
3. What advantage does a four-ply tweed effect yarn have over a flock or flake yarn? ______________________________

****

ANSWERS QUICK QUIZ 5

1. Yarn 5; four
2. by combining different colored singles yarns into a four-ply yarn
3. The ply yarn has higher abrasion resistance; in a flock or flake yarn, the color tufts are subject to abrasion.

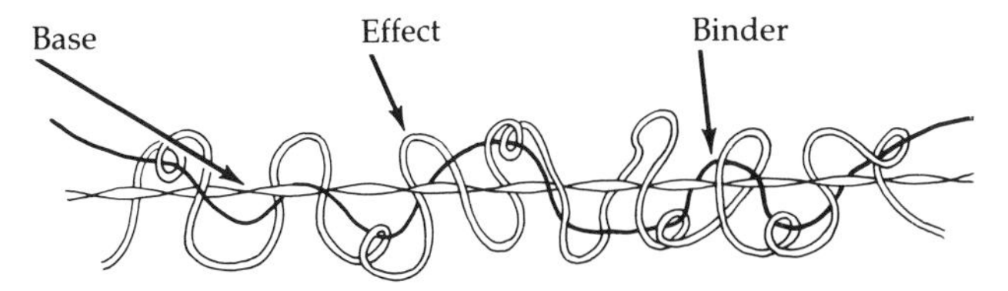

Figure 3.20. Complex ply yarns typically are made of base, effect and binder yarns.

single. Most complex ply yarns contain these three components. The base or core determines the length and stability of the yarn; the effect yarn contributes the special esthetic component desired; and the binder yarn holds the effect in place around the core.

The bouclé yarn diagrammed contains single yarns for each of the three components; however, ply yarns may be used to create the core effect or binder yarns in producing complex yarns. Definitions of various complex yarns have evolved, and the nature and handling of the effect yarn frequently determines the name of the complex yarn.

Bouclé and Loop Yarns

The major difference between bouclé and loop yarns is that the effect yarn in a loop yarn forms obvious loops on the yarn surface; in bouclé, the effect yarn is tighter and loops are not obvious. In addition, the loops on a bouclé are closer together, while the loops on loop yarns are larger and farther apart. Of the three components discussed which are typically used to form complex ply yarns—the base yarn, sometimes called the core, usually imparts strength to the final yarn.

The fancier and larger the effect yarn, the less durable the resulting fabric will be. There is more opportunity for it to abrade, snag and be damaged.

Ratiné or Gimp Yarns

Other complex ply yarns with special loop effects include ratiné or gimp yarns. They are similar to bouclé and loop yarns previously discussed, which have loops added by the effect yarn. A distinguishing characteristic of ratiné or gimp yarns is that their loops are much closer together than in the bouclé yarns.

Ratiné yarns are characterized by a rough surface created through two twisting operations. First, the regular yarn spinning occurs, then twisting in the opposite direction establishes the desired esthetic or looped effect.

Nub and Knot Yarns

The terms *nub or spot* and *knot or knop* are often used interchangeably, but there are minor differences. These yarns are typically made of base and effect yarns, with no tie or binder required.

A nub or spot yarn is produced when the base yarn is held stationary while the effect yarn is wrapped around it several times to build an enlarged segment. The enlarged nub or spot is often secure enough so that no third ply or binder yarn is needed to hold the effect yarn in place.

The knot or knop yarn is produced in much the same way, except that brightly colored fibers are frequently added to the enlarged knot. Thus, knot yarns are also typically two-ply yarns consisting of base or core and effect yarns.

Seed Yarns

Seed yarns resemble nub and knot yarns in that they also are made of base and effect components often with no binder. The difference is in the size and shape of the enlarged segment.

Yarn 1 is a seed yarn. Compare it with the nub and knot yarns 6 and 7. The "seed" formed by the effect yarn is smaller than the enlarged segment in the sample 7 knot yarn.

Chenille Yarns

These yarns resemble a hairy caterpillar—in fact, chenille is the French word for caterpillar! To create fabrics made of chenille yarns, a special fabric is constructed and then cut into narrow lengthwise strips that serve as chenille yarns. They are used to create special esthetic textures; you may recall seeing chenille fabrics that have vertical or horizontal fuzzy strips.

Metallic Yarns

Another type of yarn that exists almost entirely for esthetic reasons is metallic. It is discussed in this section because of its primary use to create novelty

QUICK QUIZ 6

1. Compare the loops in yarns 2 and 8. Based on closeness of loops, identify which is the bouclé: ____________; and which is the ratiné ____________.
2. Name the types in yarns 6: ____________________, and 7:____________________.
3. By matching, indicate which complex yarns are typically made of three singles or plies, and those usually made of only base and effect yarns.

1. ____ seed		a. base and effect yarns
2. ____ bouclé		b. base, effect and binder yarns
3. ____ loop		
4. ____ knot		
5. ____ nub		
6. ____ ratiné		

4. Which of the yarn samples is a chenille yarn? ____________ How did you identify it? ____________________
5. Name some textile products that are associated with chenille effects: ____________________

****

ANSWERS QUICK QUIZ 6

1. bouclé - 2; ratiné - 8
2. 6-nub (or spot); 7-knot (or knop)
3. 1-a; 2-b; 3-b; 4-a; 5-a; 6-b
4. 9; by the short protruding fibers that resemble a catepillar
5. bedspreads, bathrobes, rugs

fabric effects, although metallics do not fit totally under the preceding definitions of complex yarns.

Metallic yarns are made using several different techniques; however, two methods are most common. In the more common process, a thin, flat piece of metallic foil is given a plastic coating similar to the polymer used in making either acetate or polyester. After the metallic foil is coated, the film is sliced into narrow strips to form metallic yarns. The second method involves dispersing a fine granular form of the metallic substance in an acetate or polyester polymer, which is then extruded in the form of a fiber filament. This second method is increasing in use.

Metallic and chenille yarns are produced for special textural effects in fabrics, and both can contribute to limitations in fabric durability or care and maintenance procedures. In terms of their manufacture, these two novelty yarn have in common that they both can be made by slicing different fabrics into strips to form either flat metallic yarns or hairy chenille yarns.

Metallic yarns made by the fabric-slicing process are flat rather than round. The flat, monofilament coated yarns are less pliable, making the resulting fabric less flexible and more stiff or harsh than fabrics made with rounded yarns. Metallic yarns are used infrequently, not only due to durability limitations, comfort or care requirements, but because of expense. Thus, you will find metallic yarns primarily in luxury fabrics designed for special esthetic decorative end uses.

When metallic yarns are selected for a fabric, they are typically used in combination with other fibers and yarns for several reasons: other yarns can increase comfort potential, making fabric less harsh and more absorbent; durability properties can be improved by using flexible, abrasion-resistant yarns; cost can be reduced by combining yarns that are less expensive; and easy-care fibers and yarns can simplify fabric maintenance.

Core Yarns

The metallic wrapped core just discussed is used for esthetic purposes. In addition to these decorative core yarns, there are those produced for specialized durability, comfort and care features.

By definition, a core yarn is one in which a base or core yarn of one kind of fiber or blend is completely covered with a second type of yarn. Refer to Figure 3.21. The *covering yarn* may be either a simple or complex yarn; the *core yarn* is a simple yarn but may be either a single or ply.

The same fibers may be used for both covering and base yarns, but typically the cover yarn is composed of quite different fibers in order that limitations of one fiber will be offset by another. In fact, one main reason for making core yarns is to combine desired characteristics of two or more fibers or yarns. For example, a rubber core may be wrapped with cotton to yield a comfortable, relatively durable and highly elastic yarn.

Yarn sample 10 is a core yarn produced primarily for decorative purposes. Examine the yarn, locate its core or base that is surrounded by a very loosely twisted cover yarn. As in your sample, the covering yarn is often so low in twist that it resembles a layer of fibers rather than a typically twisted yarn. However, it performs the basic purpose of covering the core yarn and creating special esthetic effects.

The core and cover yarn in yarn sample 10 both contribute to yarn serviceability. The base or core is a two ply yarn that adds strength and stability to the yarn. The cover yarn contributes to comfort in cold weather through its bulk density, which traps air and contributes to thermal retention.

Core yarns are used frequently where high degrees of stretch are desired, so elastomeric fibers like spandex and rubber are especially suitable choices for the base yarn. Elastomerics are more often found in the base or core rather than as a covering yarn, because they are used for their elongation and elastic recovery rather than for esthetic, care or comfort considerations. They are best covered with an outer cover yarn selected for other properties. Also, the heat sensitivity of rubber can be compensated for by a heat resistant covering. The dye affinity problems associated with the elastomerics become no problem

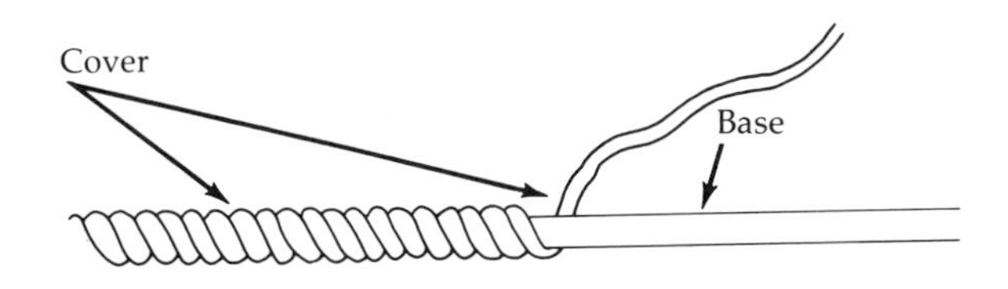

Figure 3.21. Core yarns contain a base or core yarn completely covered with a second yarn.

when the cover yarn is absorbent and takes dye readily.

Core spinning is one method of making elastomeric core yarns. It involves stretching a filament core of spandex or rubber during spinning, while the cover yarn is tightly wound around the stretch core. After the core is wrapped, the core is allowed to return to its normal length; this causes the outer cover to bunch up, providing for built in stretch or elasticity during use.

The degrees of stretch designed into fabrics vary considerably, and two terms are used to describe the relative amount that a fabric will stretch in use. The first is *comfort stretch* and is applied to fabrics containing enough give for normal wearing comfort. This stretch potential is generally under 30 percent. Consumer items for which this relatively low stretching ability is acceptable include dress shirts or slacks, blouses, skirt, dresses, blazers or any garments worn for routine, low activity purposes.

Power or action stretch describes fabrics with far greater stretch potential, normally between 30 and 50 percent, though it is possible to build in as much as 200 percent stretch. Power stretch products have extra holding power to help maintain desired shape when worn or used under excessive stretching conditions. Some items subjected to such excessive conditions include swimwear, athletic uniforms, foundation garments, and ski wear.

While it would be possible to process yarns of 100 percent spandex, most stretch yarns include only three to ten percent. This low amount of spandex, or even rubber, creates yarns with adequate stretch for most end uses, whether comfort or action stretch is needed.

Stretch can be built into yarns by using several processes. In addition to the creation of power or action stretch through the spinning of core yarns, another method involves combining spandex with some other fiber in the drawing stage. A simple, single yarn is produced from the drawn, blended sliver. These yarns are quite appropriately called *intimate blend* stretch yarns.

Yarn texturization is yet another way to add stretch to textile products. Also, as discussed in the next section, texturization is used to increase thermal retention and bulk density, decrease high luster, create irregular surfaces for certain design effects, and increase cohesiveness for yarn spinning.

QUICK QUIZ 7

1. A core yarn, by definition, has a base or core yarn of one kind of fiber or blend and a second ________________ yarn.
2. Core yarns combine desired characteristics of two or more fibers or yarns; often ______________ fibers are incorporated to add ______________ or ______________ stretch to textile products.
3. Elastomeric core yarns can be made through ____________ spinning or by combining spandex with other fibers during the drawing stage, called ______________ blending.

****

ANSWERS QUICK QUIZ 7

1. covering or cover
2. elastomeric (spandex); comfort or action/power
3. core; intimate

Special Effects Through Yarn Texturization

Most fibers tend to be straight and relatively smooth, except for animal hair fibers, like wool, whose crimp contributes natural textured effects. When smoothness is not desired, texturization is needed to create less luster on manufactured filaments, for more cohesiveness for yarn processing, or special esthetic effects. Yarns are also textured to impart bulk, stretch or a combination of both to textile products. Textured yarns may be simple or complex types.

Texturization can be done as a part of the filament or tow manufacturing processes, or it can be applied to filament fibers at a later time. Textured yarns may be made from filament-length fibers, or from filament tow which will be cut into staple length for making staple yarns. When textured staple fibers are used, the yarns are made by any of the standard yarn manufacturing processes designed to handle staple fibers.

Texturizing most often consists of distorting or altering a filament's shape, and then heat setting the resulting change to make the new form relatively permanent. The exact form the textured yarn takes depends on the method used to make it—yarns are texturized by a variety of processes including *crimping, coiling, curling* and *looping*.

Textured yarns, found in a wide variety of fabrics, are very versatile for creating esthetic and comfort features for many end uses. Because filaments are textured by processes that involve heatsetting, a majority of textured yarns are made from thermoplastic fibers. The type of texturizing process chosen depends on the specific characteristics desired in the yarn—i.e. stretch, bulk or a special esthetic effect. In Figure 3.22 are shown selected types of textured yarns which cannot be identified by inspection.

Figure 3.23 shows that textured yarn can be extended considerably. In fact, textured yarns are frequently used in apparel to provide comfort stretch, with an elongation potential up to 30 percent. However, when action or power stretch with an elongation potential of 30 to 50 percent or more is required, elastomeric core yarns are used.

Several of the complex or simple yarn types discussed previously can be simulated through texturizing. A textured yarn made by the loop process can be designed to have esthetic effects similar to such complex yarns as the loop, bouclé, and possibly ratiné, if loops are very small.

A crimped or crinkled textured yarn can be made to resemble a very high-twist crepe yarn, and crepe effect yarns are frequently made of texturized synthetic filaments. Fabric 20 is polyester fabric textured to resemble a crepe construction. It is not a true crepe, however, because it was not made from highly twisted yarns.

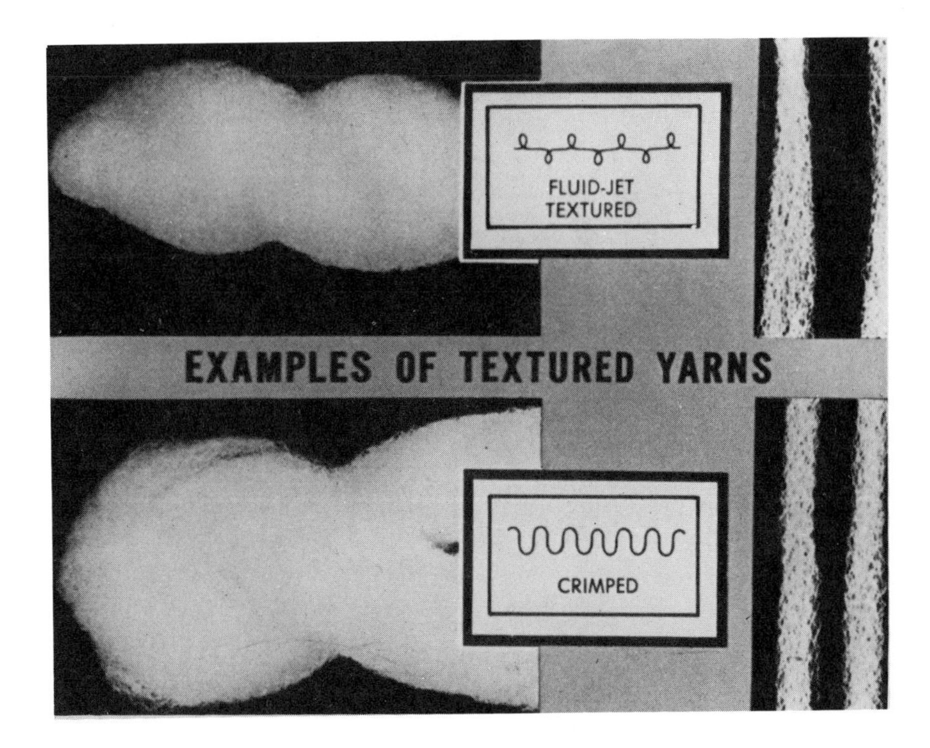

Figure 3.22. Two of the several yarn texturing processes are shown. *(courtesy Man-made Fiber Producers Association, Inc.)*

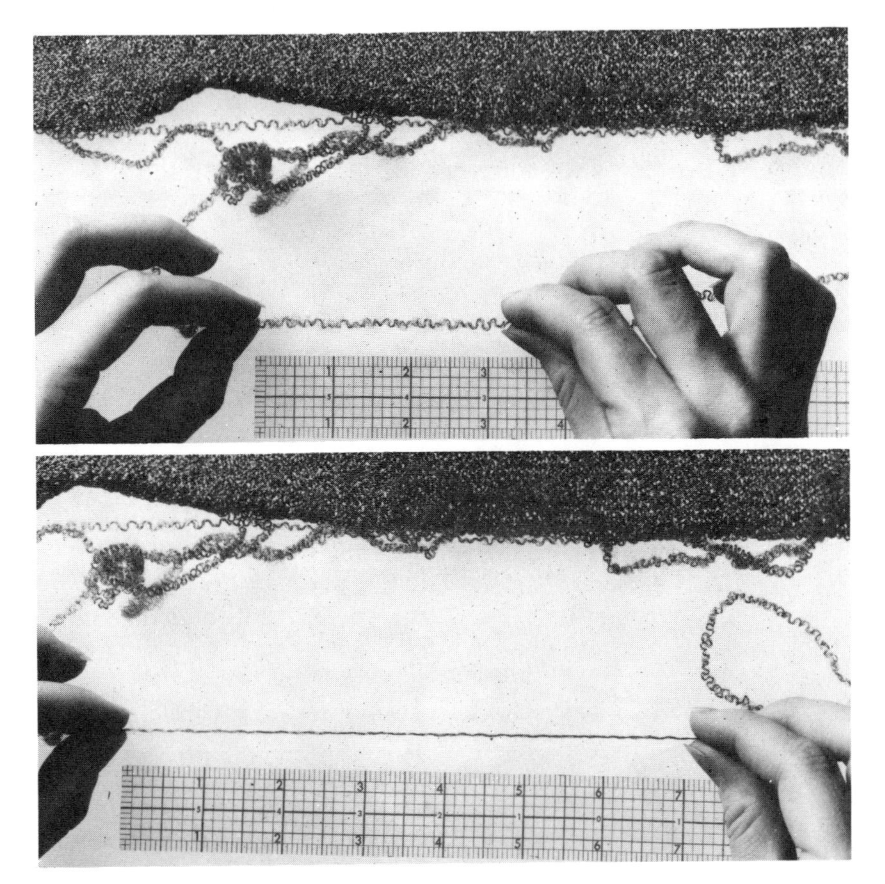

Figure 3.23. Textured yarns have built in stretch potential; note how much this crimped yarn stretches.

REVIEW

1. List the traditional steps in spinning yarns from staple fibers:
   a. ______
   b. ______
   c. ______
   d. ______
   e. ______
   f. ______
   g. ______
   h. ______

2. Which process is used only for a select group of yarns? ______
   Why? ______
   ______

3. Yarn processing steps for wool fibers are similar to those used for cotton and other staple fibers. Wool yarn made *without* a process comparable to combing is called ______ yarn, while wool yarn processed to include combing is called ______ yarn.

4. When a yarn is made of only one type of fiber, for example nylon, it is called a ______.

5. List three reasons for making blend yarns:
   ______
   ______
   ______

6. If two singles are combined into one yarn, the result is a ______ yarn.

7. If three singles are combined into one yarn, the result is a ______ yarn.

8. If two, two-ply yarns are combined into one yarn, the result is a ______ yarn.

9. Match the yarn type with its appropriate description.

___	1. crepe	a. thick-and-thin yarn
___	2. slub	b. tightly twisted yarn that appears kinked or crinkled
___	3. tweed/flake	c. complex yarn composed of core, binder and effect yarns
___	4. core	d. fuzzy, caterpillar-like yarn
___	5. bouclé	e. base yarn surrounded completely by another yarn
___	6. chenille	f. loosely twisted yarn with short colored fibers

10. Name two major characteristics of textured yarns: ______
    ______
    What usually holds in the textured effect? ______

11. List three advantages textured yarns contribute to fabrics:
    ______
    ______

12. What is the difference in stretch potential between power and comfort stretch?
____

13. What is the principal justification for most complex yarns?
____

14. The number of filaments in a multifilament yarn is determined by the ____________ ____________.

15. As manufactured fibers are extruded from the spinneret, they are either combined and twisted into ____________ yarn, or cut into short lengths for spinning into ____________ yarn.

16. The procedure for making many textile fabrics involves five steps: fiber processing, followed by spinning into ____________, which are structured into ____________, to which are added finishes, color and designs.

****

REVIEW ANSWERS:

1. a. opening and blending of fibers
   b. formation of picker lap
   c. carding
   d. drawing
   e. combing
   f. finisher drawing
   g. roving
   h. spinning
2. combing; because this straightening is needed only when extra smooth and uniform yarns are required
3. woolen; worsted
4. single fiber component yarn
5. Any three, Blending:
   maximizes the advantages of more than one fiber;
   improves the spinnability of some fibers;
   neutralizes the disadvantages of a single fiber;
   contributes to more economical manufacturing of yarn products;
   improves the versatility of the completed yarn or fabric.
6. two-ply
7. three-ply or multi-ply
8. cord
9. 1-b; 2-a; 3-f; 4-e; 5-c; 6-d
10. bulk and/or stretch, rough or crimped surface; heat setting processes
11. Any three: increased stretch, increased thermal retention, increased bulk with little weight, special esthetic effects
12. comfort stretch potential is usually under 30 percent; power stretch is usually 50 percent or more
13. esthetic appearance
14. number of holes in the spinneret
15. multifilament; staple
16. yarns; fabrics

UNIT THREE

Part III

Serviceability Characteristics

In this section the various yarn structures are discussed in terms of the serviceability concepts of durability, comfort, care and esthetic appearance. The serviceability of any yarn depends on fiber content used, as well as type of yarn construction.

The amount of twist given a yarn is an important consideration in determining serviceability properties. Up to a point, the greater the amount of twist the stronger the yarn; beyond an optimum level, twist tends to reduce fiber tenacity and make a yarn somewhat brittle or less flexible. Another major factor involved in determining serviceability of yarn is its type—*single, ply* or *cord* —and class—either *simple* or *complex*.

Figure 3.24 shows the difference in twist required for yarns made from staple and filament fibers. Because of the length of filament fibers, the amount of twist required is considerably less than for staple fibers. The long length of the filaments helps hold the fibers together; whereas staple fibers require the higher twist to form yarns.

A high degree of twist works well when spinning short, staple-length fibers which have enough *cohesiveness* or *spinning quality* to adhere together. However, if a yarn is twisted too tightly so the pulling force of the twist exceeds the tenacity of the fiber, the yarn will break or rupture.

The amount of abrasion a yarn can take is influenced, in part, by fiber content. The nature of yarn construction, especially in complex or novelty yarns, also influences abrasion—novelty effects are often loose and very susceptible to abrasion and snagging. Plus, they are uneven in load-bearing capacity, which decreases abrasion resistance. The effect ply in a bouclé or loop yarn has a tendency to abrade or snag. Another yarn which is susceptible to abrasive forces is the slub yarn because of the intermittent areas of reduced twist that produce thick, loose areas which snag and abrade easily. The relative amount of abrasion resistance necessary for serviceability depends on the planned end use.

Elongation and elastic recovery are very high in core yarns that contain elastomeric fibers. The absolute amount of abrasion resistance characteristic of elastomeric core yarns depends on both fiber and yarn structure used in the cover yarn.

Absorbency is an important property affecting fabric comfort, and yarn twist and type affect the potential for moisture absorption in a textile product. Low twist yarns contribute to increased fabric absorbency because they have more surface area available for absorption than high twist yarns.

In addition to low twist, texturization processes can also be applied to thermoplastic filaments to increase their absorbency and comfort. One advantage of a textured yarn over a smooth filament yarn is that the filaments in a textured yarn do not pack; this can contribute to increased absorbency because the spaces between the filaments permit the yarn to absorb moisture or pass it through the fabric for evaporation. Thus, in warm weather, texturization can increase comfort of yarns made from low absorbency fibers in the man-made noncellulosic or synthetic categories.

Textured yarns also can increase thermal retention when engineered with dead air spaces which provide excellent insulation and heat retention; thus, textured yarns can contribute to either warm

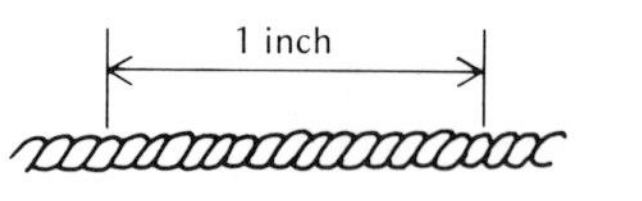

Figure 3.24. Suggested optimum twist for filament yarns is three to five turns per inch, compared with ten to 20 turns per inch for staple yarns.

or cold weather comfort. When they are texturized to allow for increased air transfer through the fabric, high thermal conductivity results; when engineered to trap air in the yarn, high thermal retention is created.

In addition to durability and comfort properties influenced by yarn structure, care procedures are likewise affected. More delicate care procedures are required when fabrics contain complex yarns.

Yarn Identification

Remove the Yarn Mounting Sheet from the back of the book and review the types of yarn you are to identify. As directed in this section, determine the yarn construction methods used in the fabric samples indicated.

To identify how a yarn is made, first remove a yarn from the fabric sample you are studying. Hold each end between your thumb and first finger of both hands; slowly untwist the yarn by turning one end of it in the opposite direction of the yarn twist. You may have to turn the yarn in both directions to see how it will untwist. Determine if the yarn is a single, ply or cord construction, and if it is made from staple- or filament-length fibers.

Using the fabric samples numbered below, pull out a yarn from one direction of the fabric and un-

---

QUICK QUIZ 8

1. The durability of a yarn is affected by what factors? ______________________
____
2. A tightly twisted yarn is somewhat (more, less) flexible than a loosely twisted one.
3. Dimensional stability is also partially dependent on the tightness or looseness of the yarn twist. The higher the degree of twist up to the optimum, the (more, less) stable the yarn because ______________________________________.
4. Why are simple yarns generally more durable than novelty yarns?
____
____
5. Name four complex yarns that can be weakened and abraded by agitation in laundering.
____
6. Why are many textured yarns likely to be damaged by high laundering, drying, or ironing temperatures? ______________________________________
____
7. What high twist yarn made from man-made cellulosic fibers, frequently rayon, can be stretched or shrunk by excessive moisture? ____________________

****

ANSWERS QUICK QUIZ 8

1. fiber content, amount of yarn twist, diameter of yarn
2. less
3. more; the fibers are compacted and adhere more securely
4. Simple yarns are typically uniform and have a higher degree of twist, which increase tenacity and abrasion resistance.
5. Any four: bouclé, slub, ratiné, loop, knot, seed, flake tweed
6. They are often made of thermoplastic fibers, and excessive heat can remove texturization.
7. crepe

---

twist it as described above; then pull out a yarn from the opposite direction of the sample fabric and repeat the process. Determine if the lengthwise and crosswise yarns are different types. Remember that two yarns of the same type can have different thicknesses and textures due to variations in degree of twist and amount of fiber used.

Simple yarn types are used more frequently in textile products than complex or novelty yarns. Most of the fabrics in your packet are, likewise, made from simple yarn types, either single or ply. Simple yarns are uniform in construction and tend to give fabrics increased durability or serviceability. They are also generally less expensive to produce.

1. Label the yarns in each sample listed as singles, ply or cord constructions.

Fabric	Yarn Type: Single, Ply or Cord
10:	________
11:	________
18:	________
19, top layer:	________
21:	________
39, top layer:	________
41:	________

   Only samples 19 and 21 of the above have different types of yarn construction in the lengthwise and crosswise directions. However, the single yarns used in samples 10, 11 and 18 differ in thickness and texture in the crosswise and lengthwise directions.

2. Recalling how single yarns are made, give two reasons for the variations: ________
   ____
   ____
   ____

3. Indicate by matching whether the following samples are made from filament-or staple-length fibers:

   ____ a. fabric 10 1. filament
   ____ b. fabric 11 2. staple
   ____ c. fabric 18
   ____ d. fabric 39, top layer

4. Fabrics 18, 39 and 41 contain different types of fibers in each single yarn; therefore these fabrics are made from ________ yarns. Fabrics 10 and 11 are made from one type of fiber, so they are structured from ________ ________ yarns.

5. Fabrics 37 and 43 have yarn construction types that create special esthetic effects. In each sample pull out one of the larger yarns. Is it simple or complex? ________
   Name the yarn type in 37: ________
   and 43: ________.

   Now cut 1½ inch by 2 inch samples from the fabrics studied above—10, 11, 18, 19, 21, 37, 39, 41 and 43. The longer direction of each small sample should be in the same direction as the long direction of the large sample, since crosswise and lengthwise yarns may differ. Attach the samples to the Yarn Mounting Sheet on the appropriate boxes for Simple Yarns.

6. Examine fabric 3, and name the complex yarns present: ________. How does construction of these yarns differ from that of the simple yarn types just studied? ________
   ____

   Attach a small sample of fabric 3 to the Yarn Mounting Sheet. Yarn types are not always easy to identify, given the smallness of the staple or filament singles used and the tightness of twist. Therefore, most of the complex yarns in the study packet are separate yarn samples to allow for easier identification.

7. As a review, take out the yarn sample card and identify each yarn type represented.
   1: ________
   2: ________
   3: ________
   4: ________
   5: ________
   6: ________
   7: ________
   8: ________
   9: ________
   10: ________

8. Pull on fabric samples 8 and 9; what characteristic related to comfort do they have in common?
   ____

9. Unravel in both directions some yarns from fabrics 8 and 9. What type of yarns were used to provide the comfort stretch? ________

10. In fabric 8, compare the crosswise and lengthwise yarns. In which direction are the 100 percent textured polyester filaments used that impart the stretch? ________

11. In sample 9, a 100 percent polyester fabric, what type of yarns were used to give two way stretch? ________________
12. Examine fabric 20. Explain how this crepe effect was created. ________________
13. Fabrics 26 and 42 are both knitted structures, and yarns are often hard to unravel from knits. Sample 42 can be unravelled, however, by pulling on its corner. What type of yarns were used to make it? ________________
    Cut small samples from the five fabrics just studied—8, 9, 20, 26, and 42. All were made with textured yarns. Attach them to the Yarn Mounting Sheet.
14. Which of these fabrics contains yarns with the highest degree of stretch? ________________
15. Fabrics may be composed of both simple and complex or novelty yarns for a variety of effects. The many types of yarns are combined in different fabric constructions to provide the consumer with varied ________________, ________________, ________________, or ________________ serviceability properties.

****

Yarn Identification Answers:

1. 10-single; 11-single; 18-single; 19-lengthwise single, crosswise ply; 21-lengthwise single, crosswise ply; 39-single; 41-single
2. amount of fiber staple or filaments used; degree of twist
3. 1-a; 2-b; 1-c; 2-d
4. blended; single fiber component
5. simple; 37-two ply; 43-cord (two, two-ply yarns are twisted together)
6. slub; the amount of twist is varied, resulting in thick and thin areas
7. 1-seed; 2-bouclé; 3-slub; 4-singles; 5-four-ply, tweed effect; 6-nub; 7-knot; 8-ratiné; 9-chenille; 10-core
8. both have comfort stretch
9. textured
10. crosswise
11. textured filament
12. through textured filaments (rather than high-twist crepe yarns)
13. textured
14. fabric 42
15. durability, comfort, care, esthetic

Posttest

1. Fabrics with rough textures are produced from ________________ yarns.
   a. woolen and worsted
   b. worsted and combed
   c. carded and woolen
   d. carded and combed

2. True crepe yarns are produced with
   a. crimped filaments.
   b. effect plies.
   c. crepe weaving.
   d. high amount of twist.

3. What property of wool fibers is beneficial for achieving a flake or tweed yarn?
   a. resiliency
   b. flexibility
   c. cohesiveness
   d. tenacity

4. The smaller and less complex the effect ply in a ratiné or bouclé yarn, the ________________ durable the resulting fabric.
   a. more
   b. less

5. What is the correct order of steps for forming a cord yarn?
   a. singles, ply, roving, cord
   b. roving, singles, ply, cord
   c. roving, ply, singles, cord
   d. ply, singles, roving, cord

6. The ________________ the yarn twist, up to an optimum point, the ________________ the yarn.
   a. less, stronger
   b. greater, stronger
   c. greater, less strong

7. Simple, singles yarns wear well in fabrics because
   a. of their uniform size.
   b. they have two or more plies.
   c. they are very tightly twisted.
   d. they are made of filament fibers.

8. You would expect a ______________ yarn to be stronger than a ______________ yarn.
   a. singles, ply
   b. ply, cord
   c. cord, ply

9. What does texturing contribute to a yarn?
   a. increased luster
   b. decreased absorbency
   c. increased stretch
   d. increased pilling

10. Which yarn would be least affected by agitation in laundering?
    a. core
    b. bouclé
    c. slub
    d. flake

11. What do natural cellulosic fibers contribute to synthetic blends in wearing apparel?
    a. increased resiliency
    b. increased absorbency
    c. increased tenacity
    d. increased abrasion resistance

12. What type of yarn is made with a base yarn that has other yarns wrapped or twisted around it?
    a. ply
    b. bouclé
    c. textured

13. What term is applied to fibers which are drawn off from the carding and combing machines during yarn processing?
    a. roving
    b. yarn
    c. sliver
    d. mat

14. What purposes in yarn spinning do carding and combing processes serve?
    a. blending fibers for combination yarns
    b. aligning staple-length fibers and removing short, fibers from lap
    c. aligning filament-length fibers and providing a smooth finish
    d. combing singles into ply yarns

15. Which of the following products would be best served by the use of low twist yarns?
    a. children's overalls
    b. towels
    c. sofa
    d. raincoat

16. An action- or power-stretch core yarn would most likely be made of which fiber combination?
    a. 80 percent spandex, 20 percent nylon
    b. 100 percent spandex
    c. 90 percent nylon, 10 percent spandex
    d. 100 percent rubber

17. What process increases the absorbency potential of synthetic yarns?
    a. combing
    b. twisting
    c. roving
    d. texturizing

18. Compared to an 18 count cotton yarn, a 35 count cotton yarn is:
    a. coarser
    b. finer

19. Which of the following yarns is always created by twisting two or more single yarns together?
    a. slub
    b. crepe
    c. ply
    d. novelty

20. Compared to a 100 percent rayon fabric, a fabric containing a 50/50 blend of rayon and polyester would have
    a. half the properties of each of the fibers.
    b. greater loss of strength when wet.
    c. increased wrinkle resistance.
    d. higher heat resistance.
    e. greater resistance to oily stains.

21. Why are multifilament yarns used more frequently than monofilament yarns in textile fabrics? Multifilament yarns have greater
    a. dimensional stability.
    b. resistance to crushing forces.
    c. flexibility.
    d. electrical conductivity.

Answers for Unit 3 Posttest:

1. c	12. b
2. d	13. c
3. c	14. b
4. a	15. b
5. b	16. c
6. b	17. d
7. a	18. b
8. c	19. c
9. c	20. c
10. a	21. c
11. b	

UNIT FOUR

Fabric Structures

In preceding units the following topics were dealt with: terminology commonly used in the discussion of textiles; fibers and their properties in relation to end use; and the construction of yarns and how yarns affect serviceability concepts. This unit is devoted to fabric structures. Procedures by which fibers and yarns are converted into fabric structures are outlined with emphasis on structures typically available to consumers.

Objectives for Unit 4

1. You will be able to identify and/or describe the fabric structures listed below, and associate selected fabric names and end uses with the structures.
   A. Woven Fabrics
      1. Basic weaves: plain, twill, satin
      2. Decorative weaves: pile, dobby, jacquard, double
      3. Triaxial
   B. Knitted Fabrics
      1. Filling knits: plain, purl, rib, pile, interlock, double
      2. Warp knits: tricot, milanese, raschel
   C. Special Structures Made from Yarns
      1. Knotted Fabrics: net, lace
      2. Braided
      3. Tufted
      4. Stitch Bond
   D. Multicomponent Fabrics
      1. Bonded and laminated fabrics: foam, adhesive, fabric-foam-fabric
      2. Quilted
   E. Nonwoven Fabrics and Special Structures
      1. Felt
      2. Bonded fiber
      3. Needle felt
      4. Spunlaced
      5. Ultrasuede®
      6. Film
2. You will be able to identify and explain the function of the basic parts of a weaving loom: warp beam, heddle, harness, reed, shuttle, battening bar, and cloth beam.
3. You will be able to relate serviceability concepts to various characteristics of fabric structure.
4. You will be able to describe appropriate care procedures for various fabric constructions.
5. You will be able to identify appropriate end uses for various fabric structures based on serviceability concepts.

Pretest

1. What fabric structure is formed by interconnecting loops?
   a. knit
   b. weave
   c. film
   d. braid

2. Which phrase describes the traditional fabric selvage?
   a. fabric bias
   b. tightly woven crosswise fabric edge
   c. tightly woven lengthwise fabric edge
   d. center fold of the fabric

3. The filling yarns in a fabric run
   a. parallel to the selvage.
   b. perpendicular to the selvage.

4. Compared to woven structures, knits are less
   a. flexible.
   b. resilient.
   c. absorbent.
   d. dimensionally stable.

5. What are the lengthwise interconnecting loops in knits called?
   a. wales
   b. courses
   c. ridges
   d. warps

6. A filling-knit structure has (high, low) crosswise stretch and (diagonal, horizontal) interconnectors.
   a. low, horizontal
   b. high, horizontal
   c. low, diagonal
   d. high, diagonal

7. Which knit is likely to "run" if snagged?
   a. warp
   b. filling

8. What are the lengthwise yarns in a woven fabric called?
   a. wales
   b. fillings
   c. courses
   d. warps

9. ______________ control the heddles, which in turn control the ______________ yarns.
   a. Reeds, filling
   b. Harnesses, warp
   c. Warp beams, warp
   d. Shuttles, filling

10. Which weave structure listed has the longest yarn floats?
    a. plain
    b. rib
    c. twill
    d. satin

11. Twill weaves are frequently stronger than basket weaves because twills
    a. are typically more loosely woven.
    b. have a plain weave pattern of interlacing.
    c. are generally woven with yarns of low twist.
    d. have a higher yarn count.

12. A satin weave is characterized by
    a. long warp-yarn floats.
    b. long filling-yarn floats.
    c. short filling-yarn floats.
    d. no yarn floats.

13. A sateen weave is characterized by
    a. long warp-yarn floats.
    b. long filling-yarn floats.
    c. short filling-yarn floats.
    d. no yarn floats.

14. Which weave structure has an extra set of yarns woven in with the basic ground cloth?
    a. novelty
    b. pile
    c. satin
    d. rib

15. How do tufted pile fabrics differ from woven pile fabrics?
    a. structure of the ground cloth
    b. depth of the pile
    c. fiber content of pile
    d. method of forming the pile

16. A dobby weave is always characterized by
    a. a balanced weave.
    b. repeating geometric patterns.
    c. elaborate floral or scroll patterns.
    d. long floating yarns.

17. Which fabric listed below is made on the jacquard loom?
    a. matelassé
    b. wafflecloth
    c. piqué
    d. birdseye

18. A crepe fabric can be achieved by any of the following *except*
    a. high twist yarns.
    b. textured yarns.
    c. special weaving patterns.
    d. slub yarns.

19. If high durability were the major consideration, which one of the following weaves would you choose?
    a. pile
    b. satin
    c. plain
    d. jacquard

20. Which weave structure would be the least abrasion-resistant?
    a. twill
    b. leno
    c. plain
    d. satin

21. High yarn counts contribute to high fabric
    a. flexibility.
    b. thermal conductivity.
    c. tenacity.
    d. absorbency.

22. If you wanted a very elastic fabric, which structure would you choose?
    a. double knit
    b. filling knit
    c. warp knit

23. Which structure yields an open mesh fabric made from yarns that have been knotted or interlooped together?
    a. net
    b. marquisette
    c. braid
    d. leno

24. What is the name of an open mesh fabric with a fancy or floral pattern?
    a. brocade
    b. lace
    c. leno
    d. piqué

25. Braid structures always have
    a. at least four yarns.
    b. more than five yarns.
    c. less than eight yarns.
    d. at least three yarns.

26. Which fabric structure is held together by fiber scales?
    a. needle felt
    b. felt
    c. bonded web
    d. malimo

27. Which fabric structure is formed through fiber entanglement by means of fine, high-pressure air or fluid jet action?
    a. malimo
    b. needle felt
    c. spunlaced
    d. bonded web

28. Which fabric structure would be the least absorbent and consequently uncomfortable to wear?
    a. unsupported film
    b. supported film
    c. fabric-foam-fabric laminate
    d. fabric-fabric laminate

29. Match the type of knit to the description of consumer need.
    1. tricot
    2. purl
    3. jersey
    4. rib

    ____ a. greatest stretch potential
    ____ b. good dimensional stability
    ____ c. high elasticity where snug fit is needed
    ____ d. crosswise structured ridges

30. Match the most desirable structure to meet the consumer need:
    1. needle felt
    2. double cloth
    3. double-faced
    4. twill
    5. melt blown

    ____ a. reversible blanket with contrasting colors
    ____ b. durable blanket suitable for camping
    ____ c. hospital protective covering
    ____ d. lowest-cost blanket
    ____ e. blanket with highest thermal retention

Answers for Unit 4 Pretest:

1. a
2. c
3. b
4. d
5. a
6. b
7. b
8. d
9. b
10. d
11. d
12. a
13. b
14. b
15. d
16. b
17. a
18. d
19. c
20. d
21. c
22. b
23. a
24. b
25. d
26. b
27. c
28. a
29. 1-b, 2-d, 3-a, 4-c
30. 1-d, 2-e, 3-a, 4-b, 5-c

UNIT FOUR

Part I

Introduction to Fabric Construction

Yarns are made into fabrics by *weaving, knitting, knotting* or *braiding.* Some fabrics, called *nonwoven* structures, are made directly from fibers rather than yarns. For the following discussion, look at the samples whose numbers are in parentheses.

Woven fabrics (2, 4, 6) are made by interlacing two or more sets of yarns at right angles in either the horizontal or vertical direction. *Filling knits* (22, 24, 42) are made by interlooping yarns horizontally, while *warp knits* (27, 27A, 28A, 29) are made by interlooping parallel yarns in a vertical direction.

Knotted fabrics are made by knotting yarns to hold them together. *Braided fabrics* are made by interlacing yarns or strips of material diagonally. *Nonwoven fabrics* (31, 33, 35) are usually made by combining fibers into a usable fabric without the intermediate yarn step. *Multicomponent* fabrics (19, 36, 36A, 39) are made by combining two or more layers of fabrics or mats of fibers.

In this unit you will learn various ways to identify the structure of fabrics used for home furnishings and apparel. For example, calling a fabric ''flannel'' identifies the fabric name, but says nothing about fiber content, yarn structure or fabric structuring methods. If a fabric is described as 100 percent wool flannel, you know that the fiber content is wool, but you still do not know about yarn and fabric structure.

When talking about a fabric, it is important to recognize that a variety of words can be used to identify it—such as the fiber type, fabric structure or name, as well as specific information about its manufacture. Most fabric names do not include specific terminology that tells the consumer about the type of yarn used, and one cannot always associate a particular fabric name with a given fiber content in a fabric structure.

Consumers are better able to evaluate fabric serviceability concepts if they can identify the *fiber, yarn* structure or fiber arrangement, *fabric* structure, and the fabric *name.* For example, if reference is made to a ''wool jersey knit,'' the fiber content is wool, the construction is knit and the fabric name is *jersey.*

In this unit you will learn to identify structure and fabric name, where appropriate, of common woven and knit fabrics. You will also study knotted and braided structures made from yarns, nonwovens made directly from fibers, and how two or more layers of fabrics are joined together to make a multicomponent fabric. You will learn to evaluate serviceability for all of these.

Comparison of Weaving and Knitting

The most common fabric construction methods presently used are knitting and weaving. These procedures will be discussed and compared first.

The untrained eye may have difficulty distinguishing between knitted and woven structures, especially in small fine constructions. In Figure 4.1, note that the knit structure consists of a series of interconnecting loops, while the woven consists of yarns interlaced at right angles.

Two basic directions are referred to in describing fabrics. In wovens, the *warp* is the long dimension when large amounts of fabric are considered; the

---

QUICK QUIZ 1

1. If a fabric is described as cotton, the reference is to ______________________.
   Has the fabric structure been identified? (yes, no)
2. If a fabric is referred to as a woven, the reference is to the ______________________.
3. When a fabric is described as *denim, corduroy* or *flannel,* the fabric ______________ is being given.

****

ANSWERS QUICK QUIZ 1

1. fiber content; no
2. construction
3. name

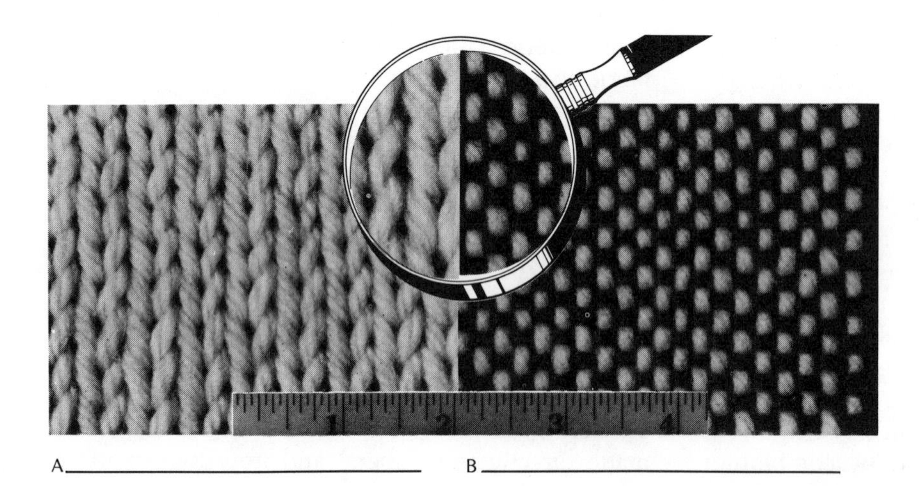

Figure 4.1. Comparison of knit and woven structures. Label each at A and B.

filling direction is the crosswise of the fabric or the shorter dimension. The traditional woven fabric has a selvage edge woven more tightly than the rest of the fabric; this edge provides stability to the fabric during processing and prevents raveling of yarns during fabric handling. However, in the interest of speed and economy of production, several newer looms do not produce a traditional selvage.

Thus, the selvage on fabrics from these nontraditional looms looks like the rest of the fabric. It lacks the characteristic tightly woven edge created from increased warp yarns, diagrammed in Figure 4.2.

The identifying terms used to describe direction in knits are *wales* for the lengthwise direction, and *courses* for the crosswise. To compare woven and knit fabric directions, remember that the *warp* of a woven and the *wale* of a knitted are both lengthwise

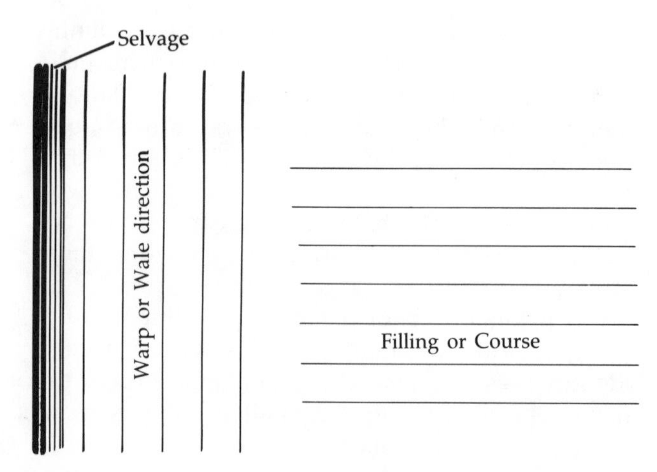

Figure 4.2. Woven and knitted fabrics have a lengthwise and crosswise direction.

QUICK QUIZ 2

1. The lengthwise direction of a woven fabric is called the ________________, and the lengthwise rows in knits are the ________________.
2. The crosswise direction of a woven fabric is referred to as the ________________, and the crosswise rows of a knit are the ________________.
3. The traditional selvage stabilizes the fabric edge and prevents yarns from raveling or pulling out. In a woven fabric with a selvage, the ________________ yarns are parallel to the selvage.
4. Knitted structures also have their fabric edges structured to be stable and not ravel. Instead of the edge being more tightly structured as in a conventional woven ________________, the knit loops themselves prevent fabric edges from raveling.
5. In order to ________________ new looms do not always produce a traditional selvage.

****

ANSWERS QUICK QUIZ 2

1. warp; wales
2. filling; courses
3. warp
4. selvage
5. increase speed and economy of production

---

terms, while the *filling* of a woven and the *course* of a knitted are crosswise descriptions.

Fabric samples used with this text are cut with the warp or wales parallel to the longer side of the sample, and the filling or courses running across. Since your small samples do not have a selvage, it would be difficult to determine the fabric direction if all the samples had not been cut and assembled in the packet with warps running in the longer direction.

Knits usually have more stretch and elastic recovery than comparable weight woven fabrics. Therefore, they tend to be more comfortable in apparel than woven fabrics. You can determine why knits are more elastic than wovens in most fabrics by recalling that weaves are formed by yarns interlaced at right angles, and knitted structures are formed by interlooping yarns. If a knit fabric is made with relatively stable yarns, it has a tendency to stretch and return to its original shape. On the other hand, a woven fabric usually does not yield as readily as a knit; but if it does yield, the structure is often distorted. Generally, the stretch and elastic recovery of knits also means that they are less dimensionally stable than wovens. However, knits tend to recover from wrinkling better than woven fabrics.

Both the weaving and knitting industries have attempted to create fabrics with characteristics different from those described above. To compete with knitted structures, the weaving industry has created wovens with increased flexibility and comfort stretch. The resulting fabrics, called *stretch wovens,* have excellent resiliency, dimensional stability and ease of care. These properties are made possible with textured yarns. Examine fabrics 8 and 9—both examples of stretch woven fabrics made from textured yarns.

Similarly, the knitting industry has designed knit structures with high levels of abrasion resistance, tensile strength and stability. These knit fabrics are successfully competing with woven fabrics in home furnishing and industrial end uses. Sample 27A is a nylon warp knit fabric that has been engineered for use in soft-sided luggage. A combination of high durability yarns, a tight knit and a coated backing provide the durability necessary for luggage fabric.

In most fabrics, however, the interconnecting loops of knitted structures elongate with use. Therefore knitting provides a means of building in comfort stretch in apparel fabrics without relying on the fiber or yarn type used. Further, if a bulky yarn is used, a structure can be knitted to contain dead air spaces,

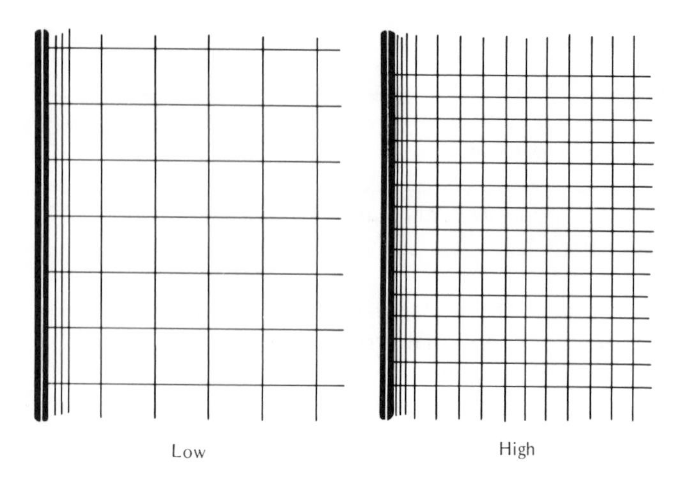

Figure 4.3. Serviceability features vary with fabrics of high and low thread counts.

which contribute to comfort during cold weather because of the fabric's high thermal retention. Such a fabric is called a *bulky knit.*

Based on the same principle of heat transfer, a loosely knitted structure made from natural cellulosic fibers yeilds a product especially suitable for warm weather wear because of its high absorbency and thermal conductivity. The extent to which a fabric is bulky depends on the size and type of the yarns and *thread count.* Thread count is the number of yarns per inch in both warp and filling directions.

The diagrams in Figure 4.3 represent woven fabrics of different thread counts constructed from interlaced warp and filling yarns. A traditionally tight-woven selvage is indicated by the lines on the left.

Now compare fabric samples 3 and 7 as to thread count. Sample 3 has a high count, compared to the top layer of sample 7 which has a low thread count. Knits likewise can be interlooped to have different degrees of tightness and looseness resulting in varied thread counts.

Knitted fabrics have become nearly as popular in apparel as woven structures for several reasons: knits can be structured for either increased thermal conductivity or thermal retention, depending on how yarns are interlooped; they allow for comfort or action stretch; they are more flexible, resilient and easier to care for than wovens; and knits allow for more moisture absorption than comparable woven structures.

Thus, knitting has become an increasingly popular method of constructing fabric over the past several decades. At the present time, it accounts for a substantial amount of fabric used for apparel and for some types of home furnishing and industrial fabrics.

A woven fabric with the same number of warp and filling yarns per inch is called a *balanced weave,* as in sample 1. An unbalanced weave has a noticeably different number of warp and filling yarns per inch. Unbalanced weaves are made by several methods—a larger diameter or bulkier yarn may be used in one direction, and smaller diameter yarns in the other. In sample 8, the textured filling yarns take up more space than the warp yarns, so there are fewer filling yarns per inch.

Another type of unbalanced weave is made by packing the yarns tightly together in one direction, while in the other direction the yarns have more space between each. Unravel a corner of sample 4; observe the unbalanced weave. Although all the yarns in the fabric are of the same diameter and type, the warps are packed tightly and have a higher warp yarn count than the filling yarns that have more space between them.

QUICK QUIZ 3

1. Woven fabrics can have varying degrees of stretch engineered in specific directions. Pull out some warp and filling yarns from samples 8 and 9. Which fabric has textured yarns in both directions? __________
2. In the other fabric, the textured yarns run in the ____________________ direction, creating ______________________ stretch.

****

ANSWERS QUICK QUIZ 3

1. 9
2. filling, crosswise

REVIEW

1. What is the difference between knitted and woven structures?
   Knitted: ____________________
   Woven: ____________________

2. Indicate by matching which fabric samples are knit and which are woven:
   ____ Sample 2 a-knit
   ____ Sample 6 b-woven
   ____ Sample 22
   ____ Sample 25
   ____ Sample 27
   ____ Sample 40
   ____ Sample 42

3. In woven fabrics, the yarns which run lengthwise are called ____________ yarns; the yarns which run crosswise are called ____________ yarns.

4. The tightly structured edge which runs parallel to the ____________ yarns is called the ____________.

5. In knitted fabrics the rows of loops that form lengthwise columns are called ____________; the rows of loops that form crosswise rows are called ____________.

6. The type of fiber, yarn structure or fiber arrangement, and fabric structure all can modify the way a fabric performs; in general, knits are selected when the following performance characteristics are important: ____________________

7. Weaves are selected when the primary performance characteristics desired are: ____________________

****

REVIEW ANSWERS:

1. Knitted structures are formed by interconnecting loops of yarn. Woven structures are formed by interlacing two or more sets of yarns at right angles.
2. 2-b; 6-b; 22-a; 25-a; 27-a; 40-b; 42-a
3. warp; filling
4. warp; selvage
5. wales; courses
6. flexibility, elongation and elastic recovery (stretch), resiliency, thermal retention or conductivity
7. dimensional stability, firmness or body

UNIT FOUR

Part II

Woven Fabrics

Recall that woven fabrics are made by interlacing two or more sets of yarns at 90° right angles. Exceptions to this rule are triaxial fabrics that are interlaced at 60° angles and three-dimensional weaves used largely for special industrial applications. Most woven structures are formed on a machine called a *loom* which has many individual parts that interlace yarns at right angles. These combine to perform the specific weaving processes.

The warp yarns, which determine fabric width, are wound around a warp beam, as shown in Figure 4.4. The warp yarns travel from the warp beam through long, flat, needlelike structures, with eyes in the center, called *heddles*. In the illustration notice that warp yarns pass through the heddles. The heddles are controlled by harnesses that raise and lower groups of heddles. The harness action causes the warp yarns to separate, allowing the filling yarns to be inserted. There must be at least two harnesses on a loom to allow for separation of the warp yarns in a manner that will permit formation of the simplest of weaves, the plain weave. Most commercial looms have more than two harnesses in order to provide for various patterns of warp yarn separation which allow for increased variety in weave patterns.

Look at the side view of the loom shown in Figure 4.5. When certain harnesses are raised, the warp yarns are separated and form a V on its side; the V is referred to as the *shed*. The device which moves back and forth through this shed is called a *shuttle* and contains the yarn which fills in the woven structure. Commercially, the warp yarns are called "ends," while the filling yarns are called "picks".

After one pass through the shed of warp yarns, two processes take place before the shuttle can be fed back through with another row of filling yarns. First, a comblike structure, called a *reed*, beats against the most recently inserted filling yarn and packs it tight against preceding yarns. The reed and shuttle are labeled in the illustration.

Second, the position of the warp yarns must be alternated to form a new shed, or the new filling yarn would pull out and not be interlaced with the previous one. Thus, the position of the harnesses which control the heddles must be changed. The alternation of the harnesses after each pass of the shuttle through the warp yarns produces the yarn interlacing at right angles which is the basic characteristic of woven structures.

The warp yarns are rolled off the warp beam and threaded through the heddles, which are attached to the harnesses. As the filling yarns are inserted and battened or packed together by the reed, the fabric is formed. The finished cloth is rolled onto the cloth beam shown in Figure 4.5.

The loom pictured in Figures 4.4 and 4.5 is a relatively simple device with four harnesses, but it provides a good example for identifying basic loom parts and the weaving process. By increasing the number of harnesses, the variation of patterns available is increased—the warp yarns can be raised and lowered in more patterns to form a greater variety of interlacing patterns.

In the textile industry, the completed fabric structure as it comes from the loom is called *cloth* or *griege* (usually pronounced "gray") *goods*. The finished fabric ready for the consumer is called, simply, fabric. While, the words *cloth* and *fabric* may be used synonymously, the term greige goods is specific to fabric as it leaves the loom prior to any additional processing or finishing. Some fabric may be ready for consumers as it leaves the looms, but this is not typical; fabric generally requires additional finishing as well as coloring.

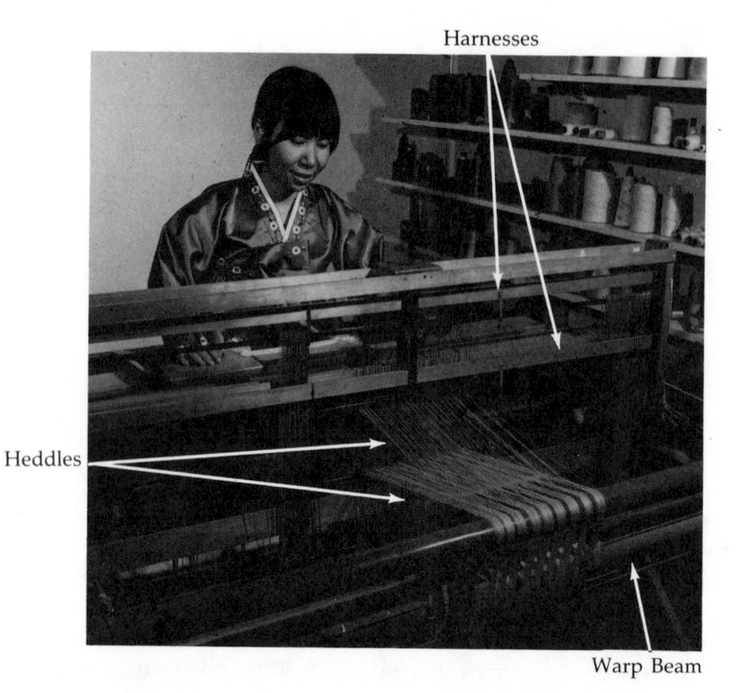

Figure 4.4. A simple manual loom is shown from the back, with the warp beam facing out.

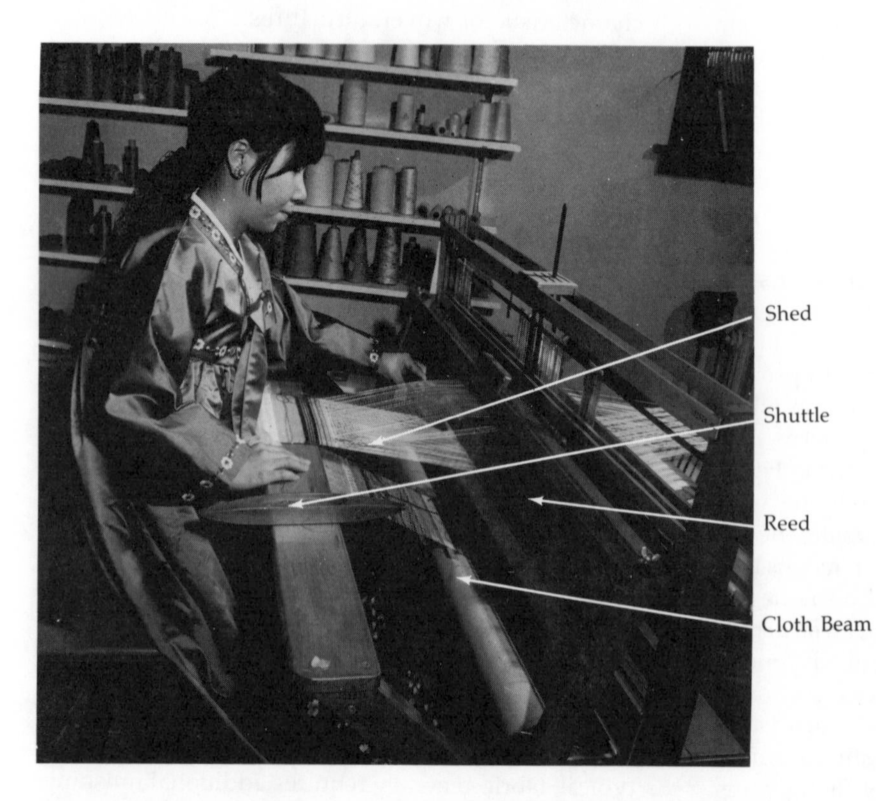

Figure 4.5. The warp yarns are positioned to form the yarn "shed," through which the filling yarns are inserted.

QUICK QUIZ 4

1. Regardless of the complexity of the basic loom, warp yarns are rolled off the ________ and threaded through the ________, which are controlled by the ________
2. After each alternation of the harnesses, the shuttle or a similar device makes a pass through the warp-yarn shed carrying a ________ which is then packed into position by the ________.
3. The completed fabric structure is then wound onto the ________.

****

ANSWERS QUICK QUIZ 4

1. warp beam; heddles; harnesses
2. filling yarn; reed
3. cloth beam

Figure 4.6. The drawing-in process used to be done by hand but now is fully automated. *(courtesy Burlington Industries, Inc.)*

Most fabric today is produced on commercial factory looms, which are much more complex and faster than the manual loom just discussed. However, the basic loom parts and processes are similar, as you can see from the commercial loom shown in Figure 4.6. It is photographed from above—observe the warp yarns at the right of the photograph before they are "drawn-in" through the flat, needlelike heddles.

To increase speed of production, modern weaving equipment has special devices instead of shuttles to carry the filling yarn across the fabric. New *shuttleless* weaving machines can turn out fabric many times faster than conventional looms. Devices such as projectiles, rapiers, water jets, or air jets may be used to insert the filling yarn through the warp yarn shed. The majority of fabrics produced in the United States are made on shuttleless looms.

Looms today are highly automated, including the production stages of filling loading, filling and warp yarn break repairs and removal of finished fabric. Computer aided design (CAD) systems are available to allow for computerized fabric design which can then be programmed into looms. Automated diagnosis of machine problems and self repair have led to

Figure 4.7. High-speed water-jet looms are seen in this weave room. The water jets add speed and efficienty to the weaving operation and also reduce noise considerably. *(courtesy Burlington Industries, Inc.)*

Figure 4.8. Modern weaving equipment in this plant uses air jets rather than shuttles to carry the filling yarn across the fabric. *(courtesy Burlington Industries, Inc.)*

almost "weaverless weaving" in modern mills where very little labor is required. Also very important is continuous quality control during weaving in which fabrics are scanned and defect locations automatically noted. The increasing importance of "Quick Response," a program linking fiber, textile and apparel manufacturers and retailers in rapid product supply systems have made quality control at each stage of textile production a necessity.

Notice the size of the weave room in Figure 4.7. Many textile mills have huge loom rooms, and when primarily shuttle-driven machines are used, the noise is thunderous. Thus, in addition to increasing production speed, shuttleless looms have the advantage of reducing noise levels to significantly improve the work environment.

Instead of the traditional selvage, many looms today produce either a fringed edge formed by the filling yarns; a heavy edge formed by catching the filling yarn end back into the interlacing pattern along the fabric edge; or an adhesive is added to the edges to secure the filling yarns where the selvage would otherwise be.

Now you are ready to study the types of interlacing patterns created on various types of looms. Basic weave structures are covered in the next section.

Basic Weaves

Three basic weave patterns—the plain, twill and satin—are used to manufacture a variety of woven fabrics produced today. The simplest type of weave requires only two harnesses and is called the *plain* weave, shown in Figure 4.9. The interlacing illustrated is formed by having the first filling yarn pass *over* the first warp, under the second, and so on. The next filling yarn goes *under* the first warp, over the second and continues this alternating over and under. This pattern is repeated throughout the fabric to create a plain weave fabric.

Plain Weave and Variations of the Plain Weave

Many of you probably have constructed plain weave fabrics using simple looms designed for making pot holders or "card" looms for weaving pillow covers or placemats. Recall the over and under patterns of interlacing you may have used.

A majority of woven fabrics are constructed by the plain-weave pattern of interlacing. The yarns can be packed tightly or loosely, and thread counts may be balanced or unbalanced. Compared to other methods of fabric construction, the plain weave is comparatively inexpensive.

Common names for plain weave fabrics include *gingham, percale, voile, muslin, organdy, batiste, chambray,* and *lawn.* Examine samples 1 and the top layer of 39 in your fabric packet to become familiar with typical, balanced yarn count, plain weave structures.

Variations of the plain weave are also produced on relatively simple looms by modifications of the weaving pattern, such as going over and under more than one yarn at a time (samples 6 and 7); by varying the closeness or looseness of the yarns (sample 2); or by varying the size, type or complexity of yarns used (sample 43). Two specific variations of the plain weave to be discussed in more detail are the *rib* and *basket* weaves.

Rib Weave

This variation is distinguished by a pattern of interlacing that creates a rib or raised section in either the warp or filling direction. This raised portion can be heavy and wide, or fine and narrow. Of the many variations between these extremes, three are

---

QUICK QUIZ 5

Whatever type of loom used for fabric production—whether it be the more traditional loom which has a __________________ to carry the filling yarn across the fabric; or a faster, quieter machine that uses ________________________ to send the yarn through the shed, the cloth or fabric that comes off the loom is specifically called __________________ goods, which generally require additional processing and finishing prior to consumer use.

****

ANSWERS QUICK QUIZ 5

shuttle; projectile, rapier, water or air jets; greige (gray)

Figure 4.9. The plain weave is the simplest, most basic weave.

shown in Figure 4.10. All of these rib weaves have unbalanced yarn counts, with the number of yarns per inch higher in one direction than the other.

When the warp yarns are interlaced as bundles or groups, the ribs will lie in the warp direction. In Figure 4.10, fabrics A and B contain noticeable ribs in the filling direction. These ribs are formed with heavier filling yarns in A, bundles of yarn in B.

A rib effect can also be created using the plain weave pattern of interlacing with yarns of the same size in both directions, but with the number of yarns in the warp considerably greater per inch than the number of yarns in the filling direction. This method was used to produce the *broadcloth* fabric shown in 4.10. Observe how the tightly packed yarns in the warp direction bunch up into a fine rib *across* the fabric as they interlace with the filling yarns.

Fine rib effects are also achieved by using in one direction yarns that are only slightly larger than those in the other direction. Study sample 4, and notice the slightly heavier filling yarns, which create fine ribs in the crosswise direction. Sample 4 is called *poplin,* the fabric name for a tightly woven, high yarn count fabric that has fine cross ribs formed by heavier filling yarns than warp yarns; and more, finer warp yarns than filling yarns.

In fact, many rib weaves are created by using larger filling yarns than warp yarns. The greater the difference in size of the yarns, the larger the rib effect. Fabric names of rib weaves produced by this method, listed in order from fine to heavy ribs, are *poplin, taffeta* (sample 5), *faille, grosgrain* and *ottoman.* All of these fabrics have ribs of varying size in the filling direction.

Poplin, frequently used in skirts and slacks, is heavier than *broadcloth,* another common rib weave fabric often used in shirts and blouses. Broadcloth has finer crosswise ridges than poplin because it is made from yarns of the same size. Recall that a high warp yarn count creates subtle ridges over the same sized filling yarns in broadcloth, whereas poplin has heavier filling than warp yarns.

Another familiar rib fabric is *grosgrain* (pronounced "*grow*-grain") ribbon. It is narrow with ribs running across the ribbon in the short dimension comparable to the filling.

The rib weave modification of the plain weave has both serviceability advantages and limitations: rib weaves provide esthetic interest, and in some cases,

Figure 4.10. Rib weave variations of plain weave.

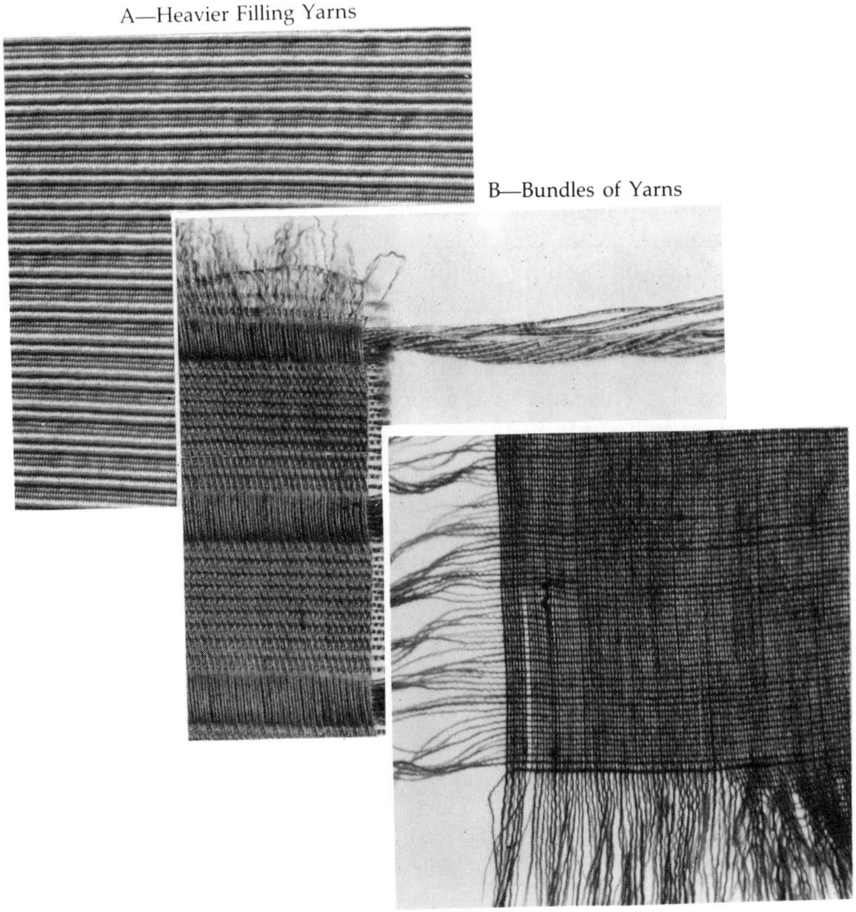

Rib Effects Achieved By:

A—Heavier Filling Yarns

B—Bundles of Yarns

C—Increased Warp Yarn Count

may be stronger than comparable plain weave fabrics. However, raised ribs are susceptible to surface abrasion; larger yarns in a fabric may abrade against finer ones. Also large, heavy ribs or bundles of yarns can decrease fabric flexibility or pliability. Care procedures have to take into account uneven wear and possible snagging and abrasion.

Basket Weave

Another modification of the plain weave is the basket weave. It is produced when two or more warp yarns interlace with one or more filling yarns; or two or more filling yarns interlace with one or more warp yarns. The yarns lie flat but interlace as units to form the basket interlacing shown in Figure 4.11.

When a yarn crosses over more than one yarn, it is said to *float* over the other yarns. If two warp yarns pass over and under one filling, a 2 × 1 basket is formed; in a 2 × 2 basket, two filling yarns float over two warp yarns then under two warp yarns. A basket weave constructed with *four* warp yarns floating over and under *two* filling yarns is described as a 4 × 2 basket construction. All yarns are laid side by side in basket weaves and do not bunch up as in certain rib structures. The flat, multiple yarns are interlaced in the same manner as one yarn in basic weave constructions.

Basket weave fabrics, like most structures, have advantages and disadvantages. They are generally softer, tend to be more flexible and have increased resiliency, when compared to similar woven plain weaves. Due to the openness of the fabrics, there is

---

QUICK QUIZ 6

1. A variety of rib modifications of the plain weave can be constructed by varying yarn size and/or yarn count. Check which method would yield the most serviceable fabric with a very large and heavy rib.
   ___ bundles of yarn
   ___ heavy single yarns
   ___ using many more warp than filling yarns
   Why is that the best method? ______________________________
   ____
2. Which of the three methods of producing rib weaves usually gives the finest rib effect?
   ____
3. Sample 5, named *taffeta,* is a ______________ weave formed by larger yarns in the ______________ direction. Compare the ribs in the taffeta and poplin (4).
   NOTE: This taffeta has a special finish applied that creates the watered appearance, which you will study in the Finishes unit.
4. Sample 3 is a plain weave fabric that has some raised effects in the filling direction created by ______________.
5. In summary, rib fabrics are formed by variations of the plain weave through unbalanced yarn counts and/or the handling or size of ______________ or ______________ yarns.

****

ANSWERS QUICK QUIZ 6

1. bundles; because large ribs could be created with multiple yarns that would share surface abrasion
2. use of more warp than filling yarns
3. rib; filling
4. slub yarns
5. warp, filling

---

also increased absorbency. However, they tend to abrade more easily than plain weaves and are not as strong as plain or rib weaves when yarn count is the same or slightly higher.

Variations of the basket weave are also found in other combinations of interlacing, such as 3 × 3, 4 × 4, 3 × 2, 4 × 3, and so on. Usually the yarns do not float over more than four yarns at a time; however, there are variations formed in an 8 × 8 structure called *monk's cloth.* It is a well known heavy and coarse balanced, basket weave fabric often brownish-white or oatmeal in color.

A popular basket weave fabric of 2 × 1 construction is *oxford cloth,* commonly used in men's and women's shirts. It has two warp yarns interlaced as one with a single filling yarn. Examine sample 6, and notice that the warp yarns are of a finer diameter than the filling yarns. The two warp yarns grouped together give the effect of having the same size as one filling yarn. Thus, the weave structure closely resembles the plain weave.

Tightly woven fabrics are not as flexible and tend to wrinkle more easily than looser weaves. Also, there is more available surface area for absorbing moisture around the loose interlacings in a basket weave than in either plain or rib weaves.

Discussion of high and low yarn counts and resulting properties are relative. For example, plain woven *burlap* has a low yarn count compared to *oxford cloth.* Generally, however, basket weaves are associated with fewer interlacings per inch than many basic plain weave fabrics.

Figure 4.11. Basket weave variations of the plain weave are described by indicating the number of yarns in the warp direction times the number of yarns in the filling direction.

In addition to the plain weave and its variations, there are two other basic weaves known as the *twill* and *satin* weaves. They are discussed next.

Twill Weave

The twill weave is recognized by parallel diagonal ridges which may be visible on both sides of the fabric, but are always visible on the face or right side. Look for the diagonal ridges on samples 8 and 9. The pattern of interlacing required to make twill weaves differs from the one used to form either the plain weave or its variations. The loom requires at least three, and usually more, harnesses so that warp yarns can be manipulated with greater variety.

The harnesses are raised and lowered in a predetermined pattern to form the twill weave construction. The simplest twill weave requires that a single warp yarn float over two filling yarns, then under one filling yarn to form 2/1 twill. The first number identifies the number of filling yarns over which the warp yarn floats, while the second number indicates the number of filling yarns under which the warp yarn passes. A twill in which a warp

---

QUICK QUIZ 7

1. A rib structure sometimes may be obtained by using two or more warp yarns as one in the interlacing pattern. How does this type of rib structure differ from the basket weave used to make oxford cloth? ______________________________
____
2. To make certain that yarns will lie flat in basket weave constructions, the fabric is (more, less) tightly interlaced than for the rib weave. As a result, basket weave fabrics tend to have (lower, higher) yarn count and durability than comparable plain or rib weaves.
3. List three advantages this lower yarn count offers the consumer:
____

****

ANSWERS QUICK QUIZ 7

1. The yarns are kept flat in the basket weave for oxford cloth, while in the rib weave they would be closely packed to produce a raised ridge.
2. less; lower
3. flexibility, resiliency, absorbency

---

yarn floats over 3 filling and under 1 is identified as a 3/1 construction; while one in which the warp yarn floats over 2 and under 2 filling is a 2/2 twill.

Examine the 2/1 twill weave illustrated in Figure 4.12; the pattern creates a definite diagonal effect on the surface of the fabric which is developed as each warp yarn, or group of warp yarns, passes over and under filling yarns in a repeating progression of interlacing. At each new pass through of the next filling yarn, it begins higher or lower than the adjacent floats. To visualize the structure clearly, use the illustration of the twill fabric in Figure 4.12. Follow a filling yarn across the fabric as it passes over one and under two warp yarns and repeats that pattern across the fabric. The next filling, either above or below, starts one yarn further in to form the diagonal line by means of the floating yarns moving one step down (or up) and over in the progressive pattern.

In Figure 4.15, notice that the direction of the twill diagonals reverses at regular intervals within the fabric; this variation is called a broken twill. The fabric shown is named herringbone, which is often used in suitings, overcoats and sportswear. Twill weaves can be varied according to the angle of the diagonal slope, the direction of the twill, and the pattern of yarn interlacing and length of floats.

The degree of diagonal angle in a twill is a guideline to the fabric durability that can be expected. The diagonal slope depends on the balance of the weave and the interlacing pattern. The greater the difference between the warp and filling yarn counts, the steeper the twill diagonals will be. Therefore, a steep twill angle is typically achieved through an unbalanced yarn count.

Figure 4.12. This twill has the warp yarns floating over two filling yarns and under one to form the parallel diagonals characteristic of twill structures.

Figure 4.13. Compare this twill with the one shown in Figure 4.12.

QUICK QUIZ 8

1. Twill floats are not limited to spans of two yarns, as you can see in Figure 4.13. How many yarns does each warp float over? ____________
2. Simple twill weaves require three harnesses; more complex twills, such as in Figure 4.14, may have as many as 15 to 18. The direction of the twill ridges is reversed from the previous illustration and the pattern is more complex. The twill diagonals run from the upper ____________ hand corner to the lower ____________ hand corner—this is called *right*-handed twill construction.

****

ANSWERS QUICK QUIZ 8

1. three
2. right; left

Figure 4.14. Novelty twills require additional harnesses to create more varied diagonal patterns.

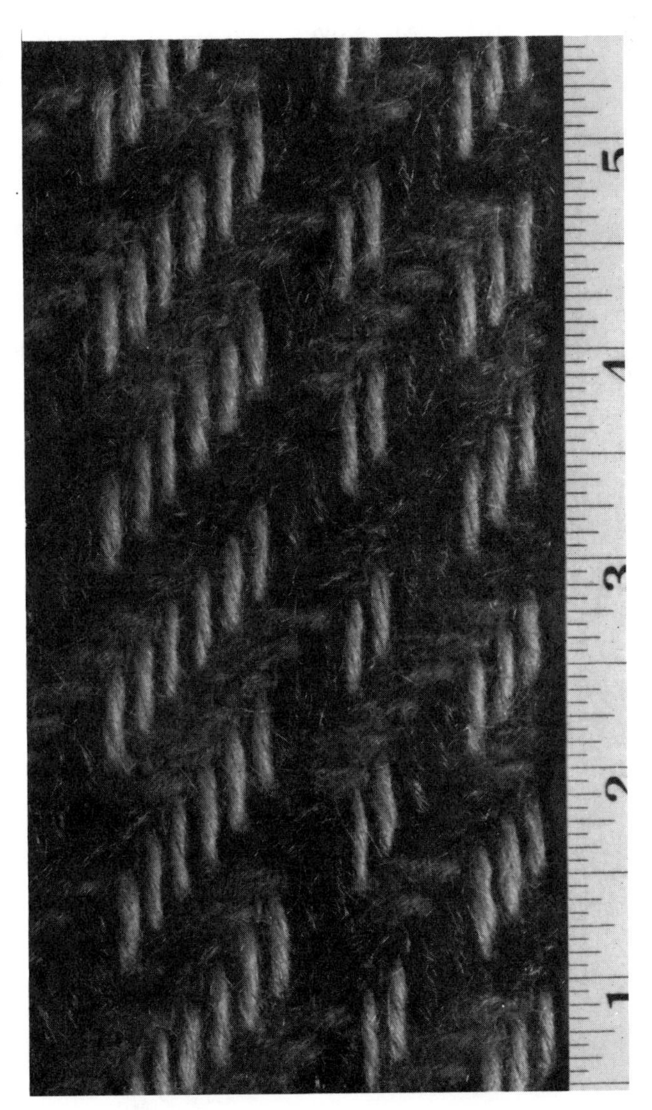

An unbalanced twill weave is generally structured with the warp direction having the higher yarn count, because the warps tend to receive greater stress than the filling. Twill fabrics with high warp yarn counts have short yarn floats and close interlacing, which contributes to increased strength. Therefore, the steeper the angle of the twill, up to the optimum point, the more durable the twill fabric will be.

Recall from the discussion of plain weave structures that balanced structures tended to be more durable than unbalanced. So you may wonder why unbalanced twill weaves with steep slopes are described as having high durability. The steep twills are more durable only when compared to other twill structures; steep twill weaves tend to have shorter floats and are, generally, more durable than twills which have low or reclining angles and lower yarn count.

While floats used in the twill and basket weaves create interesting design patterns, they can decrease fabric durability. When long floats are used, yarns tend to snag fairly easily. Since twill weaves also generally have higher yarn twist, in addition to shorter yarn floats, than basket weaves, a twill structure usually has more abrasion resistance than a basket weave.

QUICK QUIZ 9

1. Compared to basket and rib weaves, twills are generally more ______________ for end uses which demand a high measure of durability.
2. Examples of twill weave fabrics include *denim, gabardine, serge, flannel* and *jean*. When you think of blue jeans and sportswear made of denim, how do you rate fabric durability? ______________________________
3. Sample 8 is a "comfort stretch" denim. While the ______________ weave contributes to durability, additional performance properties result from the use of ______________ yarns.
4. Compare the expected durability between a twill and a plain weave of *comparable* fiber content, yarn count and yarn twist.
____
____
____
5. However, twill woven fabrics are seldom comparable to plain weaves in fiber content, yarn count and yarn twist. Thus, greater measures of durability are typically associated with twill weaves over plain weaves. Check the factors that contribute to high measures of durability for twill fabrics.

___ a. high yarn count	___ f. short yarn floats
___ b. low yarn count per inch	___ g. high number of interlacing per inch
___ c. high yarn twist	___ h. low number of interlacings per inch
___ d. low yarn twist per inch	___ i. repeating progression of floats
___ e. long yarn floats	___ j. random floats

****

ANSWER QUICK QUIZ 9

1. serviceable
2. highy durable—good strength and abrasion resistance
3. twill; texture
4. The plain weave tends to be more durable because of its balanced structure and interlacing pattern that lacks the floating yarns of twill structures.
5. a, c, f, g, i

Figure 4.15. Broken twill weave: herringbone.

Satin Weave

The third basic weave is the *satin* structure, which is characterized by long, warp floats on the front of the fabric. These floats are caught under single filling yarns as far apart as possible for the particular interlacing pattern. Parallel yarns do not interlace at touching points, as seen in Figure 4.16; this reduces the possibility of diagonal lines such as those observable in twill weave fabrics.

The satin weave is not an especially durable structure because its long floats snag easily. Since the purpose of the satin weave is to create sheen by the long warp floats, yarn selection is important. Filament yarns of low twist are most appropriately used

Figure 4.16. Satin weaves are distinguished by long, floating warp yarns on the fabric face or front.

to create the esthetic objective of satin structure; their smooth surfaces, unbroken by high twists, reflect more light and are more lustrous than staple yarns.

The length of the floats in a satin weave is dependent on the number of harnesses on the loom and the thread count; in general, the lower the thread count the longer the floats. A satin weave is structured with each warp yarn floating over at least four filling yarns, but the float span may be considerably higher.

Satin structures are usually unbalanced weaves, with more warp than filling yarns per inch to create the surface sheen. Consequently, yarns in the warp are packed tightly and create a smooth surface without any filling yarns showing on the fabric face. Examine sample 10, a typical satin structure.

A variation of the satin weave, the *sateen* weave, is illustrated in Figure 4.17. The *sateen* structure differs from the satin structure in that filling yarns float on the sateen fabric face rather than the warp yarns, as in the satin weave.

Sateen weaves are generally not as shiny as satins because they typically are made of staple-length fibers such as cotton. In fact, a familiar fabric made with the sateen structure is called cotton sateen. The sheen of a sateen weave is in the filling rather than in the warp direction of the satin weave.

Satin weaves are usually made of manufactured filament fibers in the man-made noncellulosic (synthetic) and man-made cellulosic categories. The low abrasion resistance and poor flexibility of mineral filaments make them unsuitable for satin structures. Although silk can be used as the shiny filament in satin weaves, its use is limited because of high cost.

Typical names of satin fabrics include *antique satin, slipper satin* and *bridal satin*. From the names of these fabrics it should be apparent that a majority of satin weaves are used for esthetic or decorative purposes. Since the satin weave is not selected for its durability, care procedures should be directed toward avoiding yarn abrasion and snagging; there should be minimal agitation during laundering.

Satin weaves can be varied by the use of different yarn types. A pebbly effect is sometimes achieved on the back of satin fabrics by using very high-twist crepe yarns in the filling direction. This variation is called *crepe-back satin* or *satin-backed crepe*. The filling yarns are exposed on the fabric back since the warp yarns float on the fabric face.

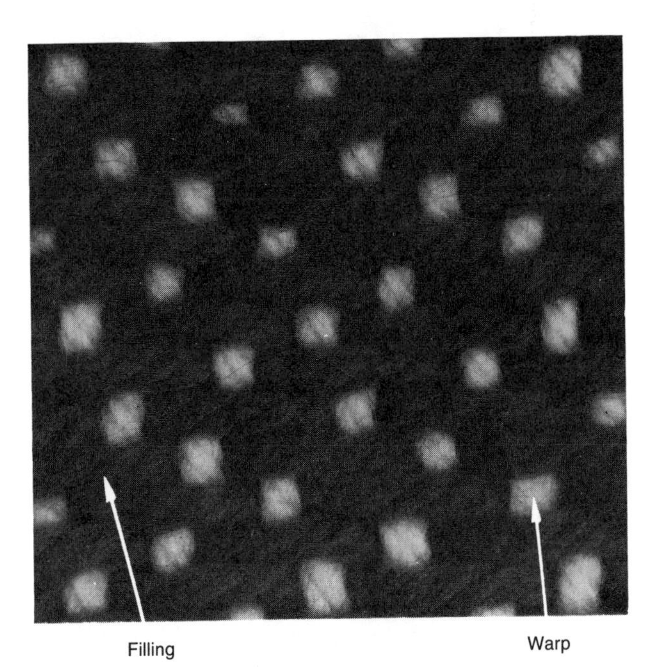

Figure 4.17. Sateen weaves have long filling-yarn floats on the fabric face.

QUICK QUIZ 10

1. Like the satin weave, the sateen weave is also an unbalanced structure. Would sateen structures have more yarns in the warp or filling direction? ________________
Why? ________________

____

2. The shiny appearance and high luster of satin weaves are accented if the filament yarns have low twist. How does the tightness or looseness of twist affect fabric durability?
____

____

****

ANSWERS QUICK QUIZ 10

1. filling; because the filling yarns are tightly packed and float on the fabric face to produce the characteristic filling sheen of sateens.
2. Low twist yarns are more subject to snagging and abrasion than high twist yarns.

In addition to providing fabrics for dressy occasions, satin fabrics are often used as coat and jacket linings. Their smooth surface increases the ease of putting on coats and reduces wear on shirts or dresses.

Review

Skill in identifying fabrics by both name and weave is a mark of the textile professional. However, in today's world of textiles, the name given a fabric by a manufacturer may differ considerably from standard names used in the past. While an expert can quickly and accurately identify the type of fabric structure, applying an appropriate fabric name may depend on such extraneous factors as current fashion, manufacturers' advertising campaigns and regional characteristics. For example, fabric 2, an unbalanced plain weave, is called batiste or shadow voile by the manufacturer; but in the past these fabric names were used for balanced, lightweight plain-woven fabrics.

Batiste and voile are both sheer, smooth fabrics made from combed yarns. Samples 1 and 2 both contain cotton, whose yarns were subjected to the combing process to increase surface smoothness.

Samples 3, 4 and 5 all have rib effects. The ribs in sample 3, the shantung, were created through the use of slub yarns; whereas the ribs in the poplin and taffeta were formed through the rib weave using slightly larger filling than warp yarns.

There is considerable difference between the two basket weave fabrics in samples 6 and 7. Not only is the interlacing pattern different, but yarn count and type of structure contribute to much of the variation. The upholstery basket fabric would have limited use due to its low yarn count if it were not stabilized by a backing layer (a fiber web to be studied in the Nonwoven section).

Samples 8 and 9 are both relatively steep twill weaves that have added stretch due to the use of textured yarns. While these twill fabrics were designed with different apparel end uses in mind, both are relatively durable due to the yarn count, type and pattern of interlacing. Also, the textured yarns contribute to comfort stretch. Generally the steeper the twill, the more durable the resulting structure; the steepest of the three twills (8, 9, 11) in the fabric packet is number 11.

Satin weaves, and the sateen variation, are characterized by long yarn floats. Sample 11, made of 100 percent cotton, has a similar texture and hand to conventional sateen; it was marketed as a cotton sateen. However, it is made with a steep twill instead of filling yarn floats in order to increase durability. Thus, this cotton sateen is another example of how manufacturers take liberties in naming and marketing fabrics. Sample 10, designed for esthetic rather than durabiity purposes, is a traditionally woven satin made with warp yarn floats of filament length fibers.

QUICK QUIZ 11

Demonstrate an understanding of the basic weave structures by diagramming the interlacing patterns characteristic of each. Use a darkened square to represent a warp yarn going over a filling yarn, and a blank square to represent a warp yarn going under a filling.

1. Fill in the diagram for a 1 × 1 plain weave interlacing in Figure 4.18. Begin the pattern by filling in the upper left hand square first to indicate a warp going over a filling yarn.
2. Represent the twill progression in Figure 4.19 by having a warp yarn float over two fillings and under one. Draw the repeating progression so that the diagonal lines emerge as the yarn floats move over one, and down on each succeeding line.
3. Diagram a typical satin weave in Figure 4.20, allowing a warp yarn to float over four filling yarns and under one. Be sure to space the points where the warp passes under so that no interlacing comes in contact with another. The "under one" interlacings should be as far apart as the 4 × 1 pattern allows: no blank spaces may touch.

NOTE: A satin weave with only four-yarn floats for the repeat pattern is the simplest type. Because of the limited number of yarns involved in the repeat pattern, it is not possible to completely eliminate the slight diagonal effect that is produced. Thus, a 4 × 1 satin weave will have a faint twill appearance.

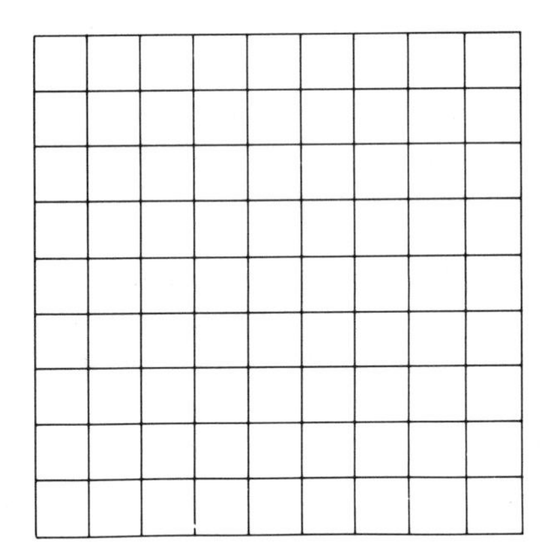

Figure 4.18. Fill in for a plain-weave pattern of interlacing.

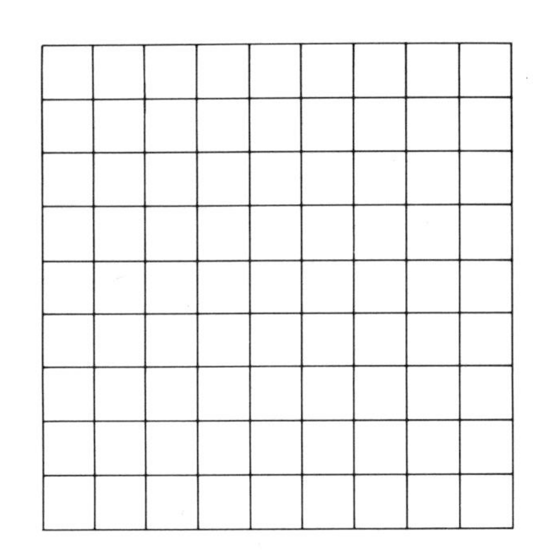

Figure 4.19. Fill in for a 2 × 1 twill-weave pattern of interlacing.

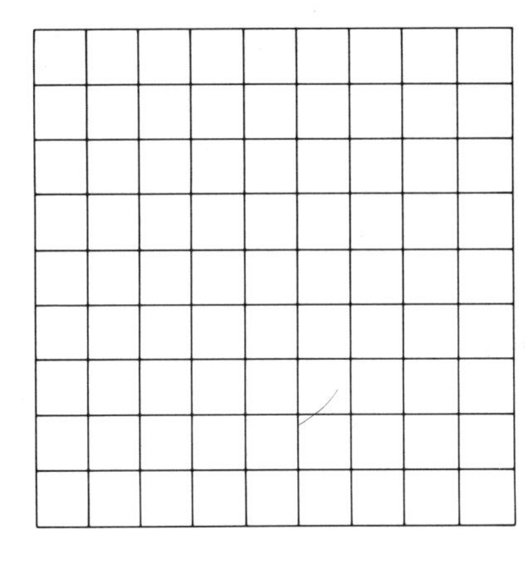

Figure 4.20. Fill in for a 4 × 1 plain-weave pattern of interlacing.

****

ANSWERS QUICK QUIZ 11

1. every other square darkened in both directions
2. diagonal wales should emerge
3. a majority of squares should be darkened to represent the floating yarns

1.

2.

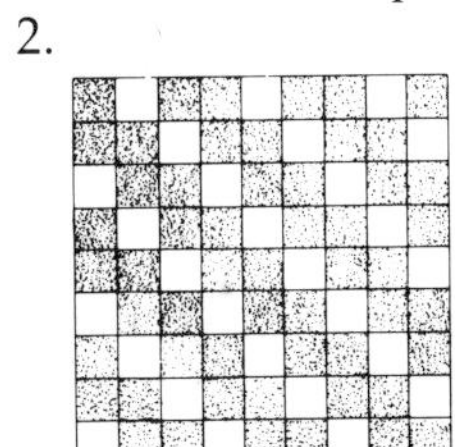

3.

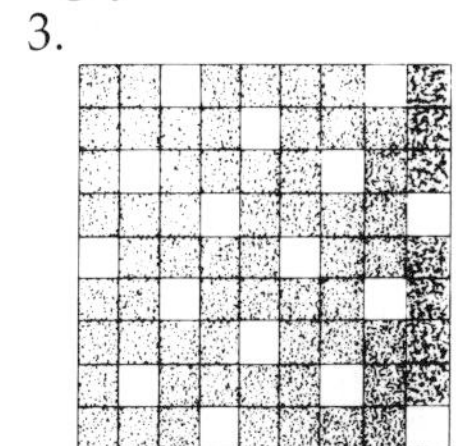

Basic Weave Identification

Identify the basic weave structures of the samples indicated below. Also try to indicate the fabric name where applicable.

1 ______________________________
2 ______________________________
3 ______________________________
4 ______________________________
5 ______________________________
6 ______________________________
7 ______________________________
8 ______________________________
9 ______________________________
10 ______________________________
11 ______________________________
40 ______________________________
41 ______________________________

Basic Weave Identification Answers:

1 - plain weave, chintz
2 - plain weave variation, batiste or shadow voile
3 - plain weave, shantung
4 - rib weave, poplin
5 - rib weave, taffeta
6 - 2/1 basket weave, oxford cloth
7 - basket weave variation, no special name—backed upholstery fabric
8 - twill weave, denim
9 - twill weave, no special name—sold under tradename Ultressa®
10 - satin weave, satin
11 - steep twill—sold as cotton sateen
40 - plain weave—sold as dotted swiss; a finish you will study later adds the dots
41 - plain weave; sold as plissé; a finish you will study adds the crinkle

Cut and attach a swatch from these samples on the Mounting Sheet for Basic Weaves. Label each in the blanks provided.

Complex or Decorative Weaves

With attachments added to create special effects, complex weaves can be structured on the same type of looms used to make plain, twill or satin weaves. Special controls and attachments modify the loom so that a variety of complex designs can be woven. These attachments provide individualized control of harnesses and, in some cases, of each warp yarn.

Pile Weaves

The first complex or novelty weave to be studied is the *pile* weave, which is made by a variety of processes. Probably the most common method is to weave an extra set of yarns into a basic ground structure. This base may be either a plain or a twill weave interlacing. In general, a high yarn count twill is preferred for the base because it tends to lock the pile yarns more securely in place; further, the special twill interlacing tends to tie in the pile yarns a little more tightly than in the plain weave interlacing.

The extra set of yarns that forms the pile is woven into the base structure in such a manner that these yarns form loops on one or both sides of the fabrics. These loops may be left as they are to form an "uncut" pile, or may be sheared to form a "cut" pile fabric.

Pile weaves are classified according to the direction of the extra set of floating yarns that forms the pile. If the extra set of yarns floats in the crosswise or filling direction, the resulting fabric is a *filling pile structure*. If the extra set floats lengthwise or in the warp direction, the fabric is a *warp pile structure*.

Filling Piles

Both *corduroy* and *velveteen* are examples of filling pile fabrics, which means that extra sets of filling yarns float over warp yarns. Traditionally, velveteen fabrics are formed so that the pile surface is uniform, while the corduroy fabric surface appears to have vertical strips of pile called wales.

Corduroy wales can be quite wide or very narrow, as in fabric 12, which is a *pinwale* corduroy. In fact, some new types of corduroy have wales so fine that they are not visible, resulting in a fabric similar in appearance to velveteen. Without an opportunity to examine such fabrics carefully under magnification, it may be impossible to determine if the fabric is a smooth-surfaced "waleless" corduroy or a velve-

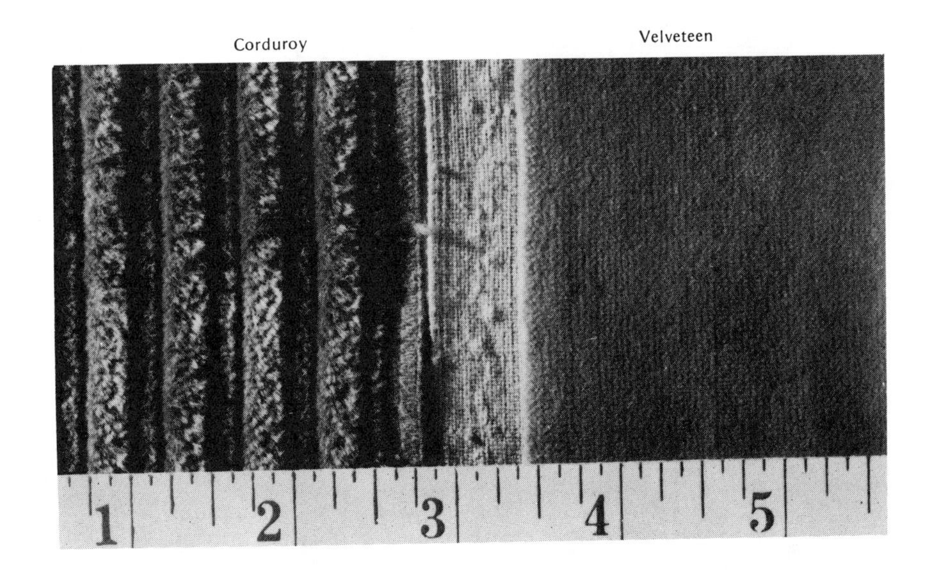

Figure 4.21. The filling yarn floats are cut to form these filling pile fabrics.

teen. The distinction between them is minimal since these waleless corduroys appear uniform in the pile formation.

The difference is in how the extra sets of filling yarns are interlaced with the base structure. In velveteen fabrics, filling floats are structured into the base fabric in an interlacing pattern that produces an overall, even pile effect when the floats are cut, as in fabric 13. In corduroy fabrics the filling floats are structured to produce wales, however fine, when cut. Nonetheless, even in velveteen structures the vertical rows of cut pile can be seen—look for them in Figure 4.21 and fabric 13.

In making velveteen fabrics it is possible to produce a deeper pile by making the floats longer. If the yarn count is high for the base fabric, the pile becomes denser than when the yarn count is low. Also the denser the pile, the more luxurious the fabric appears. The type of interlacing will also affect the depth of the pile and the density.

Figure 4.22 is a diagram representing how a filling pile structure is formed with an extra set of filling yarns floating over and under the single set of warp yarns. The woven ground structure in the diagram is a plain weave.

Figure 4.23 shows a filling pile corduroy fabric as it appears with the extra filling yarns floating on the fabric surface before the floats are cut to form the pile.

The extra sets of yarns in a filling pile construction generally float over at least three warp yarns, and sometimes as many as five or seven. The length of the float will determine the depth of the pile and its thickness, density, or closeness.

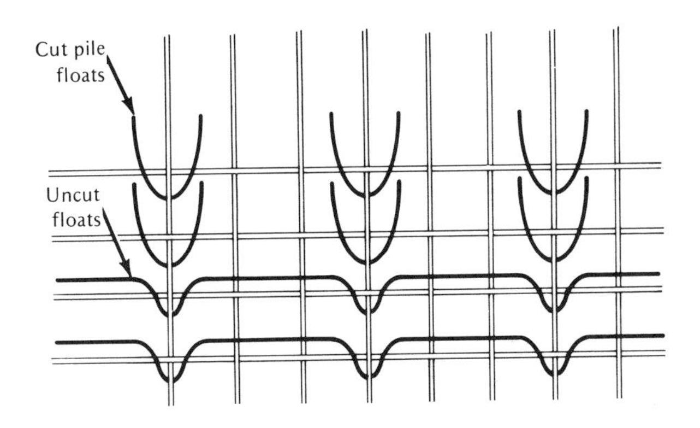

Figure 4.22. The extra filling yarns float over several warp yarns to form lengthwise columns, observable when the floats are cut.

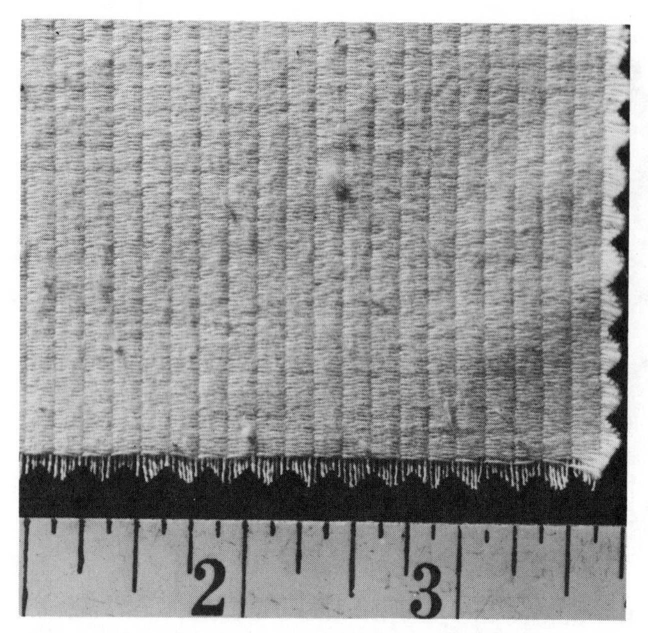

Figure 4.23. These uncut filling floats will yield the characteristic verticle pile corduroy wales when cut.

Warp Piles

Fabrics made with extra sets of warp yarns include *velvet, velour, terry cloth* and certain *carpet* fabrics. Compare the diagrams of filling and warp pile structures in Figures 4.22 and 4.24. From the illustrations, observe that the *cut filling pile* runs in the lengthwise direction, and the cut warp pile runs crosswise. Velvet and velveteen are structured by different pile methods—velvet is a warp pile, while velveteen is a filling pile—but they can be "look-alike" fabrics, as seen in Figure 4.25. They are frequently as difficult to tell apart as the filling pile fabrics studied previously—the new types of wale-less corduroy and velveteen. However velvet is typically a richer, shinier fabric than velveteen; this

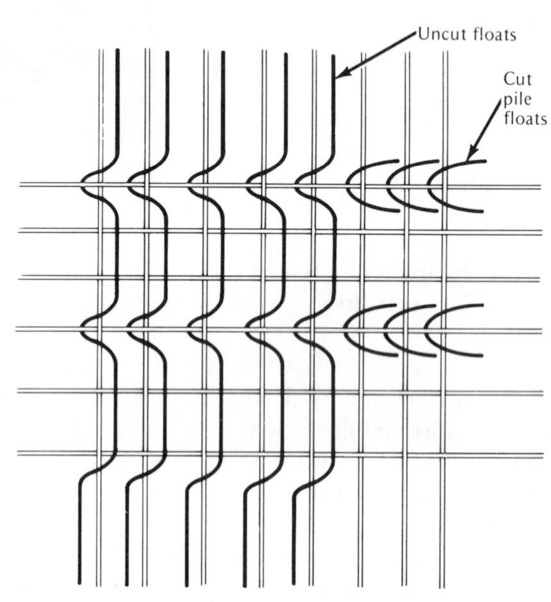

Figure 4.24. The extra warp yarns float over several filling yarns to form crosswise rows, visible when the floats are cut.

QUICK QUIZ 12

1. Name two examples of filling pile fabrics:
   ______________ and ______________
2. In these fabrics, the extra set of yarns floats in the ______________ direction; the floats span the ______________ yarns.
3. Although the filling pile is formed by floats in the filling direction, after they have been cut the pile appears in columns in the warp direction. Therefore, the raised, cut pile wales on corduroy appear in the ______________ direction.

****

ANSWERS QUICK QUIZ 12

1. corduroy, velveteen
2. filling; warp
3. warp

Figure 4.25. Notice the lengthwise breaks in velveteen, a filling pile; and crosswise breaks in velvet, a warp pile.

is because velvets tend to be made from yarns of filament-length fibers and velveteen from staple-fiber yarn.

Terry pile, perhaps the best known warp pile, is easily recognized as the typical structure used for bath towels. It is constructed with two sets of warp yarns in such a way that loops are formed on both sides of the fabric. The *slack tension* method of warp pile weaving is used in making terry cloth, which means that the extra set of warp yarns that form the looped pile is held without any tension during the weaving, while the ground yarns are maintained with the taut or tight tension required for regular basic weave constructions.

Terry cloth is often found on the market with uncut loops on both sides of the fabric; however, terry fabric is available that has had one side of the pile cut and sheared to look very much like velvet or velveteen. Such fabrics are frequently called *velour toweling* or *terry velour*. See Figure 4.26.

Terry fabric has been successful in toweling because the loops hold moisture and increase fabric absorbency; but if loops are cut on one side, absorbency is decreased. Further, the cut loops tend to produce lint which can be a problem in some end uses. Sheared velour towels then, are created mostly for their esthetic value because some consumers like their appearance.

If a pile fabric is to be marketed as an uncut or looped pile, it needs no surface cutting or shearing. Cut pile fabrics, on the other hand, must go through an extra step for the loops to be cut. One cutting process uses a special machine with guides which raise the floats so that revolving knives can roll over the fabric surface and cut the raised floats. This method is used for velveteen and corduroy.

The most common system for constructing cut, warp-pile fabrics is the double cloth method, in which two fabrics are woven simultaneously, one above the other. Study the diagram of this method in Figure 4.27. The extra set of warp yarns that joins the two fabric layers is cut after the fabrics are woven.

In the double-cloth method of constructing cut, warp pile fabrics, the depth of the pile is determined, in part, by the length of the warp yarns that secure the two layers of fabric together. Also, pile depth is influenced by final shearing which evens the pile surface.

Figure 4.26. Terry toweling is structured with an extra set of warp yarns, which may be cut (B), or left uncut (A).

QUICK QUIZ 13

1. Velvet and velveteen can be manufactured into fabrics suitable for either apparel or home furnishings, especially upholstery material. Examine fabrics 13 and 14. Identify which is the velvet and which the velveteen; also indicate whether these samples were manufactured for apparel or home furnishings.
   13 ______________________ 14 ______________________
2. Fabric 15 is a (warp, filling) pile fabric made by the ______________ tension method of weaving an extra set of ______________ yarns into a structured ground cloth.
3. Examine both sides of sample 15 and determine if it is a (cut, uncut) terry cloth. What is the fabric name? ______________________

****

ANSWERS QUICK QUIZ 13

1. 13—velveteen for apparel; 14—velvet for home furnishings, upholstery
2. warp; slack; warp
3. cut; terry velour

Cut, warp pile fabrics can also be formed from single fabric layers by using the *over wire* method. As the name suggests, the extra set of warp pile yarns passes over a wire during the weaving process; the wire has a knife edge which cuts the pile loops as the wire is removed.

Cut pile carpets made by warp pile weaving are formed by the over wire method. The wires can be interlaced in a warp pile weave in such a way that only certain warp loops are cut when the wires are removed. Thus, sculptured designs, which give raised and lowered effects, can be made by combining both cut and uncut areas in the same carpet. In both filling and warp pile weaves, the extra set of yarns which forms the pile is interlaced into the ground or base cloth as an integral part of the total fabric.

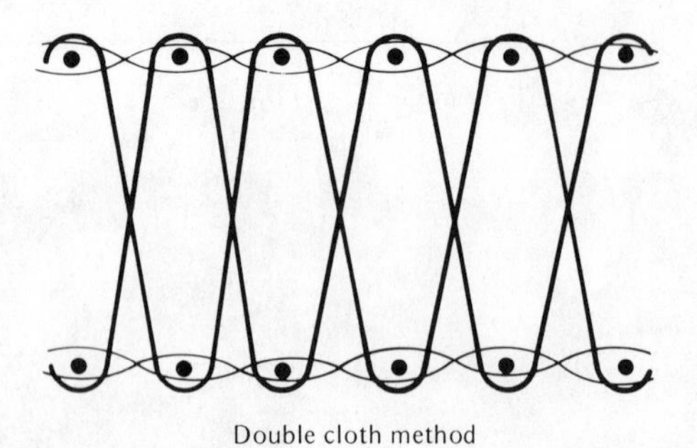

Figure 4.27. Two fabric layers are joined with an extra set of warp yarns, which are cut to form two separate warp pile fabrics. Velvet is often made by this process.

Novelty or Figure Weaves

There are many ways to produce fabrics with structured-in special appearance characteristics. Pile weaving, just discussed, is one technique. Other patterns can be built into the fabric by various devices that control the shedding or separation of the warp yarns so that special designs are actually an integral part of the weave. Two major techniques used for weaving in a pattern are the *dobby* and *jacquard* processes.

Dobby Weaves

Dobby structures are recognized by geometric patterns as shown in Figure 4.28. They are produced on the same basic type of loom used for weaving plain, twill and satin structures.

The differences are the increased number of harnesses on the dobby loom and the addition of special controls, often called dobby attachments. These raise and lower the harnesses in such a way that designs are structured into the fabric during weaving. The dobby weave is characterized by relatively small geometric designs, as in fabrics 16 and 17.

One type of loom with a dobby control attachment is pictured in Figure 4.29. The special punched plastic template controls the raising and lowering of the harnesses so that a design is formed during the interlacing of the warp and filling yarns. In this illustration, contrasting filling and warp yarns add

Figure 4.28. Repeating geometric designs are characteristic of dobby weaves.

to the geometric design being formed—look at the fabric coming onto the cloth beam.

Certain designs formed by dobby attachments have become standardized and carry specific fabric names. *Birdseye* is identified by a repeating geometric pattern composed of a diamond shape with a dot in the center. Roller towels found in commercial dispensers are often birdseye fabric. Because the design is geometric and small, it is clearly identifiable as a dobby weave.

Wafflecloth and *piqué* are two other fabric names for established dobby weaves. Wafflecloth, sometimes known as the honeycomb variation, is composed of small repeating squares; as in fabric 17. Piqué is recognized by repeated cords or vertical wales. Corduroy and selected rib weaves also are generally recognized by vertical wales, but they are not created on dobby looms.

Jacquard Weaves

Extremely complex and large designs can be woven into fabrics with a jacquard attachments, pictured in Figure 4.30. The complex unit and set of punch cards makes it possible to control *each* warp yarn on an individual basis. Looms with these special attachments are frequently called jacquard looms; however, the loom itself does not differ

Figure 4.29. Notice the rotating template which controls the geometric designs formed by the loom. *(courtesy American Textile Manufacturers Institute, Inc.)*

Figure 4.30. The elaborate head and set of punch cards at the top of the looms control the intricate patterned designs typical of jacquard structures. *(courtesy American Textile Manufacturers Institute, Inc.)*

greatly from other type of looms. The difference is in the control of the warp yarns, which allows for considerably more complex designs than in dobby fabrics.

Today, many dobby and jacquard designs are achieved through the use of electronic dobby and jacquard attachments which are integrated with computer-aided design (CAD) systems. Computer designed patterns are programmed into dobby or jacquard looms for automatic design formation.

Designs in jacquard fabrics are typically much larger, more complex and ornate, as pictured in Figure 4.31, than any of the dobby constructions. Thus, jacquard fabrics are usually more expensive to produce.

Two jacquard fabrics familiar to many consumers are named *brocade* and *damask*. Used for apparel and home furnishings, both have ornate, woven-in designs or patterns. Brocade fabrics have a surface that is slightly raised in certain areas, giving depth and additional character to the fabric. In contrast, damask is flat and smooth with a design that is usually reversible.

As you can tell from fabric 18, brocades are typically formal, dressy fabrics. In apparel, they are used primarily for evening clothing which does not require high measures of durability. But upholstery brocades are expected to be reasonably durable for the end use, so they are usually heavier than apparel brocades. Also, the selection of durable fibers from the synthetic category, such as nylon, helps provide adequate abrasion resistance for the raised design areas.

Damask is a popular choice for table coverings. They range from the formal ''white on white'' patterns to the more casual red and white damasks popular in Italian restaurants. Two reasons why damask is suitable for tablecloths is that it can be reversed during use between launderings, and the smooth fabric lies flat and provides an even base for tableware.

Figure 4.31. The jacquard structure is characterized by complex, large designs.

Other types of fabrics woven on looms with jacquard attachments include *tapestry, brocatelle* and *matelassé.* These are special jacquard weaves composed of two distinct fabric layers that are interwoven at various design points, as in sample 19, a matelassé. These ''double cloth'' fabrics are discussed in the section on double weaves.

The loom used for making such fabrics requires two warp beams and a method for inserting two different sets of filling yarns. The design is controlled by the same type of punch card mechanism

---

QUICK QUIZ 14

1. Examine fabric 18 and give its name: ____________________
2. What is the probable end use: ____________________

****

ANSWERS QUICK QUIZ 14

1. brocade
2. home furnishings, upholstery or draperies

---

used for single-layered jacquard fabrics. Due to the increased complexity of structuring these double cloths, they are more expensive than most single-layered jacquards. The second layers are usually supportive fabrics, like fabric 19, which means that they are not reversible. Other features might make them worth the expense: the second layer provides support and stability for the top, decorative layer; and increased warmth results from the combined layers. Both jacquard and dobby woven designs are usually produced by a combination of two or more basic weaves. Therefore, in certain complex designs parallel diagonal lines or ridges characteristic of twill weaves are noticeable. Also present may be long floating yarns characteristic of satin or sateen weaves, as well as over-one, under-one interlacing of plain weaves.

Crepe Effects

Fabrics which are characterized by crinkled or pebbly textures on their surface are referred to as crepes. You have studied two ways that yarns are used to achieve crepe effects: the use of highly twisted crepe yarns in a plain weave or modified plain weave construction, and the use of special textured yarns.

Still another method is the *crepe weave*, which is a complex variation of the plain weave. Because it produces a special type of appearance, it is included here under novelty or figure weaves. However, a *true crepe* is a fabric woven with a plain weave pattern of interlacing, using crepe yarns which are highly twisted.

Figure 4.32 depicts an enlarged sample of the crepe weave which can be used to produce a crepe effect. Fabric A does not have the crinkled, pebbly appearance associated with crepes because of the large yarns used to show the structure. But, the crepe appearance in Fabric B, made with normal-size yarns, was created by random interlacing that yielded floats in no set pattern.

Crepe weaves require more random floating than is possible on a basic, two-harness loom. Thus, they are structured on a dobby loom to allow enough variety of harness action to produce the "no set" pattern of interlacing. A crepe weave is not a true dobby structure, however, because the loom is not programmed to produce the special repeating geometric design characteristic of dobby weaves.

Since the interlacing pattern is random rather than in a repeating progression, diagonals are not formed as in twill weaves. Because the floats usually span only two or three yarns, the smooth surfaces that are characteristic of satin or sateen weaves do not result. Typically, satin structures are made from filament-length fibers with a low amount of twist; crepe fabrics are structured from either textured filaments or staple-length fibers with a comparatively high amount of twist.

In addition to the crepe weave, another woven crepe effect can be achieved. *Seersucker* is a fabric with a "crepe" effect in stripes of varying widths, achieved through *slack-tension weaving*. In Figure 4.33 the fabric appears smooth in some areas and crinkled like a crepe in others.

This effect is produced by carefully controlling the tension of the warp yarns. In areas where the crin-

Figure 4.32. Crepe weaves are formed using a random pattern of yarn interlacing.

Figure 4.33. Seersucker has woven-in crinkled areas produced through special tension control.

kled effect occurs, the tension on the warp is slack and less than on the warps that create the smooth areas of the fabric. These slack tension warp yarns yield the crinkled areas upon insertion of filling yarns.

Figure 4.34 illustrates the use of textured yarns as yet another way to produce a crepe effect. When interlaced in a *plain weave* pattern, these textured yarns give an effect almost identical to the one obtained in crepes made with high-twist crepe yarns.

However, special yarns designed for crepe effects are not always interlaced in the plain weave pattern. In fabric 20, a crepe studied earlier made from textured yarns, the interlacing pattern is random. Thus, a crepe weave was used to help create the crepe effect.

Leno Structures

The leno weave is produced by a modification of the basic interlacing associated with regular weaves. This modification involves the crossing of a pair of warp yarns between each filling yarn, as illustrated in Figure 4.35.

Thus, the warp yarns are handled in pairs as a single unit. The pairs of yarns are parallel to each other as in any type of traditional weaving, but the

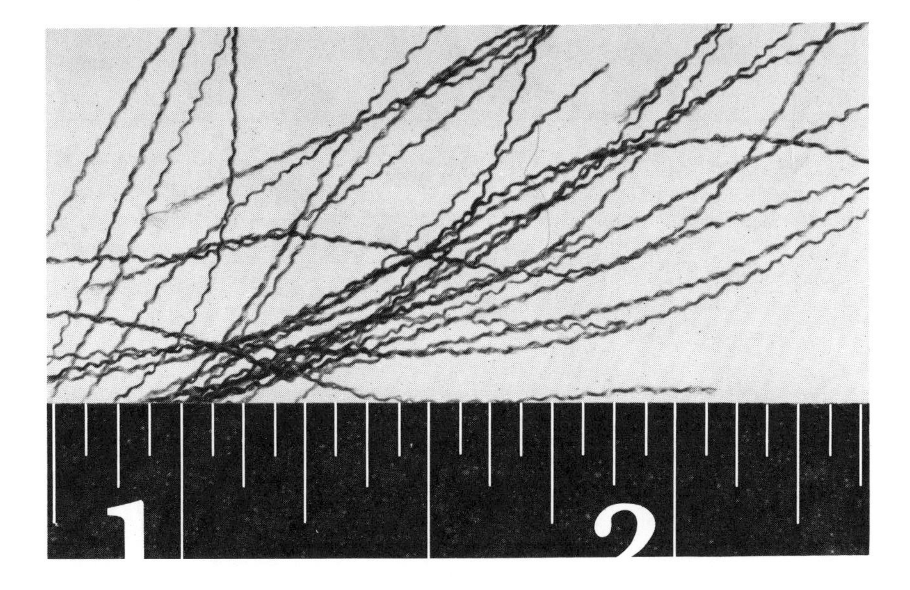

Figure 4.34. Thermoplastic yarns may be heat-set to yield crepelike texturization.

QUICK QUIZ 15

1. Match the three methods of achieving crepe fabrics to the advantage of each method:

____ 1. use of highly twisted crepe yarns	a. yields crepe effects that can be heat set
____ 2. use of crepe weave	b. produces "true crepe" with springy, lively hand
____ 3. use of textured yarns	c. results in structurally stable crepes

NOTE: A fourth method for achieving a crepe effect is through finishing, discussed in Unit V.

2. Examine fabric 21 and give the fabric name: ________________.
3. Are the crinkled effects (temporary or permanent)? Why? ________________ ________________

****

ANSWERS QUICK QUIZ 15

1. 1-b; 2-c; 3-a
2. seersucker
3. permanent, because they are woven in

---

yarns within each pair are not parallel when they crisscross one another between each filling. The leno weave produces very open fabrics with good stability; the intertwining of the warp yarns holds the filling yarns in place, so that they do not slip easily and create fabric distortion.

Figure 4.35. In a leno structure, two warp yarns are crossed before each filling-yarn insert. Generally, an open structure results.

The dual warp yarns used in making leno fabrics contribute to the strength of these sheer, open structures. Leno fabrics are especially suitable for the openwork mesh bags sometimes used to package fruits and vegetables. The strong open structure allows air circulation and also bears the weight of the contents. When leno structures are made for maximum durability, it is probable that cord yarns are used for fabric strength.

In selected end uses, leno structures are chosen for esthetic as well as durability reasons. For example, a common leno fabric used for sheer window curtains is *marquisette*. Figure 4.36 is an enlarged photograph of marquisette, which is frequently made of polyester fibers. Leno fabrics are good choices for window curtains because they allow air circulation when windows are open; the pairs of warp yarns contribute to stability in the lengthwise direction and the weave can create decorative or esthetic effects.

Modifications of leno weaves are used to produce special decorative fabrics for window coverings; sample 43 is a drapery fabric in which certain warp yarns are twisted before the filling yarns are inserted. Thus, the leno structure creates the characteristic open spaces, while still contributing to stability in the warp direction. As is often the case with novelty fabrics, several weaves are combined in sample 43—long warp yarn floats create a diagonal

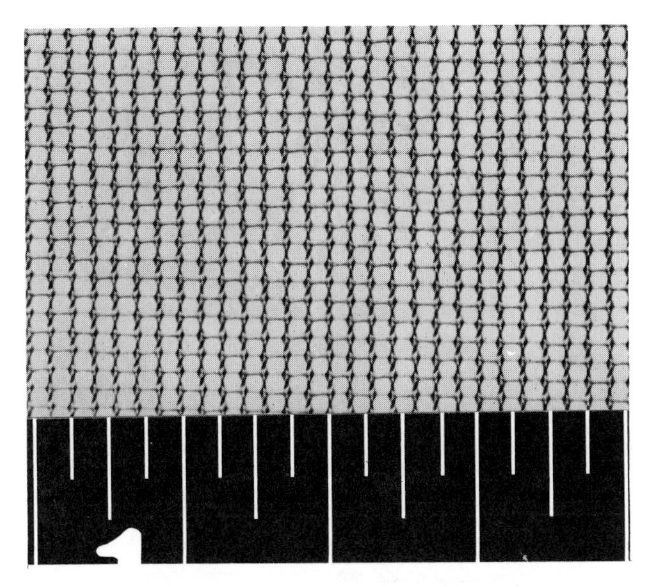

Figure 4.36. Marquisette, a sheer curtain fabric, is a typical leno structure.

twill effect, and between the novelty stripes a plain weave is used.

Double Weaves

A variety of fabrics on the market fits into the category of double cloth. A distinguishing feature among them is the number of sets of yarns used in the fabric construction. Three-, four- and five-yarn double cloths are discussed here.

A *true double cloth* has at least *five sets of yarns*, as shown in Figure 4.37. One set of warps interlaces with one set of filling yarns; a second set of warps interlaces with a second set of fillings; and the fifth set, usually of warps, interlaces between the other sets to join the two layers of fabric. The five-yarn double cloth used in making certain warp pile structures such as velvet is designed to be cut apart, resulting in two cut pile fabrics. When a five yarn double cloth is structured to remain joined, a two-layered, reversible fabric results.

While the appearance on each side may differ, fabrics made from several sets of yarn are generally reversible and designed for use in coats and suits. Such fabrics are expensive, considering the materials and time required for construction. Some features that make five-yarn double cloths worth the extra cost include: increased heat retention between layers, increased fabric stability, and greater esthetic variations due to their reversible nature.

Another type of double cloth is made using *four sets of yarns*; two examples discussed under jacquard weaves were *matelassé* and *brocatelle*. These structures have two sets of warp yarns and two sets of filling yarns interlaced by jacquard controls that produce large, complex floral or scroll designs which are not reversible. Four-yarn double cloths are often woven to be reversible; but in matelassé and brocatelle, the second set of warp and filling yarns is designed to provide support or stability to the top decorative layer.

When matelassé and brocatelle are woven in upholstery weight, the second layer reduces stretching and increases durability for the end use. This supportive layer also helps create the raised design effects characteristic of these double-woven jacquards. In matelassé, the back layer is woven tighter than the front so that the front layer forms wrinkles or puckers on the fabric face. In brocatelle, the back layer is structured to support raised designs similar to those in the single-layer jacquard fabric named brocade.

Figure 4.37. In five-yarn double cloth, two separately woven fabrics are joined with a fifth set of yarns.

Four-yarn double cloths designed to be reversible are illustrated in Figure 4.38. Notice in the fabric that periodic interlacing of one set of warp and filling yarns reverses with the other set of warp and filling yarns; this reversal of the weaving pattern joins the two layers together. In the illustration, you can determine that the reversal takes place and the fabric layers are joined where the colors change.

Another type of double cloth made using *three sets of yarns* is more accurately called *double faced* or *backed fabric*. This type has either two sets of warp and one of filling yarns, or two sets of filling and one of warp yarns. The third set of yarns is interlaced so that it forms the surface of the fabric. When the extra set of warp yarns forms the surface, the fabric is called *warp faced*; if the filling set forms the surface, the fabric is *filling faced*.

Double-faced fabrics are usually designed to be reversible; typical end uses include blankets and satin ribbons. Frequently the fabric is made so that the two sides are of different colors. For example, in a filling-faced fabric, blue could be obtained on one side and a gold on the reverse side by having one set of filling yarns blue and the second set gold.

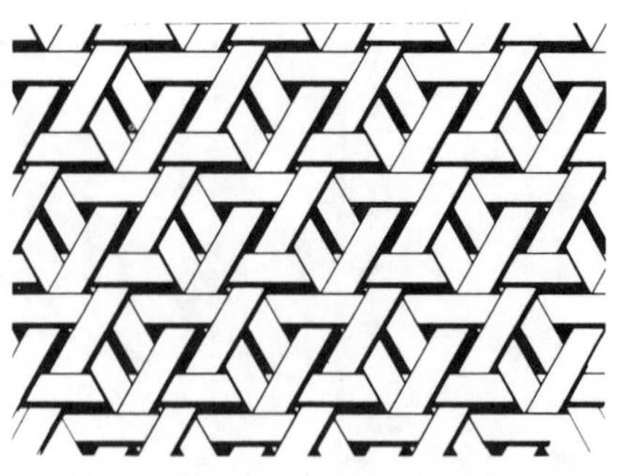

Warp yarn at 60° to filling yarn

Warp yarn at 60° to filling yarn

Figure 4.39. Examine the 60° angles of triaxial woven constructions.

Figure 4.38. Two sets of warp yarns and two sets of filling yarns interlace to form this four-yarn double woven structure; no fifth set of yarns is present.

Triaxial Weaves

Most woven structures are made with sets of yarn interlacing at right angles to each other. A more recent process interlaces three sets of yarns at 60° instead of the standard 90° right angle typical of biaxial woven fabrics. This fabric is identified as a triaxial structure, compared to regular weaves which are biaxial. Advantages of triaxial fabrics include high tear strength, multidirectional stability, uniform give in all directions, and reduced production costs.

The potential of triaxial weaving is increasing yearly. Of particular advantage is the consistent stability of these fabrics in all directions; there is no greater stretch in the bias direction than in any other direction, which is not true of regular biaxial weave fabrics. A variety of pattern designs is possible, and yarn slippage is reduced to a minimum. Some end uses where the stability characteristics of this weave are advantageous include draperies, upholstery, special apparel items that require stability without extra stretch, plus a variety of industrial uses.

REVIEW

1. By combining different kinds of fibers and yarns, an infinite number of designs is possible in double-woven fabrics. Additional modifications can be made through the use of special finishing processes. Double-woven structures are frequently used where reversible fabrics or additional warmth is desired, or where stability is needed in the ultimate end use. When a desired effect can be achieved by a single layer fabric, however, that method is generally preferred because ________________________.

2. A single-layered fabric structured to be as heavy as a double woven has to contain heavier yarns in order to match the thickness and weight of a double weave. In such instances, what advantage would the double woven have over the single layer of comparable thickness and weight? ________________________

3. A double weave tends to be (warmer, less warm) than a single-layer fabric of the same thickness. Why? ________________________

4. Explain the difference between these fabrics:
   a. filling faced ________________________

   ____

   b. warp faced ________________________

   ____

5. List two guidelines that help identify double-woven fabrics:

   ____

   ____

6. Recall that some pile-woven fabrics are made with three sets of yarns. How does their construction differ from double-faced fabrics made with three sets of yarns? __________

   ____

   ____

****

REVIEW ANSWERS:

1. single weaves are considerably less costly and bulky
2. the double weave has greater flexibility and pliability
3. warmer; the double weave traps air within each layer and between the two layers
4. a—two sets of filling yarns and one set of warp yarns are used; filling yarns form the surface on both sides of the fabric

   b—two sets of warp yarns and one set of filling are used; warp yarns form the surfaces
5. Any two: Two layers of fabric can be separated into two distinct fabrics by cutting the yarns that hold them together.

   Two distinct layers of fabric (which may or may not be separated into two fabrics) are visible.

   At least two layers or sets of yarn in either the warp or filling direction are present.
6. The extra set of yarns in the pile construction is interlaced to form loops on the surface. In a double-backed or faced fabric, all three sets of yarns are interlaced to form relatively flat reversible structures.

Complex, Decorative Weave Identification

1. Identify the fabric structure and name, where applicable, of the samples indicated below:
   12 ____________________
   13 ____________________
   14 ____________________
   15 ____________________
   16 ____________________
   17 ____________________
   18 ____________________
   19 ____________________
   20 ____________________
   21 ____________________
   43 ____________________

   Cut a swatch from each of these samples. Attach and label them in the appropriate block on the Mounting Sheet for Complex or Decorative Weave Fabrics.
2. Samples 12, 13, 14 and 15 are all pile weaves made with ____________________

   ____

   How does their serviceability compare to fabrics made from plain or twill basic weave structures?

   ____
3. However, the pile weaves in your swatch set—the corduroy, velveteen, velvet upholstery and terry velour—all appear to be relatively durable for their special end uses due to their (high, low) yarn counts and uniform (cut, uncut) pile surfaces. Thus, snagging and abrasion probably (would, would not) be a problem.
4. Four of the complex weaves in your fabric set were made on a dobby loom. Name the two samples with repeating geometric patterns that are typical of dobby weaves: ____________ and ____________. Name the two other samples not identified as dobby weaves that require a dobby loom to create the interlacing pattern: ____________ and ____________.
5. Compare the serviceability of the two jacquard-woven home furnishing fabrics—18 and 19:

   ____

   ____

   ____

   ____
6. Fiber content is very important in determining serviceability, as well as yarn and fabric structures. Further evaluate the serviceability of the jacquard fabrics, given these fiber contents: the brocade is 56% olefin, 44% nylon; the matelassé is 35% olefin, 35% polyester, 30% rayon.

   ____

   ____

   ____

   ____
7. Dobby-woven fabrics 16, 17 and the seersucker, 21, are all made of polyester/cotton combinations. Considering the fiber content and woven-in designs, evaluate durability and care procedures. ____________________

   ____

Complex Decorative Weave Identification Answers:

1. 12 - filling pile weave, corduroy
   13 - filling pile weave, velveteen
   14 - warp pile weave, velvet upholstery
   15 - warp pile weave, sheared terry or terry velour
   16 - dobby weave, no special name
   17 - dobby weave, wafflecloth
   18 - jacquard weave, brocade upholstery
   19 - jacquard double cloth, matelassé upholstery
   20 - crepe weave, crepe
   21 - slack-tension weave, seersucker
   43 - combination of weaves—plain, twill, leno
2. an extra set of warp or filling yarns
   Generally plain and twill woven fabrics are more durable and less susceptible to surface abrasion than pile weaves.
3. high; cut; would not
4. 16 - dobby
   17 - wafflecloth
   20 - crepe
   21 - seersucker
5. The single-layered brocade does not have the supportive layer of the matelassé. Also, the raised brocade pattern might be more susceptible to overall abrasion and snagging, though the stuffer yarns on the matelassé also would be subject to abrasion.
6. The nylon in the brocade would increase abrasion resistance over the matelassé that has 30% rayon. The olefin in both samples would help decrease static electricity.
7. All fabrics would be durable and easy care for most apparel uses.

UNIT FOUR

Part III

Knitted Structures

In Part 1 of this unit a general comparison of woven and knitted structures was presented, followed by descriptions of weaving processes and woven fabrics. This section includes discussions of knitting processes and knitted fabrics.

Knit fabrics are formed by interlooping yarns, also described as the interconnecting of yarns through a series of loops. Knitted fabrics are classified as *warp* or *filling* knits according to the direction in which the loops are interconnected. Just as interlaced warp yarns run in the lengthwise or vertical direction of woven fabrics, the yarns in warp knits are interlooped in the lengthwise direction. Likewise, the yarns in the filling knit process are interlooped in the crosswise direction.

Frequently it is very difficult to tell the difference between warp and filling knits. Only repeated experience will provide the basic skill required to distinguish between the two types of knit structures. However, careful attention to the diagrams provided and study of the fabric samples will help identify warp and filling knits.

In both warp and filling knits, vertical rows are referred to as *wales;* crosswise rows are called *courses*. Wales typically are observable on the face of knitted fabrics, and courses are usually more clearly seen on the reverse side of knitted structures. However, sometimes knit fabrics are printed on the back or course side, so ignore the location of applied designs in studying knits and determining the front and back.

Examine the diagrams of knit structures in Figure 4.40. While warp and filling knits have wales and courses, the direction in which the loops are interconnected distinguishes between the two.

Visualize in the diagram that the loops in the filling knit are interconnected in the crosswise or course direction. In the warp structure, the rows of wales are interlooped in the vertical direction with diagonal interconnectors, which are drawn with a heavier yarn in the warp diagram. The wales will always be parallel to the lengthwise direction of the knit, while the courses will always be parallel to the crosswise direction.

To better understand the difference in warp and filling knits, compare knitted samples 22 and 27. On the front of both samples are vertical ridges called wales. On the back of the fabrics are observable courses.

Since the fronts or faces of these samples both have wales observable in the warp direction, additional guides are needed to help distinguish between warp and filling knits. So, examine the *back* sides of these samples, where a difference in the knitted courses is evident. An identifying feature in determining warp knits is the diagonal or V shaped interconnectors observable on the back of fabric 27.

Because many warp knits are very fine structures, it is difficult to see the way in which the yarns are looped to form the V interconnections, but a magnifying glass should help. In fabric 22 there are no diagonal interconnectors, and the courses are interlooped straight across the fabric horizontally—thus it is identified as a filling knit.

Whether a knit is a warp or filling knit also may be determined by looking at the front of the fabric. On the wale side of these warp and filling knits, pull the fabrics in the crosswise direction. A different construction of the interlooping *between* the vertical ridges is apparent. When sample 22 is stretched in the crosswise direction, horizontal interconnectors are visible between the wales. These interconnectors are formed by the crosswise interlooping of yarn which is characteristic of filling knits.

QUICK QUIZ 16

1. In samples 27 and 27A, which are ______________ knits, what type of yarn connectors are visible on the reverse side of the fabrics? ______________
2. In addition to determining the differences between warp and filling knits by diagonal or horizontal interconnectors, the degree of stretch can sometimes be compared. Pull samples 22 and 27 in the crosswise direction; which has more stretch? ______________ Now attempt to stretch samples 27A. Why does this warp knit exhibit little crosswise stretch? ______________
3. Extend sample 23 in the crosswise direction and notice the diagonal interconnectors between the wales, which means it is a (warp, filling) knit.
4. Normally, warp knits have less give in the crosswise direction than does sample 23, but the use of the ______________ crepe-effect yarns contributed to increased stretch.
5. Samples 28 and 28A are warp knits with special finishes applied to the (wale, course) side. The reverse side of the fabric has wales with ______________ interconnectors.
6. The wale or course side of knitted fabrics may have finishes or designs applied, so the ''right'' side for the end use does not determine structurally the top side of a knit, which characteristically has visible ______________.

* * * *

ANSWERS QUICK QUIZ 2

1. warp; diagonal interconnectors
2. 22; it has been manufactured for use in luggage
3. warp
4. textured
5. course; diagonal
6. wales

There is generallly more stretch along the courses than the wales in both knits; and filling knits usually have more given than warp knits. However, there are variations in degree of stretch in both warp and filling knits based on the openness or coarseness of the knit, the tightness or looseness of interlooping and

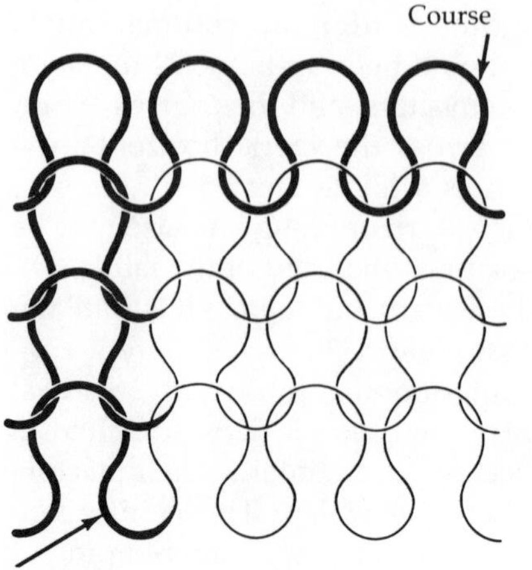

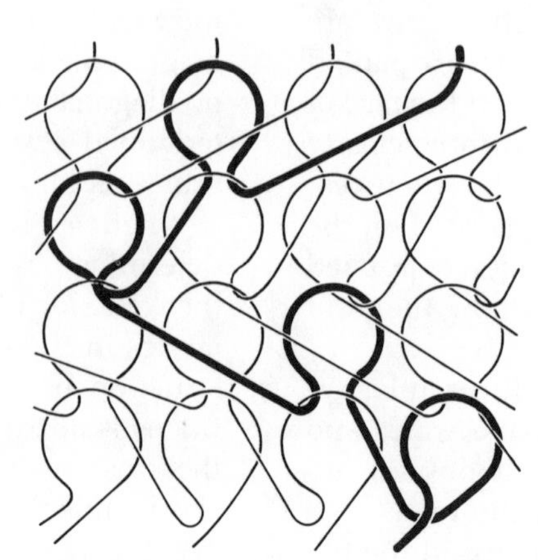

Figure 4.40. Knitted fabrics are classified according to the direction in which the yarns are interlooped.

the yarn type. For example, textured yarns are often used for specialized purposes, such as to increase stretch and produce the crepe effect in sample 23.

Warp and Filling Knit Construction

In order to control the direction in which knits are interlooped, different types of equipment are used to form warp and filling knit fabrics. Warp knits are made on knitting frames that have a large beam with as many yarns as needed for the desired width of the final fabric. These frames produce warp knits in a *flat* form, similar to woven fabrics.

Filling knits may be made on knitting frames that produce either *flat* or *circular* fabrics. A majority of filling knits are produced today on circular knitting frames due to their speed of production. A circle or tube of fabric formed may be slit as part of finishing processes, or left in tubular form for certain end uses.

Knit fabrics can vary considerably in the closeness of the loops. Some knits are structured with the loops spaced sufficiently apart to yield a fabric that appears loose and open; other fabrics may be knitted with the loops very closely packed to provide a compact, tight structure.

Although filling knits may be made by either hand or machine, a warp knit can be made only by machine. Warp knitting uses many parallel warp yarns, each carried by yarn guides in the lengthwise direction of the fabric. These yarns are picked up by needles to form the interconnecting loops which create the wales or lengthwise rows of loops on the front of the fabric.

Refer to Figure 4.41 for an illustration of a warp knitting frame. Since warp knits require separate yarn guides and needles for each yarn in order to form each vertical wale in the fabric, hand knitting methods cannot produce warp knits.

In the hand method of making filling knits, generally one yarn is used to knit left and right across the

Figure 4.41. Warp-knitting machines have many needles that interloop yarns vertically, yielding flat fabrics with straight edges. *(courtesy Collins and Aikman)*

fabric. However, commercial machines for constructing filling knits have multiple yarns and needles which knit interconnecting loops simultaneously in the horizontal direction. In Figure 4.42 the many cones of yarn used in making filling knits are shown. This knitting machine makes a filling knit into circular fabric, instead of the flat method used in hand knitting. As the yarns from the different cones of yarn are interlooped, the fabric is actually built in a spiral form rather than a true row by row horizontal manner.

Compare the circular knitting machines in Figures 4.42 and 4.43. The cones of spun yarn that feed into the machine shown in 4.42 require more yarn processing prior to the knitting process than the slivers used in the filling knit machine in 4.43.

Knitting is a fast way to construct fabrics. While some authorities say warp knitting is faster than filling knitting, this is debatable and probably no longer true with modern machinery. Regardless, knitting by either method is faster than weaving. Filling knits are more economical to produce than warp knits, since their overhead costs are less. Designs or patterns may be changed quickly using the electronic design controls on circular knitting frames. The latest of computerized knitting machines has allowed for higher speeds and flexibility for multipurpose applications. Modern knitting machinery is capable of producing elaborate combinations of patterns and textures with greater speed and accuracy. Very rapid fabric production is possible with the new large diameter circular and warp machines. As with woven fabrics, computers have allowed for integrated manufacturing from design to finished product.

Major end uses of filling, circular-knit fabrics in their tubular form include hosiery, T-shirts, some sweaters, and swimsuits. In such end uses, the fabric tube can be knitted to size so that no seams are required to connect fabric edges.

Since a warp knit is constructed by parallel yarns being interlooped in a vertical direction down the length of a fabric, rather than being built up horizontally, warp knits are always constructed flat. Thus, warp knits must be cut and sewn to form textile products such as wearing apparel.

Warp knits are generally more closely knitted than filling knits; and they have less elongation potential, especially in the lengthwise direction due to the diagonal or V interconnectors. Warp knits, therefore, are desirable structures when dimensional stability is an important property in the end use.

Full fashioned knits are a special kind of filling knit structure made by flat rather than circular knitting methods. The process permits accurate shaping of garments by adding and dropping knit stitches. Full

Figure 4.42. Circular knitting machines produce filling knit fabrics *(courtesy American Textile Manufacturers Institute, Inc.)*

Figure 4.43. A 12-feed sliver circular knit machine creates bulky knits of loosely twisted yarns *(courtesy Collins & Aikman)*

fashioned knitting is the most expensive type of knitting because it is time consuming. In a full fashioned sweater the fashion marks where stitches have been added or dropped are visible in the shoulder and armhole areas. These fashion marks are shown in Figure 4.44.

Types of Filling Knits

Several types of stitches are used to make filling knit fabrics. The *plain stitch,* also called jersey or stockinette, *purl, rib, float* or *miss* and *tuck* stitches are the most common. In addition to making a vari-

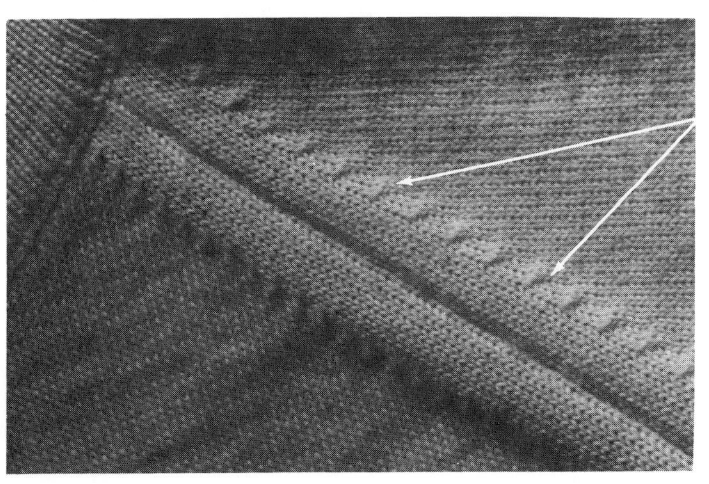

Figure 4.44. Full-fashion marks are distinctive characteristics of flat filling knits that have been knitted to shape, rather than cut and sewn.

ety of designs with these stitches, it is also possible to double knit fabrics and to create surface loops that form a pile.

Plain

The plain knitting stitch is frequently called the stockinette stitch, and fabrics made with it are called jersey knits. Unfortunately, in recent years the term jersey has been applied to knits of either warp or filling construction when the fabrics are relatively smooth and do not have a knit-in design. Sample 22 is a filling knit made using the plain or stockinette stitch; the fabric is appropriately sold as *jersey*. This lightweight, plain filling knit fabric has obvious wales on the front or face of the fabric and courses on the back or reverse side.

Figure 4.45 is an example of a plain filling knit which has much larger knitted stitches than the jersey fabric sample. The knit structure depicted is the type and size frequently used in sweaters, particularly when hand knitted. Notice the definite front and back of the plain knit structure—only wales are apparent on the front of the fabric, and only courses on the back or reverse side.

In the filling knit structure, a filling yarn travels back and forth across the fabric, serving as the anchor to which all the wales are interlooped in vertical columns. Consequently, when a yarn is snagged, a column of wales above and below the snag tends to pull out, leaving a gap called a run. Women's hosiery is a familiar filling knit product that is associated with runs.

Knits of more complex structures, and most warp knits, do not run easily. Plain filling knits of cohesive fibers such as wool are not as subject to forming runs as fabrics made of smooth fibers like nylon filaments.

Purl

A variation of the plain knit stitch is the *purl* stitch, which yields fabrics with the same appearance on both front and back. In Figure 4.46 notice that only courses are visible on either side. This structure has the advantage of producing a fabric that is reversible, but it is more costly to manufacture since the process requires more time than plain knitting.

A purl effect can be achieved in a garment with a plain filling knit by constructing the garment with the back or course side of the filling knit as the right side. When a plain knit is reversed to show the courses, the knitted fabric is called a *false purl*, as shown in Figure 4.47. A purl effect achieved by this method, rather than with a true purl knit, has the advantage of being less expensive to produce.

Rib

When knit stitches and purl stitches are alternated in each row, a *rib* effect results. On the front of the completed fabric alternating wales and courses are visible. Figure 4.48 illustrates the resulting rib knit structure. Where stitches were knitted, raised wales are seen; where purled, courses are clearly appar-

Figure 4.45. Plain or stockinette stitch, filling knit.

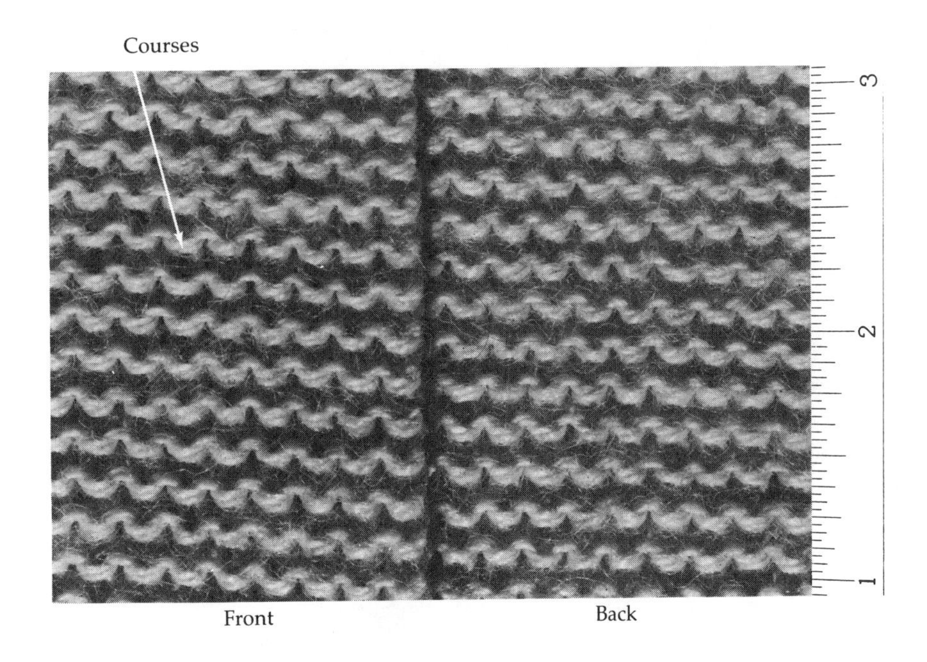

Figure 4.46. Purl Stitch, filling knit.

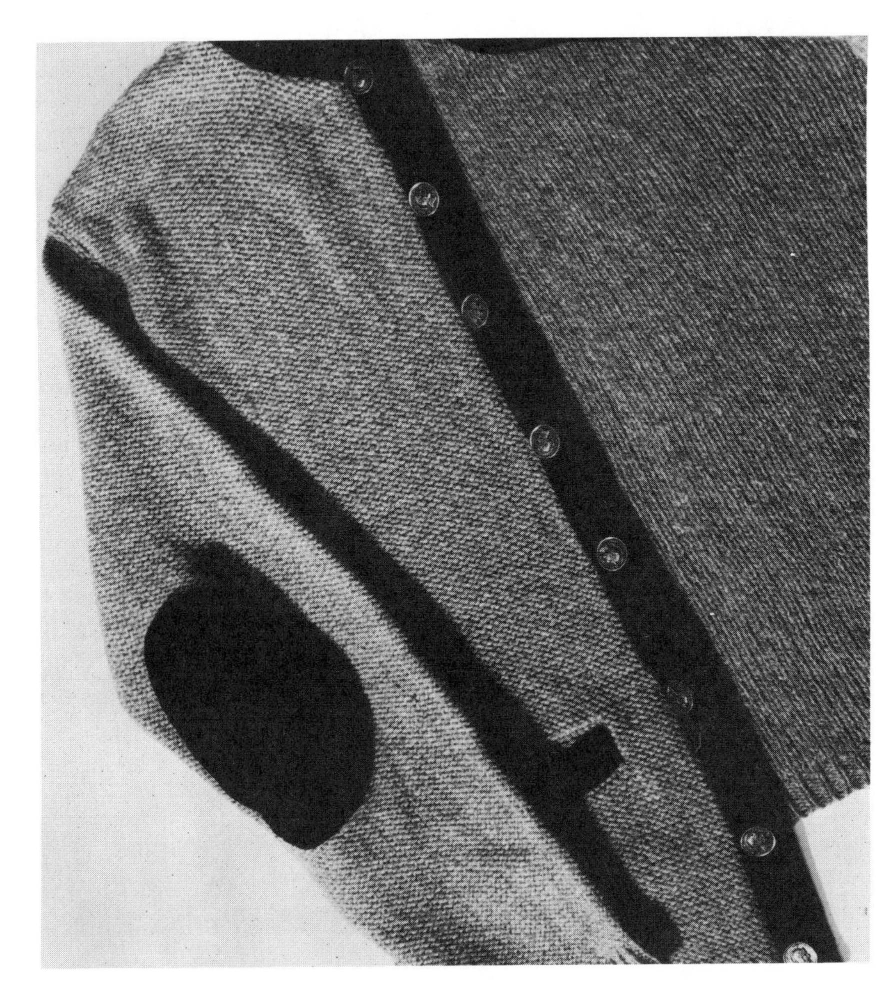

Figure 4.47. False purl, filling knit.

Front

Back

Figure 4.48. Rib stitch, filling knit.

ent. The raised and lowered areas reverse themselves, so that a raised wale on one side forms obvious courses on the other side. Therefore, rib knit structures are usually reversible.

Rib knits are especially elastic and have high dimensional stability, which means that the knit is likely to return to its original shape and size after being stretched. They are commonly used around the cuffs, collars and waists of garments to help maintain the shape at these areas of stress. Because rib knits require more yarn per square area than plain filling knits, they have increased dead air spaces, which tend to contribute to high thermal retention.

Pile

While pile knits are formed in various ways, *an extra set of yarns* is always knitted into the fabric to form the pile. These fabrics are knitted in such a manner as to form loops on one or both sides of the fabric. This extra set of yarns contributes to both special esthetic effects and to increased thermal retention. The loops may be cut or left uncut. Brushing may raise the pile or the pile may be flattened to produce special effects. A pile filling knit, with cut loops that have been brushed, is pictured in Figure 4.49. The structure is often used to make "fake fur" fabrics, such as the one shown.

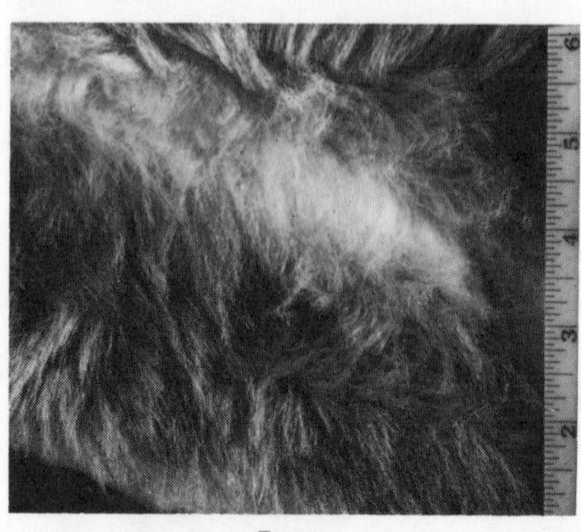

Front

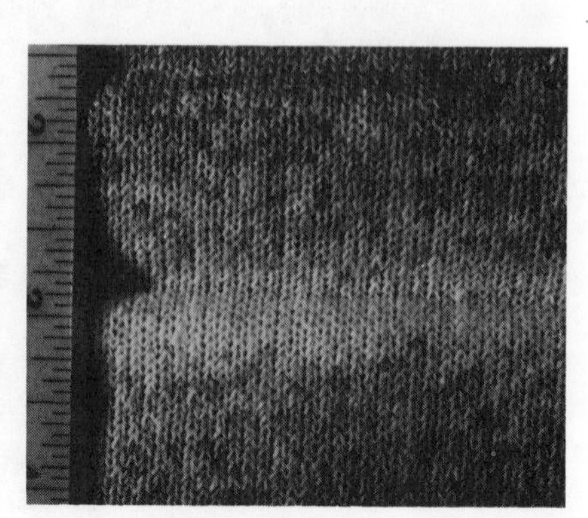

Back

Figure 4.49. Pile filling knit.

Interlock and Double

Many filling knits on the market are interlock or double knit constructions made on circular knitting frames equipped with two sets of needles. The yarns are interlooped in such a way that two layers of fabric are formed, but the two layers are securely integrated.

Interlock knits are made with two sets of yarns *completely interknitted* or interlooped as in samples 24 and 42. Double knits, however, may be formed so some areas have separate layers of fabric while other areas are intermeshed. These systems are commonly used with jacquard controls to form patterned jacquard knits as shown in Figure 4.50.

Compare double knit 25 made using jacquard controls with interlock knits 24 and 42. While none of the fabrics can have their layers separated, the two sets of knitted yarns are more integrated in the interlock fabrics than in the double knit. Also, interlock knits tend to have a finer, smoother surface than conventional double knits and, look somewhat similar to the wale side of a single, filling knit jersey.

Figure 4.50. A jacquard attachment was used on the knitting machine to form this patterned double knit.

Types of Warp Knits

Warp knitting requires many warp yarns, each one controlled by a separate needle that forms interconnecting loops vertically down the fabric. Fabrics can be knitted with great speed on warp knitting frames. Warp knits can be identified by the manner in which the yarns diagonally interconnect between wales.

There are three basic variations of warp knitting named according to the type of machinery used: *tricot* (pronounced tree *ko'*); *milanese* (pronounced mill an *eez'*); and *raschel* (pronounced rah *shell'*). Although these warp knits are made on different machines by varying techniques, they have the common characteristic of being made from loops interconnected lengthwise.

Tricot

Samples 27, 27A, 28 and 28A already discussed are examples of tricot fabrics. Samples 28 and 28A are tricots with texture added to their surfaces by special finishing processes. As you can readily see, however, yarn type and coloring techniques can allow for very different esthetics although the knit structure is the same. Sample 28, often referred to as brushed tricot, would be used for women's and children's sleepwear. Sample 28A, however, would be used in leather-like high fashion apparel. The plain, flat texture of sample 27 is typical of many tricot fabrics. This type of tricot is encountered frequently in knit apparel, especially women's lingerie. The warp knit structure of sample 27A had been engineered for the high durability needed in luggage fabric.

Figure 4.51 is an enlarged photograph of a tricot knit. On the front of the fabric the wales are visible, and on the reverse side the diagonally interconnected courses can be seen. These warp knits generally have less stretchability than filling knits and are, therefore, more dimensionally stable.

The tricot knitting machine shown earlier in Figure 4.41 is versatile in that it can produce plain or patterned fabrics in light-to-medium fabric thicknesses. Fabric thickness is determined by the diameter of the yarns used and by the number of loops per inch. Tricot machinery is best suited to relatively fine yarns of uniform diameter. This is why most tricot fabrics are made of nontextured filaments which yield light-to-medium weight fabrics of relatively smooth textures.

QUICK QUIZ 17

1. In many pile knits the pile surface covers the underlying knitted structure, as can be determined by examining fabric 26. However, on the reverse side of the pile, the _______________ of this filling knit are visible.
2. Pile knits are made in a variety of different weights, from heavy fake furs and coat linings that contribute to high __________________________ to such lighter weight fabrics as 26.
3. Distinguish between an interlock knit like 24 and a single knit jersey like 22. ____
4. Double knits are usually reverisble. Often one side will be relatively plain while the reverse side may be patterned, as in Figure 4.50 and sample 25. The novelty effects in these knits were created on a knitting machine with a ____________________________ attachment.
5. Interlock knits have a disadvantage not found in regular double knits in that they have a tendency to run. Pull samples 25 and 42 in the crosswise direction and observe that 42, the ____________________________________ knit, runs. Both interlock and double knits have interconnecting loops in the horizontal directions and are, therefore, structured as ______________________________ knits.
6. Double knits, often made with _______________________________ yarns, are sometimes popular in slacks, suits, coats and jackets since they have more body and ____________ ____________________ than plain filling knits.
7. Compare samples 22 and 25. Which has more stability? ____________ Why? ____________ ____

****

ANSWERS QUICK QUIZ 17

1. wales
2. thermal retention
3. Wales show on both sides of the interlock knit, and the courses of each layer face inside the structure; courses are visible on the reverse side of a plain filling knit.
4. jacquard
5. interlock; filling
6. textured; dimensional stability
7. 25; because it is a double knit and 22 is a single filling knit.

Wale side Course side

Figure 4.51. Tricot warp knit.

Milanese

Although fabrics made on milanese machines appear similar to those made on tricot machines, the processes are different and fabric characteristics vary. Milanese knits are smoother and more regular in structure and elasticity than tricot knits, yet considerably more tricot than milanese fabric is produced. This is because the complex machinery required for milanese knitting is forcing milanese fabrics off the market owing to the high cost of production.

Milanese knits are identified by a pronounced diagonal effect, usually produced in contrasting colors to show a diamond pattern, as in Figure 4.52. Like the tricot knit, the front of the fabric is composed predominantly of wales and the back of courses.

Figure 4.52. Milanese warp knit.

Raschel

Raschel knits, fast growing in popularity, are quite competitive with tricot and many filling knits. Fabrics made on raschel knitting machines vary from open structures such as crochet effects, knitted laces and knitted nets, to closed structures with compact and close knit stitches. Compare the knit labeled B in Figure 4.53, with more compact fabrics A and C.

Figure 4.53. Raschel warp knits.

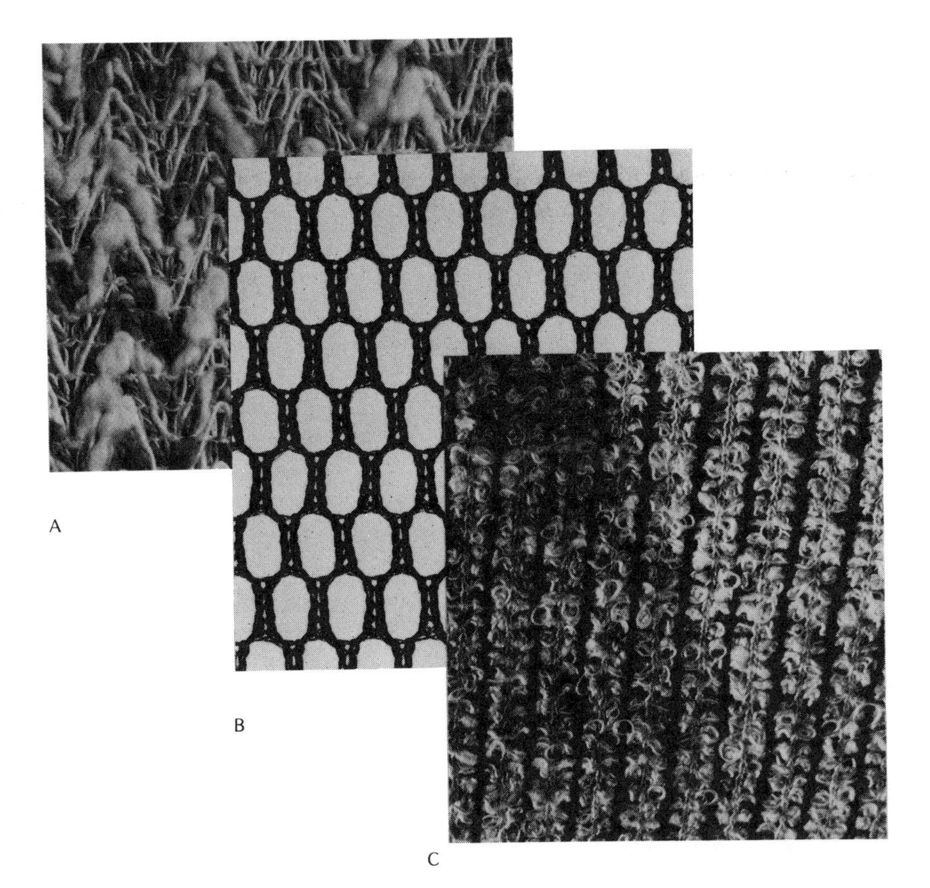

Raschel structures may also have yarns laid in, as fabric A, to increase fabric thickness and design effects. In contrast to tricot machines which are designed to handle fine and uniform yarns, raschel equipment is suited to different yarn types and sizes, including many novelty or complex yarns.

Fabrics 29 and 30 are made on raschel equipment. Both fabrics resemble other structures: sample 29 is a knitted version of woven eyelet embroidery, and sample 30 is a knitted rather than traditional knotted lace.

Warp and Filling Knit Identification

1. Without looking back to the previous discussion, identify the following samples as to warp or filling knit and give the knitted variation or fabric name.
   22 ____________________
   23 ____________________
   24 ____________________
   25 ____________________
   25A ____________________
   26 ____________________
   27 ____________________
   27A ____________________
   28 ____________________
   28A ____________________
   29 ____________________
   30 ____________________
   42 ____________________
   Cut a swatch from each; attach and label on the Knitted Structures Mounting Sheet
2. Match the identifying characteristics of these filling knit structures:
   ___ 1. plain or stockinette
   ___ 2. purl
   ___ 3. rib
   ___ 4. pile
   ___ 5. false purl
   a. wales not visible on either side, but courses easily seen on both sides; reversible
   b. extra set of yarns is drawn up through the courses side of a filling knit
   c. courses used as front side of a plain knit
   d. wales predominate on front, courses on back
   e. alternating wales and courses on both sides; reversible

Warp and Filling Knit Identification Answers:

1. 22-filling, plain or stockinette stitch, jersey
   23-warp, tricot, crepe effect
   24-filling, interlock
   25-filling, double knit, jacquard attachment
   25A-warp, tricot
   26-filling, pile knit
   27-warp, tricot
   27A-warp, tricot
   28-warp, tricot
   28A-warp, tricot
   29-warp, raschel
   30-warp, raschel lace
   42-filling, interlock
2. 1-d; 2-a; 3-e; 4-b; 5-c

REVIEW:

1. Match the parts of the loom to the appropriate description of each:
   ___ 1. warp beam
   ___ 2. heddles
   ___ 3. harnesses
   ___ 4. shuttle
   ___ 5. reed
   ___ 6. cloth beam
   a. inserts filling yarns through warp yarn shed
   b. holds the structured fabric
   c. stores the continuous lengths of warp yarns
   d. raise and lower groups of heddles
   e. bats the filling yarns after each pass of the shuttle; compacts fabric
   f. flat, needlelike devices through which warp yarns are passed

---

2. Match the basic weaves and their variations to the most accurate description of the structure:

____ 1. plain
____ 2. rib
____ 3. basket
____ 4. twill
____ 5. left-hand twill
____ 6. right-hand twill
____ 7. broken twill
____ 8. satin
____ 9. sateen

a. groups of warp yarns float over and under one or more filling yarns
b. alternating left and right diagonal lines form V's
c. smooth surface formed by long, warp-yarn floats
d. simple over-one, under-one interlacing pattern
e. long, filling-yarn floats form fabric face
f. unbalanced yarn counts create raised ridges
g. parallel diagonal lines are visible
h. diagonal lines run from upper left to lower right
i. diagonal lines run from upper right to lower left

3. Match the complex or decorative weave structures listed below with their appropriate descriptions:

____ 1. pile
____ 2. dobby
____ 3. jacquard
____ 4. filling pile
____ 5. crepe
____ 6. leno
____ 7. warp pile

a. structure formed by using textured yarns for crepe effect
b. open weave formed by crossed pairs of warp yarns
c. woven-in complex floral or scroll patterns
d. extra set of yarns creates loops in a basic ground cloth
e. extra set of warp yarns used in a basic ground cloth
f. woven-in, usually small repeating geometric designs
g. extra set of filling yarns creates loops
h. dobby loom used to create random floats, without geometric repeats

4. What is the difference between double-woven cloth and double knit?
double woven: ____________________

____

double knit: ____________________

____

5. Dress weight matelassé double cloth is sometimes referred to as matelassé crepe; heavier weight matelassé fabrics are produced for drapery and upholstery end uses. Brocade also is made in dress and upholstery weights. Which upholstery weight fabric would be more serviceable for home furnishings? (matelassé, brocade) Why? ____________________

____

____

6. What main durability property does a double knit contribute to a knit fabric?

____

****

REVIEW ANSWERS:

1. 1-c; 2-f; 3-d; 4-a; 5-e; 6-b
2. 1-d; 2-f; 3-a; 4-g; 5-h; 6-i; 7-b; 8-c; 9-e
3. 1-d; 2-f; 3-c; 4-g; 5-h; 6-b; 7-e
4. Double-woven cloth is made with four or five sets of yarns interlaced so that two distinct layers of fabric are interconnected. Double knits are made with two sets of interconnecting loops that securely interknit the yarns.
5. matelassé; its supportive layer increases fabric stability and holds taut the raised portions on the surface layer. Raised designs on single-layered brocade are subject to greater abrasive wear.
6. increased dimensional stability

UNIT FOUR

Part IV

Serviceability of Knit and Woven Structures

In this section, knit and woven structures are discussed in terms of their respective serviceability considerations. Remember that serviceability characteristics depend not only on the method of fabric construction, but also on fiber, yarn type, finishes, and color or design applications. To give maximum serviceability, fabrics should be chosen with the ultimate end use in mind.

Since the warp yarns are stronger than the filling yarns in most woven fabrics, products should be made so that the warp yarns are placed in the position subject to the greatest pulling or stretching forces. Therefore, in products such as coats and draperies, warp yarns are most serviceable when parallel to the vertical or lengthwise direction.

In evaluating durability characteristics of a fabric, a fundamental factor is the number of yarns available to "share the load". Higher thread counts produce more durable fabrics than lower thread counts. The higher the thread count, the more tightly packed the yarns in a fabric, and the firmer the textile product. Yet there are trade-offs.

Internal stresses in very high thread count fabrics decrease fabric flexibility and contribute to wrinkling; plus, tightly packed yarns present less surface area for absorbency. Thus fabrics with high thread counts usually are less absorbent, resilient and flexible than low thread count fabrics made from comparable fibers and yarns.

Abrasion resistance is also affected by thread count, as well as surface smoothness. High counts, up to an optimum level for the end use, tend to be more resistant than loosely structured fabrics to yarn snagging and surface abrasion. However, a plain weave, regardless of thread count, is more abrasion resistant than a rib weave, because there are more total yarns to share the load and spread the abrasive action. In a rib weave the raised ribs receive most of the wear.

Basket, twill and satin weaves all have yarns floating over two or more yarns in the opposite direction. The longer the yarn floats, the greater the likelihood of possible wear damage from abrasion and snagging of floats. The short floats in twill weaves are interlaced in a repeating progression that moves the floats over and down in each succeeding row. The diagonal pattern which results is more resistant to abrasion than the long filling floats of sateen, or the long warp floats of satin floats of sateen, or the long warp floats of satin fabrics.

Fiber content, thread count, length of yarn floats, and dimensional quality of the pattern affect resistance to abrasion. All are important when evaluating the durability characteristics of weaves which have woven-in patterns such as the dobby and jacquard. Fabrics made on looms with jacquard attachments are less resistant to abrasion than conventionally woven fabrics because the uneven yarn surfaces formed by the large, complex jacquard designs tend to wear unevenly.

Fabrics with excessively high thread counts also have reduced absorbency, which influences comfort. There is decreased surface are and decreased air transmission created by the tightly packed yarns. However, such characteristics may be desirable for apparel as windbreakers or raincoats.

Although knits tend to yield easily under pulling forces, their interlooped structure causes them to recover from wrinkling more easily than woven structures, contributing to high fabric resiliency. Knits can be interlooped in many different ways, resulting in great variability in knitted garments. By interlooping the yarns to create high bulk, dead air spaces can be formed in the structure. These air spaces hold heat and contribute to the comfort property of thermal retention or warmth. On the other hand, loose, open interlooped yarns increase the yarn surface area, thereby increasing fabric absorbency which contributes to body cooling and high thermal conductivity.

Since synthetic fibers have low absorbency, they may be somewhat uncomfortable unless made into fabrics that provide the necessary absorbency for comfort. The type of construction required to make synthetic fiber fabrics comfortable in warm weather is an open or loosely structured knit with spaces between and among the loops to allow for adequate air transmission.

Fibers noted for warmth such as animal hair or wool, and yarns which have been textured for bulk are often used for cold weather knit wear. Animal hair fibers trap air and hold heat against the body; bulk yarns do much the same thing by helping to hold air and preventing rapid cooling.

One technique that increases the warmth of both woven and knit fabrics is adding extra yarns to form a *pile* structure. A deep pile is desirable for apparel in cold climates because it provides insulative properties which hold body heat and prevent outside cold from penetrating the fabric.

The use of knit fabrics increased considerably

QUICK QUIZ 18

1. Which of the basic weaves is typically most durable? ________________
   Why? ____
   ____
2. Of these basic weave structures made with floating yarns, which is the least durable?
   ________________ Why? ________________________________
   ____
3. When resiliency is desired for ease of care and absorbency for comfort, would a comparably woven or knit fabric be preferred? ____________ Why? ____________
   ____
4. A woven structure would likely be selected over a knit when good ____________
   ________________ is required.
5. The higher the yarn count for the base fabric and the closer the yarns used in forming the pile, the (more, less) dense and warm fabric.
6. In pile structures, the first signs of wear are seen on the (pile, ground) yarns, which are susceptible to ________________________________

****

ANSWERS QUICK QUIZ 18

1. twill; they generally have shorter yarn floats, higher yarn twist and more tightly packed yarns than plain or satin weaves
2. satin; they generally are made from loosely twisted yarns (in order to enhance the shine of the fabric surface), which snag easily both from the length of floats and low twist
3. knit; interlooped structures tend to recover from wrinkling better than interlaced wovens and more surface area is available for absorbency
4. dimensional stability
5. more
6. pile; abrasion, rubbing or snagging

---

in recent years. While woven fabrics have returned to popularity, knit structures still find widespread use since they are generally more comfortable and easier to care for than woven fabrics. Both knit and woven structures can be modified to produce special serviceability features.

Crepe fabrics can be made with extremely highly twisted yarns, through the use of textured yarns, or by means of a crepe weave modification. Fabrics made with true crepe yarns tend to shrink when cared for because of the extremely high twist, but they may be restretched to shape by ironing. Crepe fabrics of thermoplastic textured yarns can be heat set for stability to provide better serviceability.

A leno structure is characterized by an open weave with a low thread count. Nevertheless, leno fabrics are strong despite their openness owing to the crossed configuration of two sets of warp yarns, which are generally plies of high twist.

Durability, comfort and care properties have been summarized relative to woven and knit structures. Yet the aesthetic features of a fabric are often of most importance to the ultimate consumer and will be the determining factor in product selection.

QUICK QUIZ 19

1. Match the weave with the appropriate description of its esthetic contribution to the fabric:

____ 1. pile	a. diagonal lines or wales
____ 2. satin	b. repeating geometric patterns
____ 3. dobby	c. complex, highly ornate designs
____ 4. plain	d. smooth, lustrous surface
____ 5. leno	e. crinkled, pebbly texture
____ 6. crepe	f. flat, uniform surface
____ 7. jacquard	g. fabric with dimensional surface made from extra set of yarns
____ 8. twill	h. open, figure eight construction

2. Check the knit structures which can create the greatest pattern interest:

____ double	____ plain
____ interlock	____ purl
____ rib	____ pile

3. If a knit were desired for aesthetic and comfort properties over a woven, and yet dimensional stability were important to the end use, which knit structures would be most serviceable? ____________________

****

ANSWERS QUICK QUIZ 19

1. 1-g; 2-d; 3-b; 4-f; 5-h; 6-e; 7-c; 8-a
2. double, rib, purl, pile
3. double, interlock, rib

REVIEW

1. Compare wovens and knits of comparable weight, fiber content and yarn structure regarding the following properties:
   a. abrasion resistance ______________________________
   ____
   b. elongation and elasticity ______________________________
   ____
   c. dimensional stability ______________________________
   ____
   d. flexibility ______________________________
   ____

2. Many variables affect the serviceability and performance of a fabric. Textile choices are determined according to all the elements which contribute to the final product, including ______________ content; ______________ type, twist and count; how the fabric is ______________; and finishing and coloring applications.

3. Generally, the more basic and uniform the knit or weave, the (more, less) durable the structure will be.

4. The more yarn surface area in a structure, the more absorbent the fabric. Excessively high yarn counts (increase, decrease) the available surface.

5. The more bulk and dead air spaces incorporated in a fabric, the ______________ it will be.

6. The lower the yarn count in a knit or weave, the (more, less) dimensionally stable it will be. The higher the yarn count in a fabric, the (more,less) resilient and (more, less) flexible.

7. The shorter the yarn floats in a fabric, the (greater, less) its durability.

8. The more uneven or irregular a fabric surface, the more wear from ______________ ______________ can be expected.

9. The method used in constructing a fabric influences the durability, comfort, care and appearance of the final product. In order to make the best selection of fabric it is important to consider the planned ______________.

****

REVIEW ANSWERS:

1. a. Neither has a decided advantage; abrasion resistance depends on the uniformity of the structure. (A rib knit or rib weave will be more subject to abrasion than a flat surface.)
   b. Knits tend to elongate and recover more than woven structures.
   c. Weaves have less give and are more dimensionally stable than knits.
   d. Knits are generally more flexible than weaves.
2. fiber, yarn, structured (woven or knitted)
3. more
4. decrease
5. warmer
6. less; less; less
7. greater
8. abrasion, rubbing, or snagging
9. end use

UNIT FOUR

Part V

Special Structures Made From Yarns

Net, lace, braid, tufted, and *stitch-bond* structures are discussed in this section.

Net

Net is an open mesh structure made from yarns which have been knotted; interlooped or heat fused. True net is made by knotting, but to cut costs, much of what is called net today is actually knitted on either tricot or raschel knitting machines. Heat fusion provides another "look alike" net but without the characteristics that result from either the knitting or knotting operations.

Figure 4.54 shows examples of nets made by three methods. Knotted nets are more durable than nets made by knitting. Knitted nets tend to lose the net structure if a yarn breaks, while knotted nets are securely tied at each junction. For high strength, as needed in hammocks and fish nets, ply or cord yarns are used.

Tulle, shown in Figure 4.54, is the name of a very fine net structure produced for decorative end uses like bridal veils, veils on hats, evening gown overlays and trimming for gowns. It can be made by any of the methods described; however, most is presently knitted.

Fused nets are generally made from thermoplastic fibers and are frequently fine and lightweight. They are becoming increasingly available and may be quite satisfactory if the fusing is well done and temperatures during use do not approach the melting temperature of the fused fibers. Fused nets are used primarily for decoration and packaging rather than for end uses where high strength is important. Possible end uses for a fused net are trims, party or restaurant decoration, bags and bottle coverings.

A newer and cheaper way of making a net is to run a thermoplastic film through cutting rollers which make small slits in the film. Then, when it is stretched across its width and heat set, a netlike structure is formed. The resulting product is used largely for bagging produce.

Lace

An open, highly decorative fabric characterized by intricate designs is often identified as a *lace*. Figure 4.55 shows a lace edging typical of that manufactured for trim on a variety of textile items.

Lace can also be made in full width fabrics for construction into special types of apparel. True lace is structured on extremely complex machines where yarns are knotted rather than interlooped at the points of intersection. Consequently, true lace is expensive to produce, which is the main reason that most laces are actually knitted raschels. The true lace structure shown in Figure 4.55 is more durable than a raschel net or lace made by knitting because it is knotted rather than interlooped on a knitting machine.

Sample 30 and the lace pictured in Figure 4.56 are representative of one of the many all-over lace fabrics produced on raschel knitting machines.

Many different patterns, types and weights of lace are available in a variety of price ranges. End uses range from delicate trims to fabrics for table coverings and bedspreads. Despite the undemanding end uses for most laces, they do pose problems

Figure 4.54. Nets can be produced by knotting, heat fusion and knitting.

Knotting

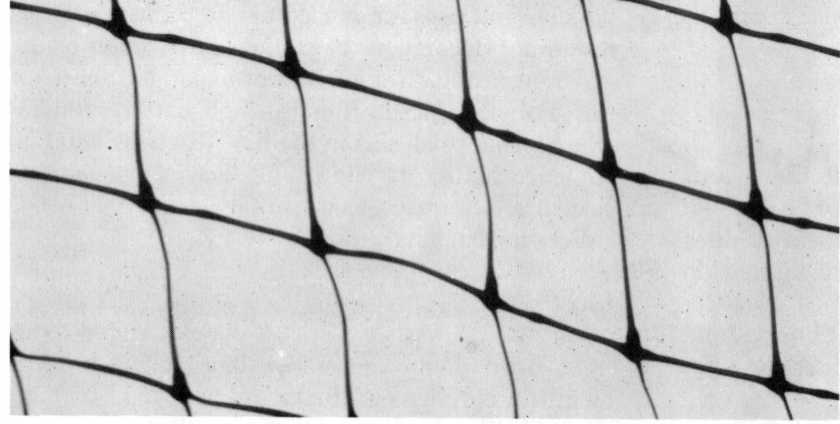

Heat fusion

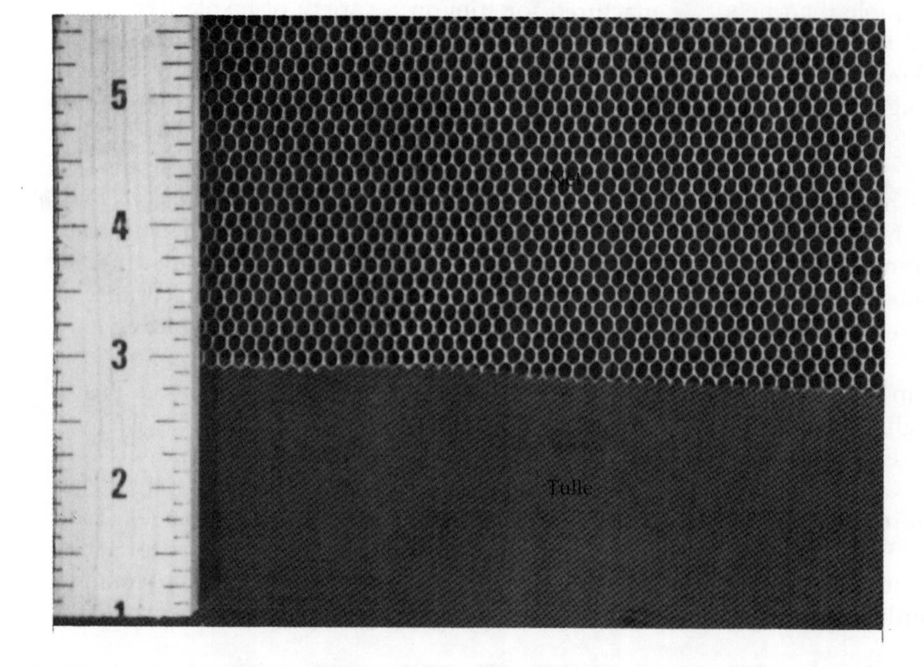

Knitting

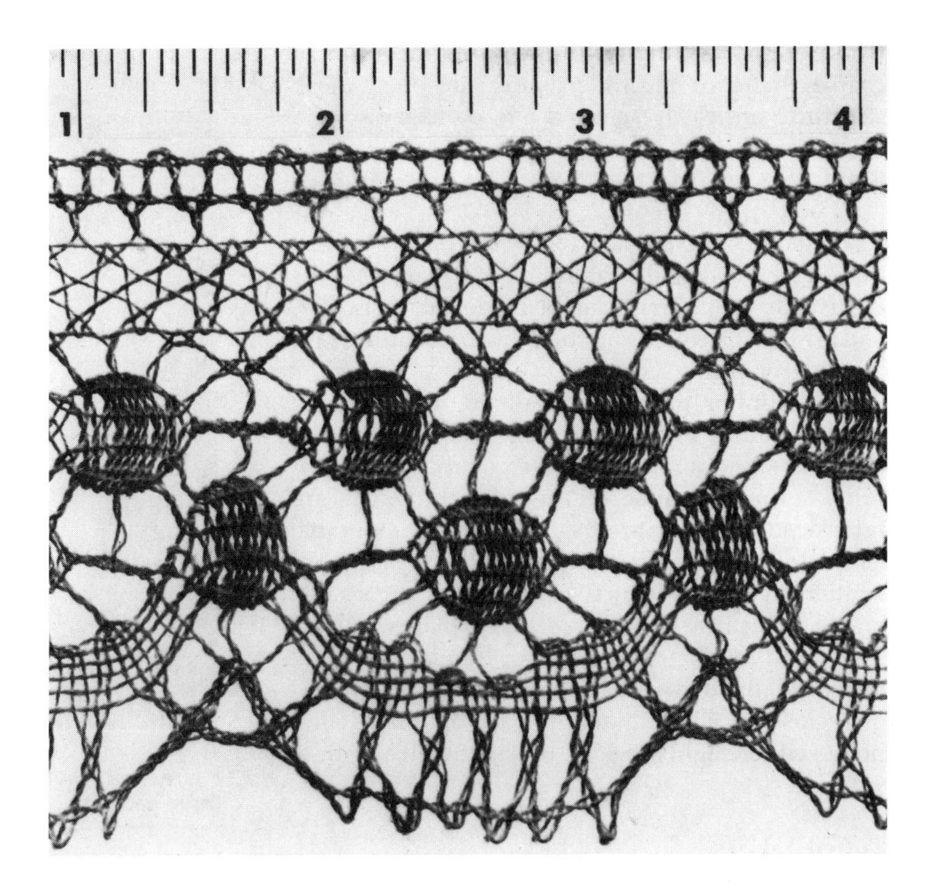

Figure 4.55. Cluny lace is a relatively durable fabric formerly made by hand, but now typically machine-made like most other laces.

Figure 4.56. A typical machine-made knitted lace.

in use and care. Whether knotted or knitted, fine or coarse, hand or machine made, the open mesh structures are easily snagged and damaged.

Braid

A braided structure is made by interlacing at least three yarns diagonally, although many more yarns are usually used. Braids can be made flat or circular, and are constructed using the same procedure as for braiding or plaiting hair.

The number of yarns involved in preparing braided fabrics depends on the width of the flat finished fabric, or the diameter of the circular or tubular one. Because of the way the yarns are interlaced to form braids, there is considerable lengthwise stretch, comparable to the bias of conventionally woven fabrics. Figure 4.57 depicts some braids, which are usually relatively narrow structures used for trim, shoelaces, or tubular industrial purposes. Braided yarns are interlaced to form angles close to 90°; the angles vary at intersecting points either slightly below, or slightly greater than 90°.

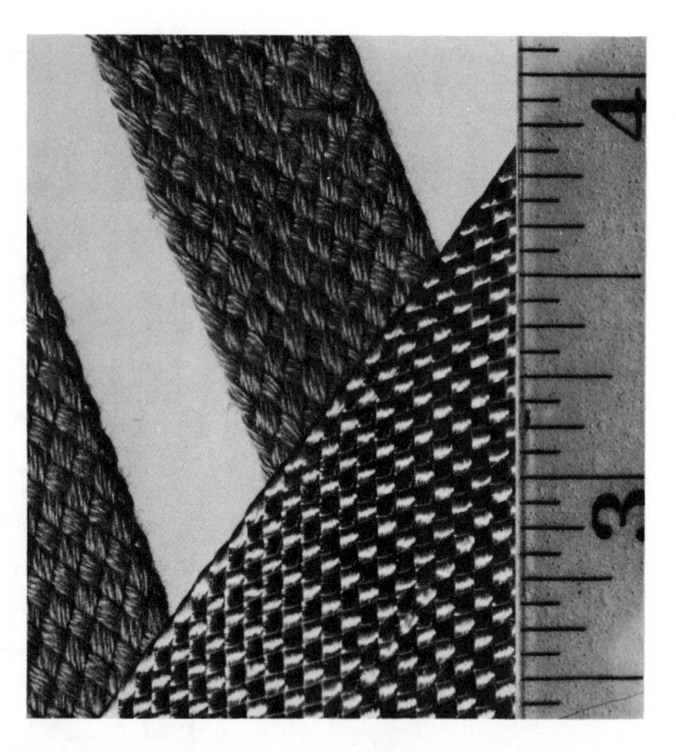

Figure 4.57. Braid structure is formed by plaiting at least three yarns together to form narrow bias structures.

Tufted Fabrics

Another special type of fabric construction is *tufting*. In this process, a woven, knitted or nonwoven

---

QUICK QUIZ 20

1. Recall the earlier discussion of triaxial weaving which also requires three sets of yarns. Compare triaxial and braided structures, explaining their differences and similarities.

___
___
___

2. Why is braid suitable for use in shoelaces and trimming the edges of textile products?

___
___

****

ANSWERS QUICK QUIZ 20

1. Triaxial weaves are interlaced at about 60° angles; braided yarns form approximately 90° angles. There is considerable lengthwise bias stretch in braided structures, whereas triaxial fabrics are stable in all directions.
2. Its lengthwise give allows shaping around an edge for trim; or stretching during shoe lace tying, which will return to size, making a secure tie.

---

base layer of fabric is fed through a machine, which actually stitches additional loops of yarn into the fabric.

The finished product has characteristics similar to pile structures, but is considerably faster and cheaper to manufacture than traditional pile weaves or pile knits. Plus, a wide variety of esthetic effects can be created on versatile tufting machines. For example, loosely twisted yarns, or even roving, can be punched through the ground cloth to yield a greater variety of effects than can be achieved with regular woven or knitted pile yarns. In bathmats, often made this way, the pile surface appears to be fibers and not yarns. Currently, over 90 percent of all carpeting is made by the tufting process. Some upholstery and numerous apparel fabrics are made by a tufted process as well.

Tufted structures can be recognized by the rows of yarns punched through on the back of the fabric that have the same appearance as rows of stitching. Figure 4.58 provides a diagram of the tufted process and an illustration of a tufted fabric.

To stabilize and prevent tufted rows from coming out, special adhesives can be applied to the back of tufted fabrics. Also, and especially for tufted carpets, a layer of sturdy, nonstretch fabric such as jute is usually sealed to the carpet back to add further protection. The increased use of tufted fabrics for carpeting stems in part from the fact that tufting is now a relatively durable process that is faster and more flexible than older, traditional methods. Tufting processes yield serviceable carpets with a wide variety of designs that can be produced quickly and at lower costs than pile woven carpets. In apparel, tufted structures also are replacing some pile wovens and knits to produce warm fabrics with high thermal retention for use in coats, coat linings, and sportswear.

Stitch-Bond Fabrics

Stitch-bond fabrics, sometimes called stitch knits, are made on machines developed in Europe during the 1960's. The fabrics produced are varied and difficult to categorize since they are not truly knits or wovens. Rather, they consist of an arrangement of yarns, slivers, and/or fiber mats held together by stitching with additional yarns. The stitching resembles a chain stitch and is very similar to the formation of knitted loops used in tricot warp knitting.

The products are often named after the machines used in their manufacture. These machines differ considerably from either weaving looms or knitting

Figure 4.58. In the tufting process, yarns are punched through a structured ground cloth rather than woven in, as with pile weaves.

QUICK QUIZ 21

1. Sample 38 is a tufted upholstery fabric. The pile fabric surface appears very similar to velvet upholstery sample 14. How can the tufted sample be identified? ______________________
____
2. Try to pull out one of the rows of tufting in sample 38. Describe the durability of this tufted structure that has an adhesive backing. ______________________
____

****

ANSWERS QUICK QUIZ 21

1. By the rows of tufts stitched on the back of sample 38; the velvet fabric has no yarns punched through.
2. The tufting is adequately secured by the adhesive; the yarn rows do not readily pull out.

frames, but have some characteristics of both. Three basic structures are made by stitch-bond methods, and the varied fabrics can be categorized into one of the three types—most commonly called *malimo*, *maliwatt*, and *malipol*.

Malimo fabrics are made from one or more layers of yarns secured with a type of chain stitch. Two layers are typically used, with one yarn layer laid in the warp direction, crossed with a second layer of yarns which is nearly perpendicular or at a slight angle, as pictured in Figure 4.59.

The two layers of yarns, comparable to warp and filling directions in regular weaving, are not interlaced. They may be placed relatively far apart with the chain stitch very much a part of the background, or the yarns can be laid close together to produce a more conventional appearing woven fabric. Study yarn placement in the malimo fabrics shown in Figure 4.59. In addition to these arrangements, yarns also can be laid in only one direction and secured by the chain stitch—a knitted-look results in these structures.

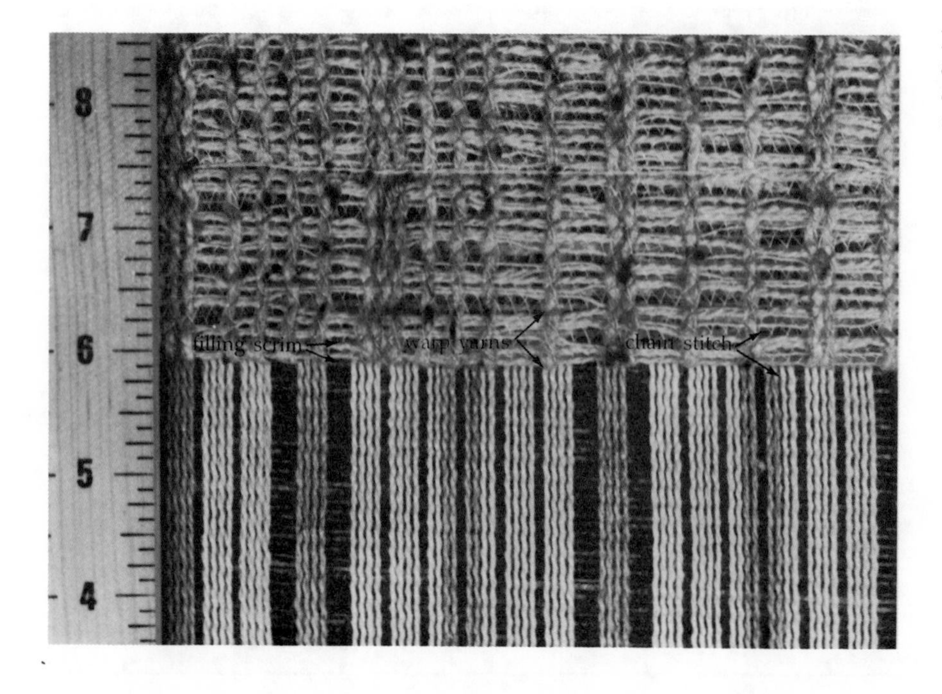

Figure 4.59. These malimo fabrics are composed of warp yarns laid on a filling scrim, with a chain stitch sewing the layers of yarns in place.

QUICK QUIZ 22

1. Examine sample 37, which is a stitch-bond fabric constructed by a combination of processes described earlier. Clearly observable on the top layer are yarns that have been laid across each other and connected by a ______________________. This process produces ______________________ stitch-bond fabrics.
2. Notice also that sample 37 has a backing layer composed of a ______________________ that is also chain stitched to the layers of yarns. This type of stitch bond is called ______________________.
3. The top layer of yarns laid in the warp direction of the sample is subject to abrasion. Pull out some of the warp yarns and untwist them. What type of yarn was used to add durability to the fabric? ______________________.
4. What is the major advantage of the stitch-bond method of manufacturing fabrics? ______________________

****

ANSWERS QUICK QUIZ 22

1. chain stitch; malimo
2. fiber web; maliwatt
3. two-ply
4. speed of production, which reduces fabric cost

A second type of stitch-bond fabric contains a mat of fibers secured by a series of chain stitches. These webs of fibers are called *maliwatt* fabrics, of which the back of sample 37 is an example.

The third method of structuring knit-sew fabrics—named *malipol*— creates products similar in appearance to pile fabrics. A mat or web of fibers, or a layer of yarns, is used as the base; the chain stitch that secures it is structured to form surface loops, which may be cut or left uncut. Observe the pile surface of the malipol fabric shown in Figure 4.60.

The durability of stitch-bond fabrics depends on such factors as the quality of the chain stitch, and the strength and abrasion resistance of the fibers used in the fabric. Other factors contributing to serviceability include type of yarn structure used,

Figure 4.60. Malipol stitch-bond fabrics have pile surfaces.

yarn count or density, and the effects of finishes and coloring methods. Draperies or curtains are examples of the most common uses of stitch-bonded fabrics in the U.S.; table coverings, napkins, blankets, coat linings and selected dress apparel are also potential end uses.

Using stitch bonding, fabrics of great variety can be produced for apparel and home furnishing end uses. However, production in the United States is limited, and stitch bonds have not become popular, perhaps because the American consumer is not willing to accept these nontraditional structures despite their versatility. One of the major reasons for the development and use of these fabrics is economic—they can be produced faster and at less cost than similar fabrics made on traditional weaving or knitting machines. Their future remains to be seen.

Special Structures Identification

1. Take out the Mounting Sheet that includes "Special Structures Made from Yarns" and determine which structures are to be mounted:
   ____
2. Identify these fabrics from your sample set; then cut, attach and label them in the designated places on the Mounting Sheet.
   30 ______________________________
   37 ______________________________
   38 ______________________________
3. Examples of a knotted net and knotted lace are not included in the swatch set due to the ______ ______________ production and resulting high ____________ of fabrics made by these methods.
4. However, these knotted structures are easy to identify when you encounter them, as are the narrow structures made from three or more interlaced yarns for end uses that require vertical give, called ______________.
5. To insure that the tufted rows do not pull out from upholstery sample 38, ______________ ______________ has been applied. What other home furnishing product is frequently made by tufting? ______________
6. Compare the serviceability of the tufted upholstery sample 38, which is 100 percent rayon; with velvet upholstery sample 14, made of 65 percent cotton and 35 percent rayon: ______________
   ____
   ____
   ____
   ____
   ____
7. For what home furnishings use would sample 37 be serviceable? ______________
   Why? ______________________________
   ____
   ____
   ____

****

Special Structures Answers:

1. lace, tufted, stitch bond
2. 30 - raschel knitted lace
   37 - stitch bond—top layer malimo; support layer maliwatt
   38 - tufted upholstery
3. time consuming; cost
4. braid
5. an adhesive backing; carpeting
6. Sample 14 is more serviceable for upholstery due to its higher yarn count and denser pile (which you can feel and see by holding both samples up to a strong light); the cotton blend allows for more abrasion resistance than the 100 percent rayon tufted sample. Also, velvet piles are structured in, compared to punched through tufted piles; however, the tufts appear to be securely anchored by the adhesive backing.
7. draperies; The stitch bond structure is adequately secured for hanging by the web backing and chain stitch, and there would not be surface abrasion on the raised warp yarns. Also, light and air can readily pass through the structure.

UNIT FOUR

Part VI

Multicomponent Fabrics

Multicomponent fabrics contain two or more layers of fabric, or a layer of fabric and a layer of fibers, or a layer of fabric and a layer of foam. They can be distinguished from double-woven and double-knitted fabrics in that the layers of fabric are made separately and then combined by an additional manufacturing step. Fabrics that may be categorized as multi-component include: *bonded* and *laminated* fabrics; *quilted* fabrics; and fabrics where the layers have been sealed by *chemical* means, *heat* or *friction*.

Bonded and Laminated Structures

Figure 4.61 diagrams the three major ways in which bonded and laminated fabrics are made. The bonding agent which fuses the layers may be a foam substance or other type of adhesive.

The usual distinction made between bonded and laminated structures is that bonded fabrics are layered fabric structures joined with an adhesive that does not add significantly to the thickness of the combined fabrics, such as the first fabric-fabric bond diagrammed. Laminated fabrics are adhered with a foam layer visible in the structure, and can be foam backed, or have foam between two fabric layers, as diagrammed. Bonded and laminated fabrics are usually designed so that one fabric is specifically identified as the face or top, while the second layer forms a lining. Thus, these fabrics are not reversible in recommended end uses.

In the adhesive bonding process, no adhesive or foam is visible between the fabric layers. The adhesive is applied to the back of the top fabric, and the liner fabric is joined to it by passing both fabrics between heated rollers. When suitable adhesive materials and processes are used to join the two layers of fabric, the resulting structure tends to have increased stability.

Despite the potential advantages of foam laminates, these products are no longer as popular in apparel as they were during the 1970's. Performance problems such as fabric separation, foam discoloration or decomposition, differential shrinkage of the two fabric layers and distortion of fabric grain lines have limited consumer acceptance of laminated fabrics for apparel. The technology is now available to overcome these problems, however. In fact, the bonding and laminating processes have opened up new markets for woven and knit fabrics in home furnishing and industrial end uses. Bonding can allow for the stabilization of fabrics that typically would not be appropriate for end uses requiring relatively rigid and durable structures. For example, the flexibility of knitted fabrics could be useful in fitting upholstery to furniture frames; however without some type of stabilization, many knit fabrics would not have the durability levels required of upholstery fabrics.

The use of bonded and laminated fabrics for apparel has been confined primarily to low cost items, despite potential advantages. The amount of time saved in constructing a garment of bonded or laminated fabric, compared to using a fabric and separate lining, lets the manufacturer keep up with fashion changes inexpensively, with the savings passed on to the consumer in the form of lower cost garments. Despite a shorter use life, less expensive bonded fabrics can be very serviceable in a culture where emphasis is on appearance and rapid style change rather than durability. High quality laminates can be produced, but the consumer must be willing to pay the price.

The consuming pubic likes easy care products. Although bonded and laminated fabrics are not always machine washable, they possess high resiliency and do not wrinkle during wear or require much ironing.

Two comfort advantages also can be realized with bonded and laminated fabrics, Sensitive skin can be protected from scratchy outer fabrics when smooth fabrics such as acetate or nylon tricot are used for the lining layer. In addition, the added fabric or foam layer increases thermal retention or warmth which contributes to cold weather comfort.

Perhaps more than any other structure studied so far, bonded and laminated fabrics require careful evaluation before purchasing.

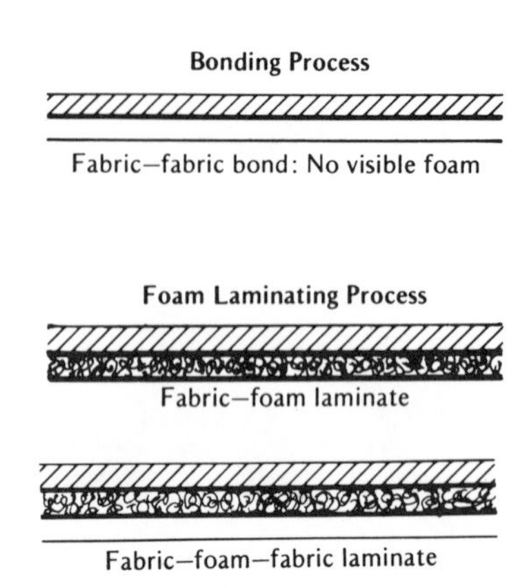

Figure 4.61. Bonded and laminated fabric layers may be joined with an adhesive or foam layer.

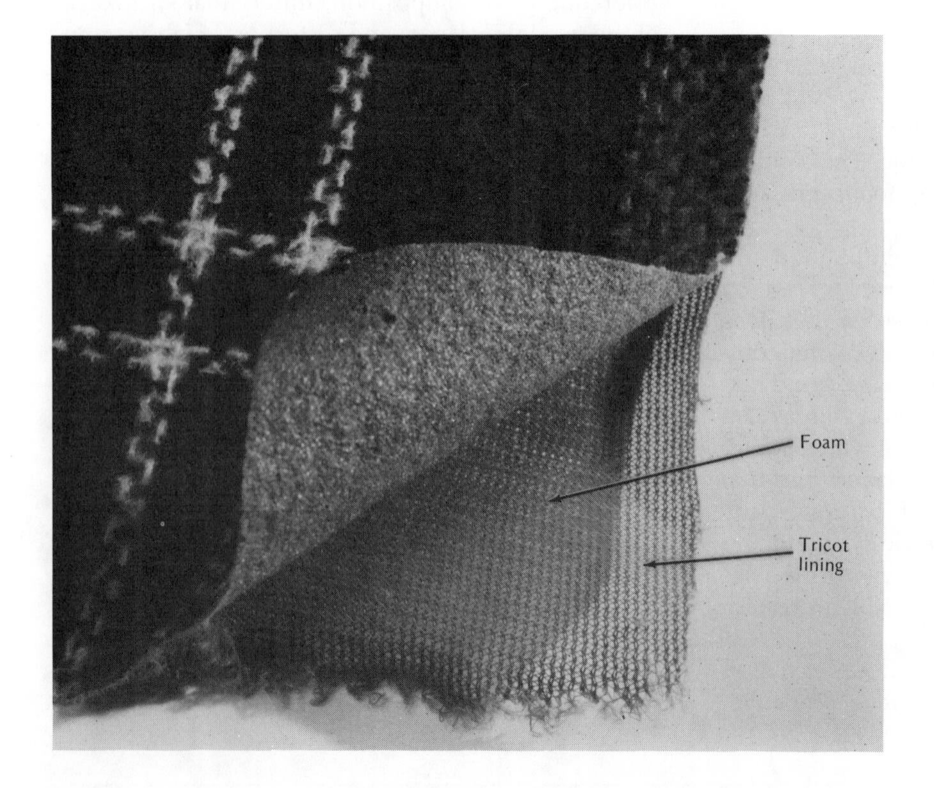

Figure 4.62. In this fabric-foam-fabric laminate, the loosely woven top fabric layer is stabilized with foam and an acetate tricot backing layer.

Quilted Fabrics

Quilted fabrics are made by stitching together two or more layers of fabric, sometimes with a web or layer of fibers between them. The stitching, which forms various designs in the fabric, may be done by machine or by hand. The size of the stitch influences durability; long stitches may break easily or be snagged. Quilted fabrics are usually designed for warmth or thermal retention. They also provide special appearance and esthetic values, and may be reversible or nonreversible.

Most quilted fabrics are heavy and have poor draping characteristics. End uses typically include home furnishings—bedspreads, comforters, upholstery, table coverings; and selected apparel—robes, skirts and vests. Other factors that affect durability include the type of fabrics used, the fiber content, and the closeness of the stitching lines.

Quilted products can exhibit these problems in care: the fabric layers may shrink differently; the fiber layer may shift and form lumps, if the design stitching lines are too far apart; or the stitching may break, resulting in layer separation. However, quilted fabrics have greater serviceability and durability than most laminated or bonded fabrics, provided fabric and stitching are of good quality.

In baby quilts, for which fabric sample 39 was designed, serviceability is probably adequate. The item is economically produced for a short-term specific end use; if offers an attractive, lightweight product with high thermal retention or warmth. But, in order to get adequate serviceability from quilted fabrics, care procedures must take into account fiber content of various layers; type of yarn and fabric structures; stitching used to join the layers; type of finishes; and color applications.

Chemical and Heat or Friction-Stitched Fabrics

Multi-layered fabrics resembling quilted structures are made using heat and friction or chemical processes instead of stitching to seal the layers, so that they look like they were stitched together. Before the layers are joined in the *chemstitch* process, one fabric is printed with a chemical in the predetermined design that then seals the layers in that pattern when heat is applied. The layers may be joined flat, or one may be sealed to the other to produce a puckered effect on the face surface when

---

QUICK QUIZ 23

1. Because the foam layer adds bulk and serves as an insulating material, the thicker the foam layer, the (more, less) thermal retention provided by the fabric.
2. Bonding can allow for the ________________ of flexible or low thread count fabrics.
3. What advantage does a bonded or laminated structure have over a comparable weight double cloth? ____
____
4. What two major advantages do double wovens and double knits have over laminated or bonded structures? ____
____

****

ANSWERS QUICK QUIZ 23

1. more
2. stabilization
3. Bonded or laminated fabrics are less expensive to produce than double wovens.
4. Double woven and knitted fabrics are structurally interconnected and therefore grain lines are aligned. They will not separate, stiffen or deteriorate during use and care.

---

one fabric layer is thermoplastic, while the other layer is not. Then, when they are laid on top of each other and subjected to heat, the portion with the thermoplastic fibers shrinks, causing the non-thermoplastic areas to pucker and create a quilted surface effect.

A newer *pinsonic* process seals two layers of fabric, usually with a layer of fibers between, with friction created by *sonic* (sound) *vibrations.* The appearance of the multi-layered fabric is similar to stitch-quilted products, as shown in Figure 4.63. At least one layer of fabric is composed of some thermoplastic fibers as in the chemstitch process. The two layers of fabric, and a central layer of fibers, are fed through a machine that has a large roll patterned with small pins extending from the surface. In Figure 4.63 the two fabrics enter the pinsonic machine from rolls at the top, right-hand side of the photograph. Sonic vibrations cause the pins to transmit adequate heat when they roll over the fabric so as to melt the thermoplastic fibers and seal the layers together; thus the name *pinsonic.* Where the pins contact the fabric, they leave an appearance of stitching.

Care of fabrics made by the chemstitch and pinsonic processes depends on fiber content used, yarn type, and fabric construction, as well as finishes applied and coloring methods. Generally, care procedures should take into account the heat sensitivity of the fabrics, which typically can be cared for as any polyester and cotton blend. While the layers are relatively securely attached, they can be fairly easily pulled apart; so minimum stress and agitation during use and care is important. In addition, product size can impose care limitations; for example, large bedspreads may not fit into home laundry equipment, so outside the home care is usually required.

While the pinsonic process has been used to make bedspreads and table placemats, other uses have been extremely limited. A major problem likely to occur with these fabrics during use and especially care is the separation of fabric layers at the sealing points. There are no thread stitches connecting the layers for additional security, and there are fewer surface areas where fabric layers are sealed in the pinsonic process. They are more likely to separate than bonded and laminated structures which are adhered throughout.

Despite the potential limitations of chemstitch and pinsonic fabrics, their use will probably continue. The processes produce quilt-like fabrics that can be made more quickly and at less cost than by conventional stitching methods.

QUICK QUIZ 24

1. List three factors that help determine the quality of the stitching used in quilted fabrics: ________________
2. Examine sample 39, which is a ______________ fabric composed of a top ______________ woven layer, a middle fiber layer, and backing layer of ______________ knit.
3. Evaluate the stitching used to hold the three layers together: ______________

****

ANSWERS QUICK QUIZ 24

1. Any three:
   size of stitch, type of stitch, fiber content, yarn type used for stitching thread, closeness of the lines of stitching, and amount of open spaces between them
2. quilted; plain; tricot or warp
3. The stitches are long and might pull out easily; the lines of stitching are far apart.

Figure 4.63. The pinsonic process joins fabric layers with sound-generated heat instead of stitching. The machine shown is a laboratory model. *(courtesy James Hunter Machine Company, Inc.)*

Multicomponent Fabric Identification

1. Six fabrics in the sample set contain two or more layers of fabric, a layer of fabric and a layer of fibers, or a layer of fabric and a layer of foam, which makes them ______________ structures.
2. Take out the Mounting Sheet which include space for Multicomponent Structures; examine samples 7, 36A and 37. The top layer of 7 is a ________ ________ weave; 37 is a ______________ structure. The top layer of 36A, a warp knit, is more difficult to identify than the structures in 7 or 37. All three fabrics are (rigid, flexible) structures that have their serviceability (increased, decreased) by the backing layer.
3. Each fabric has a supporting layer that consists of ____________ . In sample 7 and 36A, the layers are (bonded, laminated) together; and in sample 37 the layers are held together by __________ .

4. Sample 34 is a nonwoven fabric that will be discussed in the next part; it is backed by a 100 percent acrylic crushed foam. It is included here for mounting as a multicomponent fabric because it ______________________________ .
5. Samples 36 and 39 both have knitted fabrics as a backing layer. The flexibility and pliability of the top vinyl layer of 36 is increased by the (filling, warp) knit backing, whereas sample 39 has a (filling, warp) backing for increased ______________ ______________ .
6. The multicomponent structures included in the samples represent different methods of fabric joining—the two layers of sample 36 are adhered by ______________________________ while sample 39 is a ______________________ structure.
7. For review, describe the structures of the samples listed below; then cut swatches from each. Attach and label them on the Multicomponent Structures Mounting Sheet.
   7: ______________________________
   36A: ______________________________
   37: ______________________________
   34: ______________________________
   36: ______________________________
   39: ______________________________

Multicomponent Fabric Identification Answers:

1. multicomponent
2. basket; stitch bond; flexible; increased
3. fiber webs; bonded (because no foam is visible); chain stitching
4. contains two layers—fiber and foam
5. filling; warp, body and stability
6. bonding; quilted
7. 7-basket weave, bonded to fiber web
   36A-warp knit bonded to fiber web
   37-stitch bond—malimo top layer with maliwatt backing, chain stitched
   34-nonwoven, backed with crushed foam layer
   36-vinyl, bonded to filling knit backing
   39-quilted structure—plain weave top layer, center fiber layer and tricot backing layer

---

REVIEW

1. Name two processes used in producing bonded and laminated fabrics: ______________________ and ______________________
2. List three problems that can be encountered with bonded and laminated structures:
   ____
   ____
3. When might bonded or laminated fabrics be preferred over a double woven or knitted fabric?
   ____
   ____
4. Firmly woven fabrics which are dimensionally stable are generally more serviceable if they are not ______________________ .
5. Among the two-layered fabric structures, reversible fabrics are generally made by the following three types of construction: ______________________ , ______________________ and ______________________ .
6. Nonreversible, two-layered structures are usually made by ______________________ or ______________________ .

****

REVIEW ANSWERS:

1. adhesive bonding, foam laminating
2. Any three:
   separation of fabric layers; off-grain lamination of the two fabric layers; differential shrinkage of the layers; foam discoloration, decomposition or crumbling, and stiffening
3. When low cost products are desired; when the same effect can be achieved by laminating or bonding; and when durability is not a major concern, i.e., when short-term fashion wear is a priority.
4. bonded or laminated
5. double weaving, double knitting, quilting
6. bonded, laminating

UNIT FOUR

Part VII

Nonwoven Fabrics Made Directly From Fibers

Weaving, knitting, knotting, braiding, tufting, and *multicomponent* fabric structuring have been discussed in this unit. Current terminology defines as nonwoven textiles those fabrics made directly from fibers or fiber solutions, and which lack the yarns necessary for weaving and knitting. Yet, knits, laces, nets and braids are actually nonwoven structures, too. However, *by definition,* they are not classified as nonwovens.

Nonwoven fabrics manufactured directly from fibers comprise some of the oldest as well was some of the newest methods of creating fabrics. Some of the newest fabrics made directly from fibers involve the manufacture of fiber webs from nonfelting fibers, and even *microfibers.* These processes include *fiber joining through the use of adhesives; special matting and bonding of thermoplastic fibers with heat or chemicals;* or *mechanical interlocking and entanglement of fibers.*

The first fabric made directly from fibers was felt, formed by the tangling or matting together of animal hair fibers, primarily wool. These original nonwovens are discussed first.

Felt Fabric Formation

Short staple, animal hair fibers with surface scales will felt when subjected to *agitation* and *pressure* in the presence of *heat* and *moisture.* Felting takes place when wool fibers shrink, entangle and mat together; the interlocking or entangling of the fibers is enhanced by the fiber scales.

Felt fabrics can vary in thickness from 1/32 of an inch to over three inches, based on how many layers of fibers are used. A wide variety of felt fabrics is produced for many different end uses. Felts have been around for a long time and can be expected to remain on the market for use in hats, bedroom slippers, pennants, and for decorative functions that do not require frequent maintenance. Felts have an advantage in garment construction over woven structures in that there is no distinct fabric grain line to follow, and neither seams nor fabric edges have to be finished because nonwovens do not ravel.

While fiber felting is deliberately induced to create nonwoven fabrics, care must be taken to avoid inadvertent felting of woven or knit wool fabrics, or additional felting and shrinkage of completed felt fabrics. Recall that certain care procedures should be followed to minimize felting in finished fabrics: avoid rapid temperature changes and water that is too hot; avoid strong alkaline substances; and make sure there is a minimum of agitation in laundering or drying.

When felts were first produced, only leftover fibers were used; today, however, high quality fibers are prepared especially for felting. The first step in forming felt is to run the fibers through a machine which removes foreign matter and aligns the fibers; this process, also used in yarn manufacture, is called carding.

After initial carding or straightening, the aligned mats of fibers are placed in alternating layers at oblique angles to each other. Then the fibers are felted by subjecting the fiber mats to hot water and an alkaline solution, to agitation, and to pressure. The result is a compressed layer of tightly matted fibers which are randomly oriented. The pressure and vibration rearrange the fibers in such a way that no warp or filling grain is discernible.

Wool is most commonly used in felt, but rayon and other inexpensive fibers are often added to the fiber mat to reduce cost. No more than 75 percent of a nonfelting fiber, however, may be combined with wool or animal hair. The nonfelting fibers lack the surface scales which are the key to fiber entanglement required in making felt fabrics by conventional felting process. Figure 4.64 depicts a felt fabric that has been pulled apart. The fabric was photographed with a light behind it so you can see the random orientation or arrangement of the fibers.

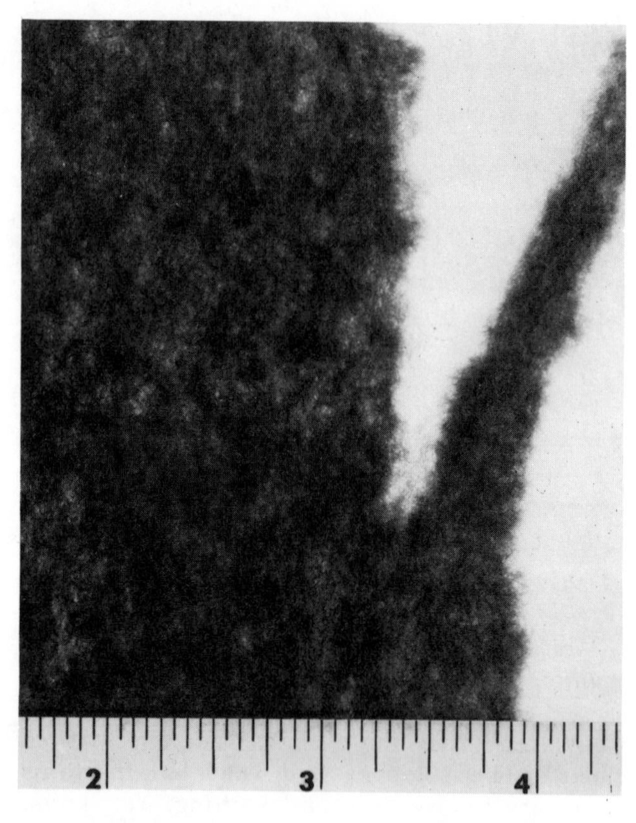

Figure 4.64. Felt fabrics are made by subjecting wool or animal-hair fibers to heat, moisture, agitation, and pressure.

Web Formation and Bonding Techniques

For the remaining discussion on other types of nonwoven fabrics made from *nonfelting fibers*, con-

---

QUICK QUIZ 25

1. Based on your knowledge of wool fiber properties, and the nonwoven fabric characteristics of felt, what are some desirable properties that can be expected from wool felt fabrics? ____
2. Garments with felt reinforcement offer several advantages to a tailor: ease of fabric shaping, lack of raveling, no grain lines, and good body; but potential problems might arise with apparel made entirely of felt. Name several: ____

****

ANSWERS QUICK QUIZ 25

1. high thermal retention or warmth; high resiliency; no warp or filling grain lines to follow in construction techniques; no warp or filling yarns to ravel
2. Felts have low fabric strength and can stretch, causing loss of garment shape. Exposed fiber ends are subject to abrasion and pilling. Felts have a relatively stiff body and do not drape well.

sideration must be given first to *web formation,* and then *bonding techniques* for adhering the fibers. Webs of nonfelting fibers are formed primarily in two ways: 1) the fibers may be randomly oriented by a process called *air layering,* or 2) the web may be carded and fibers laid in parallel or perpendicular positions.

When fibers are carded and placed parallel to the length of the fabric, there is more strength in the lengthwise direction; however the strength of the fabric in the cross direction is much lower. If layers of parallel fibers are alternated and *crosslaid* so that adjacent layers are at angles to each other, there will be equal strength in both directions of the fabric. However, crosslaid webs are expensive to produce and have not proven significantly better than random webs except in products in which balanced properties are essential. Thus, random fiber webs made by air layering have become the most frequently produced; they provide adequate serviceability for most planned end uses.

Then, based on fiber content and end product desired, the manufacturer will bond the formed web using adhesives, heat, chemicals or mechanical fiber entanglement. A *spray adhesive* may be used with nonfelting fibers or *thermal bonding* can be used to secure the fiber web when thermoplastic fibers are used in the webs. For example, nylon and polyester bonded fiber webs can be securely sealed with the thermal bonding method, while rayon webs require the use of some type of adhesive.

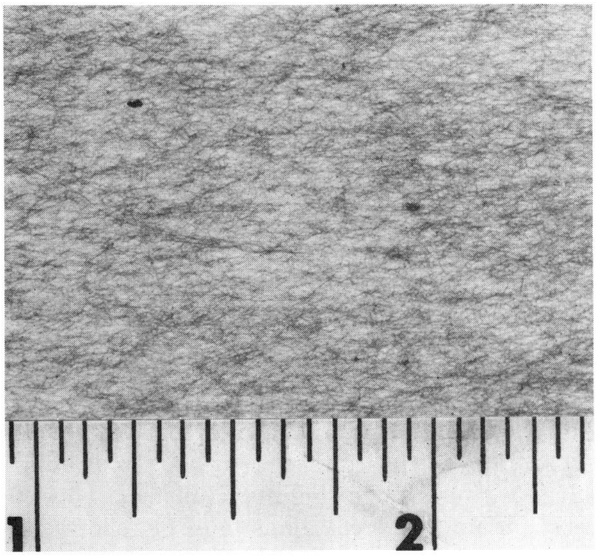

When *chemical bonding* is used to join fiber webs, an appropriate solvent is applied that softens the fibers on the surfaces so they can be fused together under pressure. Also, different types of fibers can be blended in a bonded fabric and then adhered as appropriate. For example, to increase the durability of acetate fibers, synthetic fibers can be blended in the web. Then either chemical bonding or thermal bonding processes could be used to secure the web. See Figure 4.65 for an example of a nonwoven fabric formed without fiber felting.

The remaining method of bonding fiber webs—*mechanical or fiber entanglement*—does not require heat, adhesives or chemicals to bond the fibers. Rather, barbed needles or jets of water entangle the fibers securely. These processes yield fabrics referred to as needle felts or spunlace.

Now that general web formation processes and bonding techniques for nonfelting fibers have been presented, a discussion of characteristics and uses of nonwoven fabrics produced by these processes follows.

Figure 4.65. In bonded fiber web structures, nonfelting fibers can be adhered by the addition of adhesives, chemicals or heat.

Bonded Fiber Fabrics

Bonded fiber web fabrics should not be confused with bonded or laminated fabrics made from two or more layers of fabric held together by adhesives or foam. These multicomponent structures were discussed earlier in this unit. To avoid possible confusion in identifying these different types of structures, use the term *bonded fiber web* when referring to a nonwoven structure made directly from nonfelting fibers and *bonded fabric* to describe a type of multicomponent structure. Remember, however, that a bonded fiber web can serve as a backing in a multicomponent structure.

Staple Fiber Webs

Bonded webs were first made of short staple waste fibers, which produced nonwovens of limited serviceability. Now high quality fibers are cleaned, carded, bleached and blended according to the serviceability specifications determined by the proposed end use of the product.

Bonded fiber fabrics first made a place for themselves in the limited use or throw-away fabric field. Disposable diapers alone have been a major success, and the medical field also uses bonded fiber fabrics for disposable gowns, bedding and other purposes. Key factors that help define the market for disposable products are low cost, short-term use, and increased sanitation for certain end uses.

Sample 31 is a 100 percent staple fiber polyester nonwoven that was *chemically bonded* with solvents. Pull on the structure and hold it up to a light source; note its durability and stability. While thermal bonding could have been used to fuse the fibers, since polyester is thermoplastic, in this example a solvent was used to achieve the desired effect. In fact, the fabric tradename is Phun Phelt™; thus this nonwoven no doubt was made with felt end uses in mind.

Since bonded fiber webs were formerly used primarily for disposable items, some people may still confuse them with disposable cellulose paper fabrics. Consumers have learned, though, that many bonded fiber fabrics are reuseable and need not be thrown away after one use. Fiber webs can also be used as backings in multicomponent fabrics such as sample 36A. In these cases, fiber webs become a part of a durable, long use fabric. Sample 36A is an upholstery fabric with a suede look made possible by supporting a warp knit with a fiber web backing.

Spunbonded Webs

These relatively durable webs are made from continuous filament synthetic fibers which have high, balanced strength properties. Their increased durability over staple fiber webs is due to random fiber orientation and efficient bonding of the continuous filament web network. Consequently, spunbonded fabrics last longer than webs made mainly of shorter and weaker cotton, rayon, or acetate staple fibers. Examine sample 32, a rayon/polyester blend Handi-wipe®. This "semi-disposable" product is intended to be used several times before it is thrown away.

Spunbonded fabrics and certain staple fiber nonwoven formed by crosslayering or air layering techniques to produce random orientation are successfully used in reusable items. *Interfacings*—those supportive fabrics placed between layers of fabric in garments—have become a successful and growing market. However, bonded fiber interfacings sometimes lack the qualities of a woven interfacing. Also, bonded fiber fabrics are frequently used as the base for tufted carpeting.

Melt-Blown Webs

To increase the comfort, flexibility, and barrier properties of nonwoven structures to an even greater extent, *microfibers* with considerable smaller deniers than staple fibers or filaments are produced using a *melt-blown process.* This heat bonding structuring method developed by Van A. Wente of the U.S. Naval Research Laboratory in 1954, and first commercialized some 20 years later, is opening up increased applications for nonwoven fabrics. The microfibers contribute to increased comfort; plus they yield fabrics which are excellent *barriers,* meaning that chemicals and other potentially dangerous substances cannot pass through the fabric.

Thus, end uses for protective clothing have great potential, such as in apparel for agricultural or industrial workers who need protection from pesticides and other chemicals; and for medical personnel who need the apparel to protect patients from bacteria transfer. To make the soft and flexible microfiber melt-blown webs more durable for apparel uses, the structures can be bonded to spun-bonded nonwoven fabrics, or staple fibers can be added to the microfiber webs for reinforcement. Surgical gowns and masks are successful end uses where a thin melt-blown fabric is bonded between two layers of spunbonded webs. The resulting products provide barrier and comfort properties important in this health-related application.

In addition to protective clothing, microfiber webs make excellent insulation in apparel. Thinsulate® is the trade name for a melt-blown fabric that incorporates polyester staple fibers into the microfiber web to increase strength and resiliency. The resulting web is a good insulator, too, since more microfibers can be placed in a given area; this leads to increased fiber surface areas and therefore more dead air space than conventional fibers and filaments. Warmth with substantially less weight results; thus, coats and jackets with Thinsulate® as the interlining material are advertised for their warmth without weight or bulk.

Even with synthetic continuous filament bonded webs, certain care precautions must be followed to preserve the fabric structure. Cleaning chemicals need to be selected that will not alter the bonding agent, and excessive heat must be avoided. Agitation should be kept to a minimum, since the struc-

QUICK QUIZ 26

1. A bonded web of fibers can be constructed to withstand laundering by selecting proper fiber content. Which fibers are especially suitable for bonded fiber webs to be laundered?
____
2. Compare samples 31 and 32; both are synthetic bonded fiber fabrics, made for different end uses by different processes. The chemically bonded fiber web, sample ________, was designed to simulate ____________; the heatset ____________ web, made from continuous filament polyester, contributes to fabric ________.
3. Melt-blown nonwovens that are made from ________ fibers are well suited in protective clothing for what two reasons? ________ and ________.

****

ANSWERS QUICK QUIZ 26

1. nylon or polyester filament fibers
2. 31, felt; spunbonded, durability or washability
3. micro; increased comfort and barrier properties

ture is a web of fibers that does not contain sturdy yarns.

Both felted and bonded fiber web structures have less flexibility than woven or knit structures of comparable thickness. This makes for more rigid fabrics that are somewhat unsatisfactory for use in draped, folded or creased applications. While the melt-blown fabrics made from microfibers are flexible, comfortable and light weight, they are generally used in conjunction with other fabric layers. Thus, careful consumer decisions are needed to evaluate when a nonwoven web is suitable for the intended use.

Needle Felts

Needle felt or needle-punched fabrics are another type of structure created directly from nonfelting fibers. *Mechanical entanglement* of the fiber web causes the fibers to bond together, creating a unified fabric.

Up to the point of web formation, needle felts go through similar fiber processing as do bonded fiber webs or felts that contain wool. The difference in bonded webs, felts, and needle felts lies in how the fibers are adhered in the web. In the production of needle-felt fabrics, the prepared fiber web passes through a needle-punch machine, sometimes called a loom, where multiple barbed needles pass through the web a sufficient number of times to entangle or bind the fiber web together. See Figure 4.66.

The fibers may be entangled only with each other; around a layer of filament yarns; or around a center, loosely woven fabric called a scrim. The scrim is formed from yarns which have been interlaced in a very open plain weave fabric of extremely low yarn count. The end use of the needle-felt fabric determines the type of central yarns or scrim, if any, that is used. Note the type of needle-felt structure in Figure 4.67.

The face of a needle-felt fabric looks similar to a bonded fiber web or a conventional felt fabric, as observable in needle-felt sample 35. There is a difference, however, which can be seen by holding the needle punch fabric up to a strong light. Notice the fiber configuration where the barbed needles have punched through the web to entangle the fibers.

Because the needle-felt process of fabric production is faster than either weaving or knitting, many blankets are made by this method. Since needle-felt blankets are constructed to be low cost items, expense-adding scrim is not generally used. Instead, the center in needle-felt blankets often includes parallel filament yarns if any central reinforcement is used.

In fact, needle-felt structures were first introduced for consumer use in rayon blankets which sold for a very low cost. However, they had poor

Figure 4.66. To form needle-felt fabrics, many barbed needles are punched back and forth through a web of fibers, causing the fibers to entangle with each other.

wearing qualities due to the fiber properties and nonwoven structure which didn't stand up well to pulling forces or abrasion. Since rayon fibers lose strength when wet, the product tended to have low

Yarn Scrim

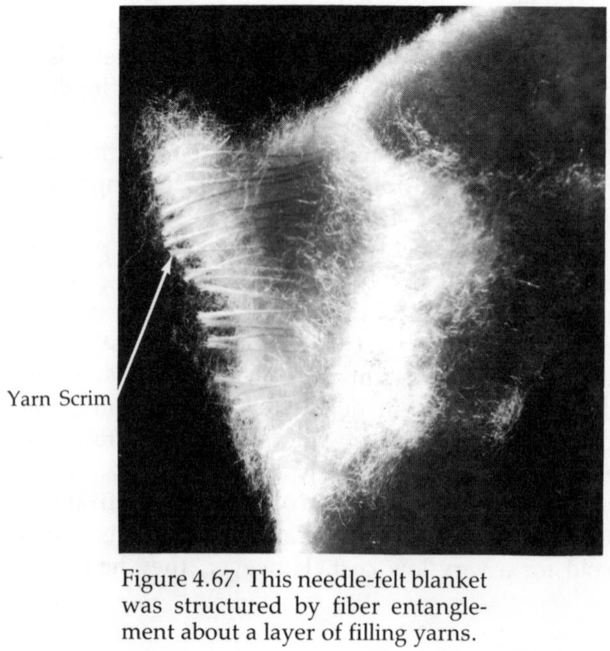

Figure 4.67. This needle-felt blanket was structured by fiber entanglement about a layer of filling yarns.

stability during laundering. Thus, more durable acrylic needle-felt blankets have proven serviceable.

Another successful end use for needle-felt structures is in low cost carpeting where a center scrim is used usually to increase fabric stability and reduce wear. Even though weaving the center scrim increases production time, the cost is still less than tufting—the major structuring method for carpets.

Needle-felt carpets were first made of inexpensive olefin fibers, which are also responsible for the success of this low-cost floor covering. The zero absorbency of olefin fibers makes them especially suitable for use in the indoor-outdoor needle-felt carpet market because the fibers do not absorb moisture or stains. But the low heat tolerance of olefins can cause a problem where high temperatures are found, such as in kitchens or under clothes dryers.

Spunlaced

In 1973 a new method of bonded fiber fabric formation, called spunlaced, was introduced to the market. This fabric structuring process, patented by DuPont, involves entanglement of a fibrous web by the action of an impinging fluid stream from jets. Since the fabric is formed from a fiber web, the

---

QUICK QUIZ 27

1. Needle-felt fabrics are used for industrial purposes, inside furniture, and on the backs of other fabrics. Sample 35 is a polypropylene needle-punch industrial fabric constructed (with, without) yarns or a scrim.
2. In determining care procedures for a needle-felt structure, the ____________ content, fiber ____________ and the type of yarn or scrim center must be evaluated.

****

ANSWERS QUICK QUIZ 27

1. without
2. fiber; arrangement or entanglement

---

structure is classified as a nonwoven fabric. Tradenames for spunlaced fabrics include Nexus and Sontara.

Jet streams of water are forced through the web structure, forming an almost woven-like fabric because the fibers interlace with each other as a result of the jet action. Examine sample 33, a 100 percent polyester spunlaced fabric, which is formed by water jet action with no binder or chemicals used.

A combination of knitted and woven fabric properties results. Spunlaced fabrics are less rigid than wovens, and somewhat stronger than knits of equivalent weight. They have lower elongation than knitted structures. The fibers in the spunlaced web have a loose bond, compared to the relatively rigid bond of fibers sealed with adhesives. Thus, spunlaced fabrics have a hand that is typically soft, subtle and lofty, as can be determined by handling sample 33.

The spunlaced webs are currently made of polyester fibers although other fibers may be used. These easy care structures can be machine washed and tumble dried. They are made in a variety of weights and styles for apparel, home furnishings and industrial purposes.

Home furnishing spunlaced fabrics are available for draperies, upholstery, domestics, and wall coverings. Industrial applications, involving the use of spunlaced fabrics in combination with other layers, include secondary carpet backings, shoe liners, coating substrates, lamination backings and plastic reinforcements. In the end use areas where high measures of durability are required, the fibrous spunlaced web is used successfully in multicomponent products.

In apparel end uses, spunlaced fabrics are suitable for robes, beach coats, lingerie and blouses. Basic spunlaced fabrics can have finishes, color and design added to produce a variety of effects. These structures are serviceable for many end uses. However, end use has not reached the level originally predicted, perhaps due to consumer preference for conventionally knit and woven fabrics.

Ultrasuede®

A nonwoven fabric that looks like suede has become very successful in high fashion apparel. Ultrasuede® has been designed to give the appearance and esthetic value of true suede. It has an important advantage over suede—it can be easily and safely washed and dried in standard laundry equipment. The fabric is expensive, however, and comparable in cost to real suede.

Ultrasuede® has good flexibility and fairly good draping qualities for a nonwoven structure. It is used in outer apparel for both men and women, including jackets, skirts, coats, suits and dresses.

Ultrasuede®, sharing some of the same properties as other nonwoven structures, has several important characteristics: the fabric is nonraveling and has no grainline; there are no visible woven or knitted properties, the fabric is stiffer than either woven or knitted fabrics; it has good thermal retention, and a firm body and hand.

QUICK QUIZ 28

1. Spunlaced fabrics can be successfully machine washed and tumble dried without shrinking or fabric damage. Of course, normal good laundry procedures should be used to avoid wrinkling of any thermoplastic fabrics. Name two procedures:
   ____
   ____
2. Compare fabrics 33 and 34. Tradenamed Nexus, both are nonwovens formed by fiber entanglement, called ______________ structures. The difference between the two is the additional ______________ layer and printed design applied to fabric 34.

****

ANSWERS QUICK QUIZ 28

1. Avoid excessively hot water and high drying termperatures; avoid overloading the machine; remove fabrics from dryer as soon as it stops.
2. spunlaced; backing or support

REVIEW

1. Differentiate among the manufacturing methods used to form the following fabrics directly from fibers:
   a. felt ______________________________
   ____
   b. spunbonded ______________________________
   ____
   c. melt blown ______________________________
   ____
   d. needle felt ______________________________
   ____
   e. spunlaced ______________________________
   ____

2. The thickest of the above nonwoven structures are usually ______________ and ______________, which have dead air spaces resulting in fabrics with high ______________.

3. The most common end uses for needle-felts at the present time are ______________ and ______________.

4. The nonwoven structure most likely to be used in disposable or limited use items is the ______________ containing rayon fibers.

5. Industrial applications and protective clothing which benefit from barrier fabrics include nonwoven layers made from ______________ by the ______________ process.

6. Generally speaking, felt, spunbonded and needle felts are not as durable as comparable woven or knit structures because they are made primarily from ______________ ______________.

****

REVIEW ANSWERS:

1. a. made from at least 25 percent wool or animal hair fibers which are held together by fiber felting action
   b. made from nonfelting, continuous filament fibers bonded by adhesives, chemicals or thermal bonding techniques
   c. microfibers held together by thermal bonding
   d. fibers mechanically entangled with each other, around yarns or a woven scrim, by barbed needles punched through the structure
   e. filament fibers entangled by high-pressure, water jet action
2. felt, needle felt; thermal retention
3. carpets, blankets
4. staple fiber web
5. microfibers; melt-blown
6. fibers rather than yarns

UNIT FOUR

Part VIII

Nonwoven Fabrics Made From Fiber Solutions

The nonwovens made from fiber solutions—called film structures—are not true textiles in that they contain no fibers or yarns. However, film products which have a fabric-like hand and drape are used in some of the same end uses as conventional fabrics. When used in textile applications, film layers are often referred to as *vinyl*. The basic material used to form textile film products is a chemical polymer generally from the *vinyon* or *polyester* generic classification.

The process consists of flowing, extruding or molding a fiber solution into a flat sheet or layer, instead of into filament form. A wide variety of thicknesses, decorative features and degrees of opaqueness and translucence can be achieved by this process. Products range from thin, lightweight rainwear and shower curtains to heavy, durable upholstery fabrics. Films can be plain and smooth, or they can have printed on or pressed in surface designs, as pictured in Figure 4.68.

Many films are joined to a woven, or knit fabric layer by the bonding or laminating process discussed previously. A film that has a fabric backing applied is known as a *supported film*, while an *unsupported film* is one not joined to another fabric layer.

Films are easy-care items especially resistant to soiling since they have no absorbency; all that is required to clean them is wiping with a damp cloth or sponge. In fact, laundering or dry cleaning is seldom needed and usually not recommended, because many plastic or vinyl films are sensitive to heat, strong alkalies and some cleaning solvents.

Since films cannot be readily laundered or dry cleaned, certain types of soil, such as perspiration, are difficult to remove. Thus, vinyls are seldom used in making apparel except as outer wear. When films are supported, there is an added problem of cleaning the supporting fabric layer; in such instances it is necessary to launder or dry clean the product with extreme caution, following precisely any care instructions provided. Apart from possible cleaning and care difficulties with vinyl films, other limitations can deter use in wearing apparel.

These include their somewhat stiff and plastic-like characteristics; the tendency to stiffen, shrink, and crack over time; and the possibility of stretching in the presence of heat or when dry cleaned. Thus, their care procedures are limited primarily to wiping with a damp cloth or very gentle laundering.

Supported vinyl films are used successfully in place of leather for luggage, handbags, and upholstery. Other items made from films include shower curtains, rain apparel, liners for baby pants, upholstery, shoes, place mats and table linens, in addition to certain industrial products. These supported vinyls are durable, do not water spot, and are resistant to most surface abrasion, except scratching or gouging.

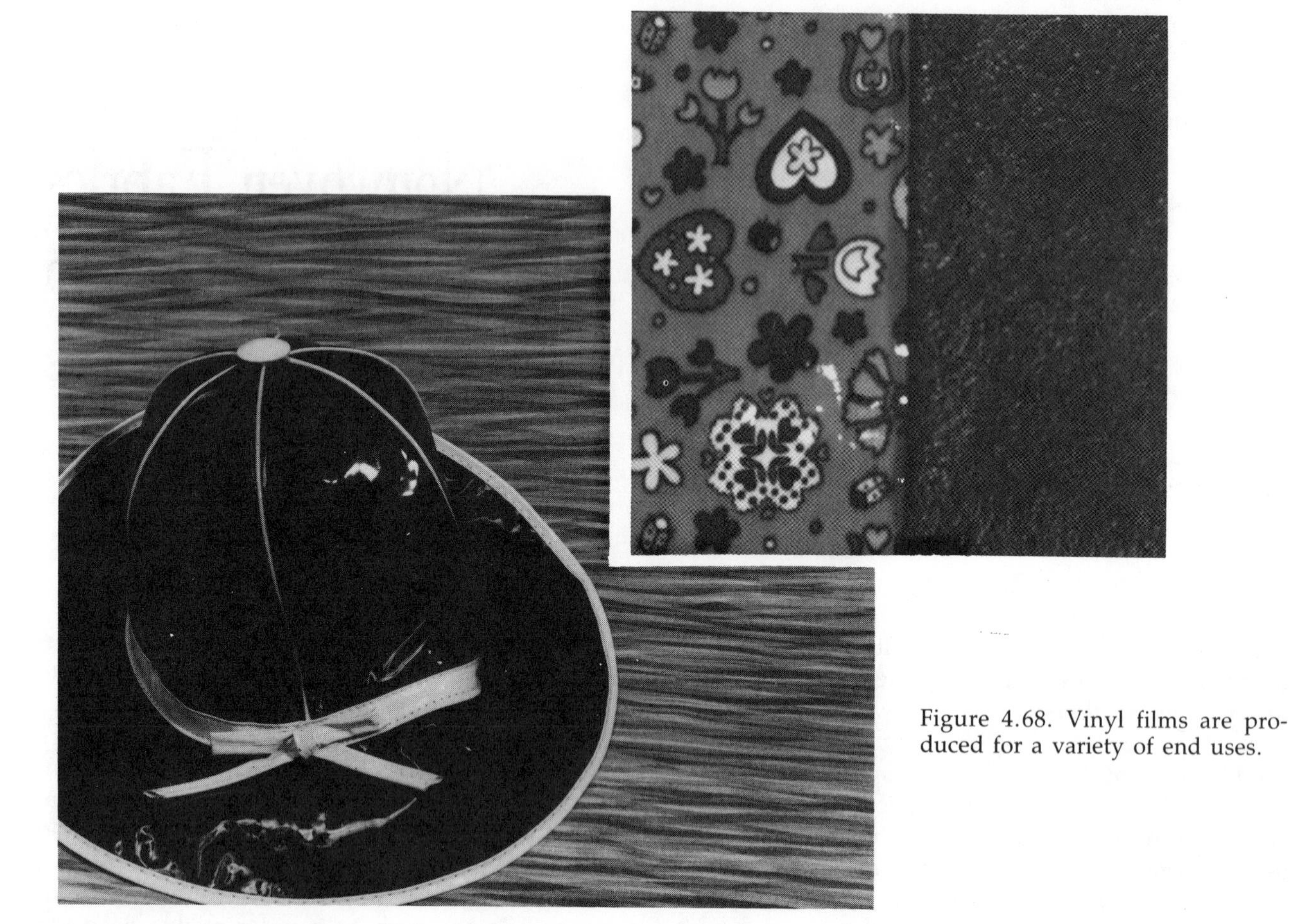

Figure 4.68. Vinyl films are produced for a variety of end uses.

QUICK QUIZ 29

1. The pressed-in pattern on the vinyl surface in Figure 4.69 is a repeating geometric design. Could the structure by described as a dobby (yes, no) Why? ______________________
____
2. In a supported vinyl film, would a knitted or a woven backing tend to give more flexibility? ______________ Why? ______________________
____
3. Can warp and filling grain lines be seen in an unsupported film? (yes, no) Why?
____

****

ANSWERS QUICK QUIZ 29

1. no; because dobby structures are woven-in repeating designs
2. knitted; because the interlooping in knits provides more give than the interlacing in wovens.
3. no; yarns are not present

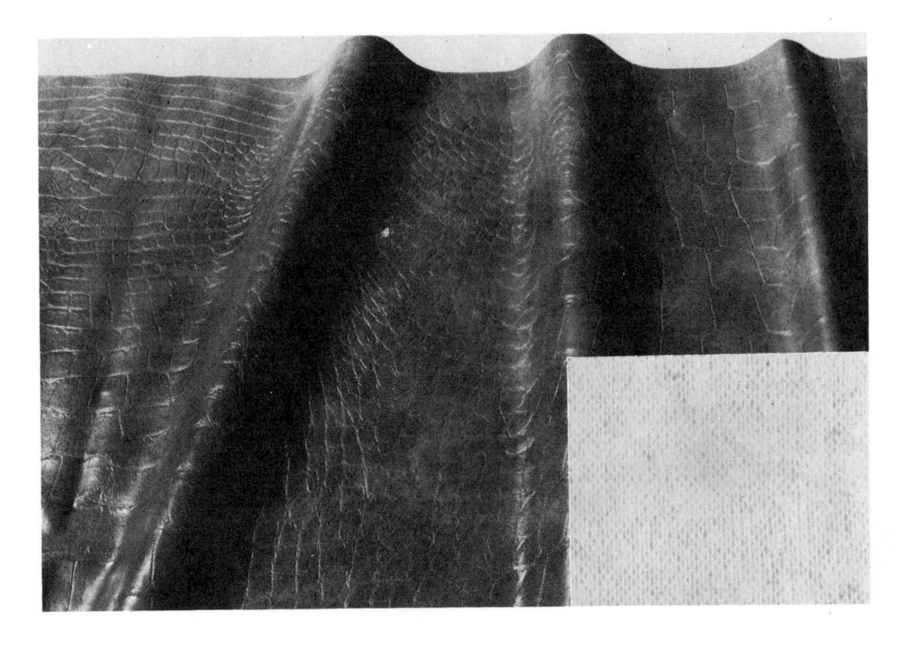

Figure 4.69. This supported film is bonded to a filling knit layer to increase fabric flexibility and yield a more fabric-like hand.

Nonwoven Fabric Identification

1. Sample 36 is a (supported, nonsupported) vinyl designed for use in ______________________ .
   Assess its serviceability: ____________________
   ____
2. Recall two generic fibers used in film form for coating metallic yarns: ____________________
   and ______________________ .
3. Samples 31 and 32 are both nonwovens made from webs of fibers. By holding the samples up to a strong light, determine which one has a more random arrangement of the fibers: ______________ .
4. Sample 32 is made of rayon and polyester fibers. In this spunbonded fabric, the fibers are joined together by ____________________________ .
5. If sample 31 were made of 100 percent rayon fibers, as are some disposable nonwovens, the fibers would have to be adhered using ______
   ____________________ .
6. Samples 33 and 34 are examples of spunlaced fabrics tradenamed Nexus®. How were these fiber webs formed? ____________________
   ____
   ____
   ____
7. Describe the structural differences between these two spunlaced samples, and indicate how serviceability is affected.
   33: ____________________________________
   ____
   ____
   34. ____________________________________
   ____
   ____
8. Sample 35 is a needle-punched fabric. How are the fibers entangled in this type of construction?
   ____
   ____
   ____
9. Why is a supported vinyl, like sample 36, often preferred to an unsupported one? __________
   ____
   ____
   ____
10. Thick supported and unsupported films are often used as furniture upholstery because they are abrasion resistant and easy to care for fabrics. Yet, there is a comfort disadvantage to vinyl upholstery because ____________________
    ____
    ____

11. Take out the Mounting Sheet for Nonwoven Structures and study the labels where samples are to be attached. Cut and mount swatches from samples 31, 32, 33, 34, 35, and 36 in the appropriate places.

****

Nonwoven Fabric Idenfication Answers:

1. supported; upholstery; durable and easy care for the end use
2. acetate, polyester
3. 31
4. thermal bonding
5. adhesives
6. pressurized jets of water entangled and bonded the fibers
7. 33—not backed, has increased drapability and flexibility over supported fabric 34
   34—the backing adds dimensional stability for certain end uses; but fabric pliability, flexibility and drape are decreased
8. by punching barbed needles through the fiber web
9. increased flexibility and more fabric-like properties; plus, greater durability with less chance of stretching than with a nonsupported film
10. it is nonabsorbent and can be sticky and uncomfortable in warm weather, and cold in cold weather
11. 31—chemically bonded Phun Phelt™
    32—spunbound Handi-Wipes®
    33—unsupported spunlaced Nexus®
    34—supported spunlaced Nexus®
    35—needle felt
    36—supported vinyl

UNIT FOUR
Part IX

Summary Review of Fabric Structures

This section concludes the unit on fabric structures, which are important elements to consider when evaluating fabric serviceability for the end use. The *structure* of the fabric explains the method by which it was made, i.e. weaving, knitting, knotting or fiber bonding. For example, if a fabric is described as a cotton wafflecloth piqué, the structure is a dobby weave, the fiber content cotton, and the fabric name wafflecloth piqué.

In determining whether to select a woven, knit, or nonwoven structure for a given end use, consider the type of wear the fabric will receive. If a product that gives with wear is desired, a knit structure is preferred; whereas an end use that requires high dimensional stability is best served by a firmly woven fabric. A woven structure with a low yarn count tends to have low dimensional stability; but it can be stabilized with the addition of another layer by a process known as lamination or bonding. Bonded and laminated fabrics usually have knit backings because they are more flexible and resilient than woven ones. The knit backing is generally a warp knit because warp interlooping is more stable than filling. Felts and bonded fiber webs are not as durable as woven fabrics in many end uses, because they are made directly from fibers rather than yarns which provide added strength.

Major characteristics and potential end uses of selected structures follow:

Satin weaves provide a lustrous surface for coat linings that slip on easily. However, abrasion resistance is low due to the long warp yarn floats.

Uncut terry pile makes an absorbent beach robe, especially when made of cotton, because the loops provide for increased surface areas to assist moisture absorption.

Nylon tufted pile is a serviceable choice for moderately priced family room carpeting that receives steady wear; the tufted rows are inexpensive to produce but adequately secured by an adhesive coating on the back. Woven pile carpets are too expensive for the majority of families.

Needle-felt structures are acceptable for inexpensive sun porch carpets because they can be produced rapidly to reduce cost; they resist weathering especially when made of olefin fibers.

Braided structures are good for flexible bindings to go around curved jacket edges because the yarns are interlaced diagonally, creating a bias which can be easily stretched and shaped.

Double-cloth fabrics might be worth the extra cost for certain esthetic gains, as in reversible coats; the structure also is a good heat insulator because air is trapped not only within each layer but also between them. Double-cloth upholstery fabrics are chosen for the increased durability, resulting from the supportive underlayer of fabric, or to allow special design effects that are held taut by the supportive layer.

Tightly woven structures, such as high yarn count nylon plain weaves, make warm windbreaker jackets that are lightweight. Warmth is increased by having a nonwoven interlining made of low density fibers or microfibers.

Double knits are serviceable knit structures for winter slacks; the double construction traps more air. Plus, they are more stable than a single knit which can sag or stretch out of shape.

Leno structures or raschel knitted nets make good laundry bags that hold the weight of clothes but still allow for good air and/or water circulation.

Seersucker is the fabric name for a woven-in, crinkled stripe structured by the slack tension weaving on a dobby loom.

Jacquard structures that are flat and reversible, like *damask*, are more serviceable than raised brocades for tablecloths.

Velvet is usually made of filament fibers to provide a luxurious, shiny pile fabric for evening wear. Velveteen is generally made of staple-length cotton fibers, producing more casual fabrics.

Leno, spunlaced, and stitch-bond structures are among the fabric structures suitable for use in openwork curtains that are relatively durable and allow high levels of light transmission.

Nonwoven structures used to be thought of as disposable fabrics of limited durability and use. Now, the newer spunlaced and melt-blown fabrics are serviceable for many specialized and re-useable purposes and are increasing in use, especially for medical, industrial, and agricultural protective apparel.

Now, the Posttest which follows will help determine how well the fabric structures have been mastered.

Posttest

1. Most woven structures are characterized by yarns
   a. cross-laid at right angles.
   b. interlaced at right angles.
   c. interlooped at right angles.
   d. braided at right angles.

2. The crosswise direction of a fabric is the ______________ direction.
   a. warp or course
   b. wale or filling
   c. selvage or filling
   d. filling or course

3. The selvage runs parallel to the
   a. warp yarns.
   b. filling yarns.
   c. bias.

4. Compared to knits, weaves generally have greater
   a. flexibility.
   b. resiliency.
   c. stability.
   d. absorbency.

5. A major disadvantage of filling knits used in hosiery is their
   a. lack of stretch.
   b. tendency to form runs.
   c. limited construction styles.
   d. low absorbency.

6. Warp knit structures typically have
   a. low crosswise stretch and diagonal interconnectors.
   b. high crosswise stretch and horizontal interconnectors.
   c. high crosswise stretch and diagonal interconnectors.
   d. low crosswise stretch and horizontal interconnectors.

7. The circular knitting process cannot be used to produce
   a. women's hosiery.
   b. men's T-shirts.
   c. double knit suits.
   d. full-fashioned sweaters.

8. Which of the knits listed below has the least crosswise stretch?
   a. purl
   b. tricot
   c. jersey
   d. rib

9. Which part of the loom interlaces the filling yarns through the warp yarns?
   a. heddle
   b. reed
   c. harness
   d. shuttle or its equivalent

10. Considering the weave balance, which fabrics listed below would wear most uniformly?
    a. gingham
    b. faille
    c. broadcloth
    d. taffeta
    e. oxford

11. Which durability property is decreased most by a rib weave?
    a. tenacity
    b. abrasion resistance
    c. flexibility
    d. resiliency

12. The low yarn count often typical of basket weaves yields fabrics with
    a. increased tenacity and increased abrasion resistance.
    b. decreased tenacity and decreased flexibility.
    c. increased resiliency and decreased flexibility.
    d. increased resiliency and increased flexibility.

13. The ______________ the angle of the twill, generally the ______________ durable the fabric.
    a. steeper, less
    b. flatter, more
    c. steeper, more

14. Twill weaves are typically stronger than basket weaves because twills tend to have
    a. lower twist yarns.
    b. higher yarn counts.
    c. random interlacing patterns.
    d. longer yarn floats.

15. In a satin weave structure, the ______________ the yarn count, the ______________ the yarn floats.
    a. lower, longer
    b. higher, longer
    c. lower, shorter

16. Sateen weaves are usually made from ______ length fibers.
    a. filament
    b. staple

17. The most frequently used construction process for cut, warp-pile fabrics is the ______________ method.
    a. over-the-wire
    b. double-cloth
    c. double-faced
    d. none of the above

18. When the loops on terry cloth are cut, how is the serviceability of towels affected?
    a. decreased absorbency
    b. increased absorbency
    c. decreased tenacity
    d. increased flexibility

19. What is the major reason that over 90 percent of the carpets produced are tufted pile?
    a. increased durability
    b. high speed, lower cost of production
    c. esthetic qualities
    d. ease of care

20. Match the fabric to its weave structure. A weave may be used more than once.

____	1. herringbone	a. leno
____	2. piqué	b. rib
____	3. marquisette	c. jacquard
____	4. damask	d. twill
____	5. gabardine	e. filling pile
____	6. corduroy	f. warp pile
____	7. monk's cloth	g. basket
____	8. velvet	h. dobby
____	9. matelassé	i. broken twill
____	10. broadcloth	
____	11. brocatelle	

21. Which weave structure is most serviceable for durable school slacks?
    a. twill
    b. dobby
    c. crepe
    d. jacquard

22. Which weave structure is most abrasion resistant for a drapery fabric?
    a. crepe
    b. leno
    c. dobby
    d. plain

23. The __________ the yarn bulk, the __________ the thermal retention.
    a. lower, higher
    b. higher, higher
    c. higher, lower

24. What is an advantage nets made by knotting have over nets made on knitting machines?
    a. less expensive to produce
    b. more durable
    c. more versatile
    d. less likely to shrink

25. Laces tend to have low
    a. elasticity.
    b. abrasion resistance.
    c. flexibility.
    d. resiliency.

26. Compared to woven structures, braids have
    a. more lengthwise stretch.
    b. less lengthwise stretch.
    c. less crosswise stretch.
    d. about the same stretch.

27. The separation of a five-yarn double cloth can yield two
    a. warp pile fabrics.
    b. corduroy pile fabrics.
    c. velveteen fabrics.
    d. filling pile fabrics.

28. What advantages do fabric-fabric bonded structures offer the consumer over double-cloth fabrics?
    a. higher thermal retention
    b. lower shrinkage potential
    c. higher dimensional stability
    d. lower cost

29. Which structure would probably provide the most comfortable waterproof raincoat?
    a. unsupported film
    b. high yarn count plain weave
    c. supported film
    d. high yarn count twill weave

30. Match the structure with its appropriate description:
    ____ 1. 100 percent polyester bonded fiber web
    ____ 2. 100 percent rayon bonded fiber web
    ____ 3. 100 percent acrylic needle felt
    ____ 4. 100 percent cotton tufted
    ____ 5. 100 percent wool felt
    ____ 6. 100 percent polyester spunlaced
    ____ 7. 50/50 rayon and cotton stitchbonded
    ____ 8. 100 percent polyester melt-blown web

    a. mechanical tangling of fibers used in structure formation
    b. adhesive required to retain structure
    c. water-jet action used for structure formation
    d. heat can be used to bond structure
    e. chain stitch used to adhere structure
    f. fiber properties necessary for structure formation
    g. micro fibers bonded by heat
    h. extra set of yarns necessary

31. Match the most appropriate carpet description to the consumer need:
    ____ 1. wool woven pile
    ____ 2. acrylic tufted pile
    ____ 3. olefin needle felt
    ____ 4. nylon tufted piled
    ____ 5. olefin tufted pile

    a. den carpeting that imitates natural fiber competitor
    b. low-cost porch carpeting
    c. durable, abrasion resistant carpet for hallway or den
    d. nonabsorbent and durable kitchen carpeting
    e. cost is no object for living room carpet of high workmanship

32. Match the drapery or curtain fabric you would select for each end use:
    ____ 1. warp pile velvet
    ____ 2. plain weave burlap
    ____ 3. jacquard brocade
    ____ 4. leno marquisette
    ____ 5. dobby cord

    a. den drapery with woven-in
    b. rich, elegant living room drapery
    c. inexpensive curtain for cabin
    d. theater stage curtain
    e. lightweight, durable curtain

33. Match each towel to the most serviceable fabric. Structures may be chosen more than once or not at all.
    ____ 1. durable beach towel
    ____ 2. decorative guest hand towel
    ____ 3. bath towel with velvet-like appearance
    ____ 4. nonlint dish towel
    ____ 5. absorbent bath towel

    a. uncut terry pile
    b. cut terry pile
    c. cut warp piled
    d. dobby birdseye toweling

34. Match the fabric to the property it would give a blouse or shirt:
    ____ 1. firmly woven broadcloth
    ____ 2. rib knit turtleneck
    ____ 3. jersey T-shirt
    ____ 4. oxford cloth
    ____ 5. spunlaced
    ____ 6. moderate yarn count, plain weave

    a. highly elastic, good shape retention
    b. good body, fine ribs
    c. soft texture, wrinkles badly if not finished
    d. smooth, uniform surface
    e. comfortable, high stretch
    f. lightweight, soft, lofty

35. Match the fabric preferred for each type of slacks described:
    ____ 1. corduroy
    ____ 2. denim
    ____ 3. poplin
    ____ 4. double knit
    ____ 5. velvet

    a. stretch jogging slacks
    b. evening slacks
    c. durable work slacks
    d. lightweight summer slacks
    e. durable slacks with dimensional pattern

36. Which woven structure is made by interlacing three sets of yarns at 60-degree angles to each other?
    a. filling pile
    b. tufted
    c. triaxial
    d. none

Answers for Unit 4 Posttest:

1. b
2. d
3. a
4. c
5. b
6. a
7. d
8. b
9. d
10. a
11. b
12. d
13. c
14. b
15. a
16. b
17. b
18. a
19. b
20. 1-i; 2-h; 3-a; 4-c; 5-d; 6-e; 7-g; 8-f; 9-c; 10-b; 11-c
21. a
22. d
23. b
24. b
25. b
26. a
27. a
28. d
29. c
30. 1-d; 2-b; 3-a; 4-h; 5-f; 6-c; 7-e; 8-g
31. 1-e; 2-a; 3-b; 4-c; 5-d
32. 1-d; 2-c; 3-b; 4-e; 5-a
33. 1-a; 2-b or d; 3-b; 4-d; 5-a
34. 1-b; 2-a; 3-e; 4-c; 5-f; 6-d
35. 1-e; 2-c; 3-d; 4-a; 5-b
36. c

UNIT FIVE

Finishes

Previous units have included information on fibers, yarns and how they are made, fabric structures and processes involved in construction, as well as resulting fabric serviceability characteristics. This section covers some of the many finishes that may be applied to a fabric. Some are applied as a basic part of fabric manufacture and are necessary to produce an acceptable product for the consumer. Other finishes are applied later to alter appearance or hand of the fabric. A third group of finishes is applied to fabrics to alter or modify their performance or function. The finishes that are a basic part of fabric manufacture are most commonly called *routine*, preparatory or general mill finishes.

Finishes that alter the appearance or hand of fabric may also modify or change its typical performance; however, the major reason for their use is to provide consumers with fabrics that look and feel the way they want them to. *Finishes that alter performance* or function include those that change care procedures, durability and performance under specific conditions. This latter group can also affect appearance and hand, changes that may be desirable or undesirable.

Finishes can be categorized as *durable* (permanent) or *temporary* (renewable); they may be *chemical* or *mechanical*; and they may be *decorative* or *functional.* The following material should help you understand the use of various finishes, their importance to consumers, problems in their maintenance or care, the way in which they alter or modify fabric behavior, and why they are used.

Objectives for Unit 5

1. You will gain an understanding of the following:
   A. Preparatory or initial finishes—singeing, desizing, scouring and bleaching; and final general mill finishes—calendering, tentering or crabbing, and inspection.
   B. Finishes that alter appearance and hand—singeing and shearing, sizing, weighting, brushing, napping and sueding; fulling, beetling, embossing, glazing, ciréing, moiréing, flocking, schreinerizing, parchmentizing, and chemical crepeing.
   C. Finishes that alter performance and/or function—abrasion resistant, antislip, shrinkage control, mercerization, absorbent, antistatic, slack mercerization, softeners, soil and stain resistant, mothproofing agents, biologically resistant, water repellent and waterproof, flame retardant, and durable press.
2. You will be able to indicate how specific finishes contribute to the various serviceability aspects of durability, comfort, care or maintenance and esthetic value.
3. Given an end use, you will be able to determine the most suitable finish or finishes to apply to a fabric.

Pretest

1. General mill or preparatory finishes are most necessary and most frequently used when processing fabrics made of fibers in the ______________ categories.
   a. natural cellulosic and man-made cellulosic
   b. synthetic and protein
   c. man-made cellulosic and synthetic
   d. natural cellulosic and protein

2. The preparatory finish called scouring is used to remove:
   a. gum from silk
   b. impurities from either cotton or wool
   c. sizing from synthetics
   d. soil from fabric

3. What is sometimes used to give the illusion of whiteness to gray goods?
   a. chlorine bleach
   b. enzyme detergents
   c. optical brighteners
   d. oxygen bleaches

4. Singeing is generally done to fabrics of ____________ length fibers.
   a. staple
   b. filament

5. A final process which makes the fabric ready for market is:
   a. singeing
   b. tentering
   c. inspection
   d. padding

6. A textured wall covering for a stairwell should be finished so that it is
   a. waterproof.
   b. weighted.
   c. flocked.
   d. abrasion resistant.

7. Abrasion resistant finishes are most likely to be used on fabrics made from
   a. natural cellulosic and man-made cellulosic fibers.
   b. natural cellulosic and protein fibers.
   c. synthetic and man-made cellulosic fibers.
   d. synthetic and protein fibers.

8. To reduce fabric slippage at seams, antislip finishes are most necessary with __________ yarns.
   a. bouclé
   b. nontextured filament
   c. textured filament
   d. slub

9. Relaxation shrinkage is more likely to occur in ____________ yarns.
   a. filling
   b. warp

10. When concerned about progressive shrinkage in normal laundering, consumers should avoid purchasing fabrics made from which of the following fibers?
    a. nylon
    b. cotton
    c. wool
    d. olefin

11. Consumers who buy mercerized cotton fabrics benefit primarily from:
    a. decreased luster
    b. added abrasion resistance
    c. increased strength
    d. reduced depth of color

12. Fume-fading resistant finishes are an advantage in drapery fabrics made of ________________ fibers.
    a. rayon
    b. acetate
    c. glass
    d. silk

13. Antistatic finishes are needed most on fabrics made from ____________ fibers.
    a. protein
    b. synthetic
    c. mineral
    d. man-made cellulosic
    e. natural cellulosic

14. Soil or stain resistant finishes decrease soiling by creating:
    a. an environment that decreases surface tension
    b. an environment that increases static charge
    c. a rough surface that hides soil
    d. a more absorbent surface

15. What property is decreased in fabrics that have a durable press finish?
    a. dimensional stability
    b. economy
    c. abrasion resistance
    d. wrinkle resistance

16. When wools are purchased with a mothproof finish, consumers ______________ need to use mothballs on them.
    a. do
    b. do not

17. Antiseptic finishes delay bacterial growth, lessen perspiration odors, and inhibit mildew growth.
    a. true
    b. false

18. Despite comfort limitations, a consumer might choose a fabric with a waterproof finish when buying ______________ ; while a fabric with a water repellent finish would be more comfortable and acceptable when buying ______________________________ .
    a. a golf jacket, an umbrella
    b. a raincoat, a tarpaulin
    c. a raincoat, a golf jacket
    d. an umbrella, a tarpaulin

19. Children's pajamas are required by law to be:
    a. antistatic.
    b. flame retardant.
    c. durable press.
    d. flameproof.

20. A consumer adds starch to a fabric as a
    a. permanent sizing.
    b. temporary desizing.
    c. temporary sizing.
    d. permanent weighting.

21. To make silk fabrics more serviceable for the consumer, they are often weighted to:
    a. decrease static charges
    b. increase tenacity
    c. decrease thermal conductivity
    d. increase body

22. Fulling is used on many wool fabrics to slightly felt the wool fibers, make the weave tight and firm, and decrease relaxation shrinkage.
    a. true
    b. false

23. What finish increases fabric loft and thermal retention?
    a. embossing
    b. napping
    c. calendering
    d. moiréing

24. Which household item is beetled to give it a lustrous surface?
    a. velvet upholstery
    b. damask tablecloth
    c. tufted carpet
    d. brocade drapery

25. As an interior designer, why would you choose fabrics with an embossed finish?
    a. to achieve a true crepe
    b. to increase durability
    c. to achieve a raised, dimensional effect
    d. to increase luster

26. Glazing is a process which polishes the fabric surface
    a. after regular calendering has been completed.
    b. instead of the regular calendering process.
    c. as an additional part of the regular calendering process.

27. Which fabric finish reflects light in a wavy or watermarked manner?
    a. moiréing
    b. embossing
    c. calendering
    d. schreinerizing

28. To keep slips from creeping up under knit skirts, a(n) ______________ finish is helpful.
    a. absorbent
    b. sized
    c. antistatic
    d. beetled

29. If you wanted to purchase an inexpensive fabric with a pilelike surface, which finish might you select?
    a. glazed
    b. flocked
    c. sized
    d. ciréd
30. Which finish can be applied to create a leather-like fabric?
    a. sueding
    b. parchmentizing
    c. beetling
    d. crepeing
    e. fulling
31. Match the appropriate finish to the consumer need it would fulfill.
    ___ 1. parchmentizing
    ___ 2. Sanforized®
    ___ 3. shearing
    ___ 4. mercerization
    ___ 5. chemical crepeing
    ___ 6. bacteriostatic
    a. plissé
    b. soft, lustrous cotton fabric
    c. carpet with even cut pile surface
    d. garment with comfort stretch
    e. product with shrinkage less than one percent
    f. stiffened, translucent effect
    g. product with shrinkage less than two percent
    h. action stretch
    i. athletic socks

Answers for Unit 5 Pretest:

1. d
2. b
3. c
4. a
5. c
6. d
7. a
8. b
9. b
10. c
11. c
12. b
13. b
14. a
15. c
16. b
17. a
18. c
19. b
20. c
21. d
22. a
23. b
24. b
25. c
26. c
27. a
28. c
29. b
30. a
31. 1-f; 2-e; 3-c; 4-b; 5-a; 6-i

UNIT FIVE

Part I

Preparatory or General Mill Finishes

The initial preparatory or routine finishes discussed in this section include: *singeing, desizing, scouring* and *bleaching*. These finishes are applied to most fabrics to improve reception of subsequent coloration and finishing. *Calendering, tentering* or *crabbing*, and *inspection* are final general mill finishes used on most fabrics to make them ready for the consumer.

Grey or greige goods is the term used to describe fabrics that have received no finishing treatments of any kind. Obviously, few grey goods are ready for the consumer market as is. Most fabrics require some processing before they are ready for sale and use.

"Grey" goods are seldom grey! The color may be ivory, white or off-white, cream, tan, occasionally somewhat grey; or colored, if fibers or yarns have been dyed prior to the fabric construction process. However, even this latter group of precolored fabrics still needs various finishing processing before being ready for consumer use.

The big business of fabric finishing may be done by the fabric manufacturer on the same site where the fabric was made, or fabrics may be shipped to *converters* who do the finishing and color application. Fabrics receive finishes based on specifications established either by the manufacturer or the purchaser.

While finishes may or may not be visible on the final product, they are applied to produce a fabric appropriate for the intended end use. Preparatory finishes remove impurities, unwanted color, and those surface irregularities that can be safely removed without damage to the fabric surface. Descriptions of these finishes are presented next, followed by a brief discussion of each.

Singeing is required on certain fabrics to remove fiber ends that protrude from the surface as a result of yarn and fabric manufacture. *Desizing, scouring* and *bleaching* are preparatory finishes that remove sizing used for weaving, and soil, color, oil and other impurities that might have accumulated during fabric construction. Scouring and bleaching are done in one basic series of operations.

Calendering is the final smoothing or flattening process that "irons" the fabric. *Tentering*, called *crabbing* for wool, is used to make certain that fabric grain is maintained accurately and that fabric dimensions are maintained during subsequent processing. *Inspection* identifies flaws and imperfections that need to be removed, corrected or repaired, or serious flaws that require discarding of the fabric or reprocessing into an acceptable product.

Singeing

Fabrics with numerous fiber ends require a special kind of heat treatment, called *singeing*, as a part of routine processing. The fabric is passed rapidly over a hot roller or between gas flames, which burn off the short fiber ends, to produce a smoother surfaced fabric. The fabric is then quenched as it moves into the next finishing process, desizing, followed by scouring and bleaching.

Desizing and Scouring

In preparation for weaving, a starchlike substance called sizing is usually applied to give warp yarns stiffness, which assists in fabric structuring and reduces the possibility of yarn breakage. The process is shown in Figure 5.1.

Removal of warp sizing is an objective of the general mill process called *desizing*. Desizing precedes *scouring* and *bleaching*—the three steps used to remove the warp sizing, natural impurities and extraneous matter in the fabric, unwanted color, machine oils, and other contaminants that might have been picked up during fabric construction. These preparatory finishing processes are most important for fabrics made of natural cellulosic and protein fibers that tend to have a great amount of discoloration as well as impurities from their growth and production. However, a majority of fabrics, including those made of man-made fibers, require some desizing, scouring and bleaching prior to further processing.

Dilute sulfuric acid can be used as a part of the scouring process in wool because it chars the organic matter upon heating and converts it to carbon, which can later be "dusted" off. This scouring process is referred to as *carbonizing* because of the chemical change to carbon. After carbonizing, wool fabrics are given a weakly acidic detergent scouring.

The gum on filaments extruded by the silkworm during spinning of the cocoon is wholly or partially removed during silk cleaning or scouring. The gum is generally removed by boiling the silk in a mild detergent solution; this method can successfully be used on silk even though it is a protein fiber, because silk, unlike wool, does not contain sulfur bridges which are broken by alkaline solutions. When the gum has been removed by this scouring process, the fabric is called *de-gummed* silk.

Bleaching

A general preparatory finish applied to whiten grey goods involves the use of hypochlorite (chlorine bleach) or hydrogen peroxide (oxygen bleach). The converter must make certain that the bleach selected is both safe to handle and safe for the fibers involved. Figure 5.2 illustrates a commercial bleaching operation.

The chemicals used in fabric converter bleaching can also include *brighteners*. These add an "invisible color" to the fabrics that reflects ultraviolet light. Optical brighteners are really a type of dye; however, they are considered bleaching agents because they make the fabric appear whiter and brighter.

QUICK QUIZ 1

1. Singeing is needed more on fabrics made of ____________ length fibers than on those made of inherently smooth filament fibers.
2. A special benefit of singeing to the finished product is the reduction of the balling up of fiber ends, called ____________.
3. Both chlorine and hydrogen peroxide (oxygen) bleaches whiten effectively. But even a dilute chlorine bleach solution damages certain fibers, especially ____________ and ____________.
4. When wool fabrics require bleaching during finishing operations, ____________ bleach is generally used.

****

ANSWERS QUICK QUIZ 1

1. staple
2. pilling
3. wool, silk
4. hydrogen peroxide (oxygen)

Figure 5.1. Warp yarns are sized on this "slasher" by immersing the entire warp in a hot starch solution. After drying, the warp yarns are wound on a warp loom beam in preparation for weaving. *(courtesy American Textile Manufacturers Institute, Inc.)*

Many detergents available to consumers also incorporate some of these whitening chemicals in them, making fabric care somewhat easier.

Bleaching or whitening is typically included in the general preparatory finishing of almost all fabrics, even when they are to be dyed dark colors. In order to make fabric color uniform and of a consistent hue and brightness, it is important that the fabric be white at the start of the process. If color or other impurities are in the fabric when it is dyed, the final color may vary and be uneven throughout the fabric.

Figure 5.2. Batch bleaching is done as a part of general mill finishing. Fabric is stacked in bins for a time to allow the bleaching chemicals to work before goods are withdrawn from the bottom of the bin. *(courtesy American Textile Manufacturers Institute, Inc.)*

To ensure the safety of fabric during bleaching, it is essential to carefully control the concentration of the bleach. In addition, solution temperature must be monitored to make certain that it does not reach too high a level so that fabric damage occurs. To remove all the bleach solution, it is important that the fabric be thoroughly rinsed.

Calendering

After the initial preparatory finishes of singeing, desizing, scouring and bleaching, fabrics are subjected to final general mill finishing processes which serve to press and straighten them. Calendering is a special kind of pressing used in general mill processing to give fabrics a smooth appearance. Fabric is passed between hot rollers under a high level of pressure, as seen in Figure 5.3. Operation of the rollers requires careful monitoring of temperature and roller pressure to suit the requirements of different types of fibers.

Calendering is performed on practically all fabrics with smooth, flat surfaces. However, when a pile or noncompacted surface is desired, the fabric is typically steamed to smooth it instead of being pressed flat by calendering rolls.

The calendering process is not used routinely on wool fabrics, since the heat and pressure would tend to crush and compact wool fabrics. Also, high temperature and pressure embrittle wool fibers and lower their resiliency. When compact and flat wool fabrics with a high sheen are desired, such as wool gabardine or broadcloth, a modified form of calendering, called pressing, is used.

Figure 5.3. In the calendering process, fabric is passed around heated rollers, under pressure, to smooth and iron it. *(courtesy American Textile Manufacturers Institute, Inc.)*

Tentering

Tentering is a mechanical straightening and drying process used to properly align fabric grain. A tenter chain or tenter frame holds the fabric between two parallel chains, as shown in Figure 5.4.

As a part of finishing woven fabrics, grain lines are pulled into position and held by the tenter clips or special pins; the marks of the clips are often observable in the selvage. The warp and filling yarns are aligned at right angles, with the warp yarns parallel to the lengthwise grain and the filling yarns perpendicular to the selvages.

Fabric is held tautly in the properly aligned position on the tentering frame while it is drying. Fabrics may be tentered several times during finishing and coloring—following bleaching, as a part of the application of functional finishes, and during the heat setting of synthetic fabrics.

There are a number of devices, such as "electronic eye" sensors, that help set the fabric on a true grain. Controls can be programmed to stop the tenter if the grain becomes distorted.

The basic finishing equipment will vary according to the type of fabric being processed; e.g., knitted fabrics are finished on machinery designed to correctly set knits and make them of the proper density or compactness. The less stable knitted fabrics, as well as nonwovens, require finishing machinery

Figure 5.4. The tentering frame aligns fabric grain and helps maintain a uniform fabric width. *(courtesy FAB-CON Machinery Development Corp.)*

that is designed and controlled to not exert excessive pressure and tension that distorts unstable structures.

Crabbing

Wool fabrics are subjected to a different straightening process from tentering, called *crabbing*. Wool is immersed in hot water, then cold water, and then passed between carefully controlled rollers so that the yarns are set at close to right angles.

Careful control is required in the crabbing process to assure a safe temperature and roller pressure that will avoid the problem of wool fiber felting. Just as in the tentering process, woven or knitted wool fabrics must be passed onto the crabbing rollers straight, in order to preserve the fabric grain line.

QUICK QUIZ 2

1. For woven fabrics to be on grain, the tenters must hold the fabric so that the warp and filling yarns are ______________________ to fabric edges.
2. The tenters establish the desired width for the fabric and pass it along the tentering frame into the heating unit, which ______________ the fabric to its uniform width.
3. The drying oven or frame serves to dry fabrics made from all types of fibers, as well as heat set ______________ fibers into their permanent shape.

****

ANSWERS QUICK QUIZ 2

1. parallel and perpendicular
2. sets and dries
3. thermoplastic or synthetic

Inspection

This discussion of the various routine or preparatory finishes has provided a general sequence of treatments applied to most fabrics. Fabric is passed through what is referred to as the finishing line, which is carefully monitored by control centers such as shown in Figure 5.5.

The machinery used to apply finishes is marvelously designed for automatic precision work; however, continual monitoring and quality control are essential to the finishing process. Inspection, as shown in Figure 5.6, is an important aspect of fabric processing. The additional cost of monitoring both the equipment and the actual fabric yields more uniform products; thus the extra investment by the manufacturer is returned by reduced rejects, increased sales, and greater consumer satisfaction.

Figure 5.5. The complexity of machinery used in finishing, often referred to as the finishing line, requires precise controls at all stages of finishing. A portion of the finishing line is shown through the window at the left of the control panel. About a mile of fabric is needed to thread through the basic finishing operation. *(courtesy American Textile Manufacturers Institute, Inc.)*

Figure 5.6 Fabric inspection. Finished fabric is carefully inspected prior to leaving the plant. (*courtesy of the American Textile Manufacturers Institute, Inc.*)

REVIEW

1. Preparatory or general mill finishes include various cleaning processes, especially important for natural cellulosic and wool products. Impurities are removed from cellulosics by ______________________; on silk fabrics ______________________ is usually necessary.

2. What finishing process is used to make the fabric surface of staple length fibers more smooth and uniform?______________________

3. What is the special purpose of singeing fabrics made of wool fibers?
____

4. To improve color purity in later dyeing and printing, practically all fabrics are ______________ during the general preparatory finishing.

5. The general mill finishes which smooth the surface and straighten the fabric grain and width are ______________________ and ______________________.

6. Match the preparatory finishing processes to the most appropriate description:

____	1. carbonizing	a. straightening process for wool fabrics
____	2. crabbing	b. ironing or smoothing process
____	3. calendering	c. process for removing impurities from wool fabrics
____	4. bleaching	d. whitening process used to prepare fabrics for color
____	5. degumming	e. sets warp and filling yarns at close to right angles
____	6. tentering	f. cleaning process for silk

****

REVIEW ANSWERS:

1. scouring; degumming
2. singeing
3. to reduce pilling
4. bleached
5. tentering, calendering
6. 1-c; 2-a; 3-b; 4-d; 5-f; 6-e

Matrix VIII below will serve as a review for the preparatory or general mill finishes just presented, and for an overview of finishes that contribute to durability discussed in the next part.

Remove Matrix VIII from the back of the text and fill in the information about the general mill finishes listed. As you work through Part 2, make notations under the durability finishes as well.

MATRIX VIII
GENERAL MILL AND DURABILITY FINISHES

Finish	Process		Typical Fiber or Structure	Purpose
	Mechanical	Chemical		
GENERAL MILL FINISHES				
Initial:				
Singeing				
Desizing				
Scouring				
Carbonizing				
Degumming				
Bleaching				
Final:				
Calendering				
Tentering				
Crabbing				
DURABILITY FINISHES				
Abrasion Resistant				
Antislip				
Relaxation Shrinkage				
Progressive or Fiber Shrinkage				
Mercerization				
Slack Mercerization				

UNIT FIVE

Part II

Finishes Contributing to Durability and Fabric Performance

An appropriate finish can increase the durability of a fabric in a number of ways. There are finishes that increase *abrasion resistance, slip resistance, shrinkage resistance* or *dimensional stability*, and *strength*. Each of these qualities is desirable when durability is an important factor in end use. These finishing processes require the use of the padding machine or padding mangle.

The padding machine shown in Figure 5.7 applies the finishing solution by a process similar to calendering, using rollers to exert pressure on the fabric. Padding differs from calendering: the padding machine has two rubber rollers; while the calendering machine, used only to press or smooth fabric, usually has a smooth heated steel roller pressing against a rubber or other resilient backup roller. Padding machines can apply liquid additive finishes as well as dye and coloring. Excess solution squeezed out of the fabric by rollers is recycled if it has sufficient finishing or coloring components left for repeated use.

The padding machine or mangle is used primarily to apply many of the special purpose finishes discussed throughout this unit. If a fabric is to receive special purpose finishes that require finishing solutions, it will pass through the padding machine and then directly to other equipment depending on the finishing processes planned. After all finishing is completed, the fabric goes through a final calendering and/or tentering process in order to have the desired appearance for consumers.

Abrasion Resistant Finishes

Fibers with low abrasion resistance like the natural and man-made cellulosics may need a specialized chemical finish if their end use will include rubbing and friction. These resins are typically applied to cotton and other cellulosics after the general preparatory finishes of scouring and bleaching. The majority of synthetic fibers, especially nylon, have high abrasion resistance and therefore do not require abrasion resistance finishes.

Antislip Finishes

An antislip finish is used to retard the natural slippage of yarns at seams in textile products, as well as within the fabric itself. Fabrics which tend to slip at the seams are generally made of smooth, filament length fibers whose slipperyness increases the probability of runs, especially in filling knit structures. Resin applied to such fabrics lessens the slippage, which can be further reduced by texturing filament yarns to make their surfaces more cohesive. Fabrics made of natural cellulosic or protein staple-length fibers are not likely to need or be given antislip finishes because their surfaces are naturally cohesive and not likely to slip easily.

Shrinkage Control Finishes

Antislip and abrasion resistant finishes are not used as frequently as shrinkage control finishes,

Figure 5.7. In the padding machine, rollers press finishing and/or dyeing solutions into the fabric. This equipment is often called the workhorse of the finishing industry. *(courtesy National Cotton Council)*

which contribute to high dimensional stability in textile products. There are two basic kinds of shrinkage the industry attempts to control: *relaxation* shrinkage, which is due to the tension created during yarn and fabric manufacturing and processing; and fiber or *progressive shrinkage* which is due to the particular nature of the fiber itself, including its molecular structure and physical characteristics. The shrinkage that may remain in a fabric after shrinkage control processes are applied is called *residual* shrinkage.

Shrinkage control is extremely important in all textile products, especially apparel, because garments are frequently laundered and may shrink to such a degree that they are no longer usable. The term *preshrunk* on a label means nothing, unless information is given on how much more shrinkage might take place during or following laundering or dry cleaning. To be certain a product will not shrink beyond the usable point, check the label for the percentage of potential shrinkage remaining in the fabric.

If a garment shrinks three percent, it is shrinking about one size. Therefore, for a garment to be serviceable, the label should state that *residual* or remaining shrinkage is guaranteed to be less than one percent during subsequent use and care. Shrinkage control is important in fabrics made from fibers in the natural and man-made cellulosic categories; from protein fibers, especially animal hair; from some types of synthetic fibers; and from highly twisted crepe yarns. Shrinkage from laundering is not generally a serious problem in most fabrics made of synthetic, thermoplastic fibers because they can be heat set during finishing to reduce or control shrinkage. However, if excessive heat is used during drying or ironing, additional shrinkage can occur.

One unfortunate aspect of consumer information concerning shrinkage is that there is no legal requirement regarding inclusion of such information on tags or labels. If information is given regarding the shrinkage characteristics of the textile product, it is offered simply because the manufacturer believes it will help sales and encourage repeat purchases. While information regarding shrinkage is important and helpful to consumers, it is presently lacking on a large proportion of merchandise on the market.

Types of Shrinkage

During the construction of any woven fabric, the stiffened, sized warp yarns are held under considerable tension. Then moisture from laundering allows the stretched and stiffened warp yarns to return to their natural dimension, creating a smaller or distorted fabric. This shrinkage, resulting from the tensions used in fabric structuring is referred to as *relaxation shrinkage.* If the resulting fabric were to be made into a garment without the benefit of relaxation shrinkage control finishing, the consumer would encounter noticeable shrinkage during the first home laundering.

For the same reason that the loom contributes to relaxation shrinkage in woven products, the knitting machine contributes to varying yarn tension in knits. Simple mechanical finishes may be used to reduce the residual shrinkage in both wovens and knits to a relatively small amount, so that the fabric size will not be markedly altered during use and care procedures such as laundering or dry cleaning.

Compared to woven structures, knits present more of a shrinkage problem because the interconnecting loops of knitted fabrics are less dimensionally stable than the interlaced yarns of woven fabrics. As a result of warp yarn relaxation after the tension of weaving is removed, a certain amount of shrinkage can be expected to occur in woven fabrics when they are exposed to moisture for the first time. The shrinkage control finishes eliminate or reduce relaxation shrinkage to a minimum before the fabric goes to the consumer.

Another problem is *fiber shrinkage* caused by changes in the molecular or physical structure of a fiber. It is sometimes called *progressive shrinkage* since, as a result of fiber felting or swelling, it can continue to occur in subsequent launderings or dry cleanings.

The major differences between the two types of shrinkage are these: relaxation shrinkage is the mechanical readjustment of tension in the fabric yarns created during fabric and yarn construction. Fiber shrinkage results from changes, such as wool felting, in the fiber itself.

The propensity of wool to felt and consequently shrink is used to advantage in finishing woven wool fabrics. They are passed through a hot solution which causes the fibers to felt slightly, thus filling in the spaces between the woven yarns. The name given to this finishing treatment is *fulling,* a finish routinely used to produce wool fabrics that appear more closely and firmly woven than nonfulled woolens.

Shrinkage Control Processes

A mechanical finishing treatment known as the *compressional shrinkage process* compensates for relaxation shrinkage in fabrics made of natural and manmade cellulosic fibers. Sanforized and Rigmel are two tradenames for the compressional shrinkage process which involves compressing fabric an amount equal to the potential shrinkage which is determined before the finishing process is begun. As yarns are shrunk in the compressional shrinkage process they are pressed more closely together, thereby increasing the thread count and strength of the fabric.

Stone, acid and garment washing are processes that also alleviate relaxation shrinkage. These processes involve washing of completed garments, sometimes with pumice stones or acids, to create a soft hand and weathered appearance.

Compressional shrinkage finishing techniques do not control fiber shrinkage in fibers which have a propensity to felt or swell in the presence of moisture. *Fiber swelling* is caused by the introduction of moisture into certain fibers which causes them to increase in thickness and decrease in length. Cotton, and especially rayon, exhibit swelling shrinkage. For example, a pair of cotton denim jeans may be extremely tight following laundering, but wearing them for a short time causes the fibers to stretch back to their original dimension. Thus, swelling shrinkage may be somewhat reversible during use.

Rayon swells more than cotton and the introduction of moisture into the rayon fiber creates an unstable situation. The fiber may swell and shorten measurably, or, if any tension is put on rayon when it is wet or damp, the swollen fibers can be stretched considerably. These characteristics create size changes or dimensional instability that is frequently observable on humid days in draperies made of rayon warp yarns. This problem has been called the *elevator property* of rayon.

Newer rayons, such as the *high-wet-modulus* (HWM) types, have less inherent fiber or swelling shrinkage. They can be stabilized just like cotton by the compressional shrinkage process which controls relaxation. Resins or cross linking chemicals may be used to stabilize regular rayons as well as other cellulosic fibers. The resins, which are designed to prevent moisture from entering the molecular structure, are impregnated into fibers during finishing to control fiber or progressive shrinkage.

The general swelling of rayon tends to occur progressively when moisture is introduced into the fiber if an appropriate resin finish has not been applied. This swelling shrinkage is a greater problem than the elevator property of rayon, which typically occurs only when the rayon fibers are subjected to tension and humidity. Recall that *residual shrinkage* is defined as the amount of shrinkage remaining in a fabric after shrinkage control treatments.

Chemical chlorination (halogenation) treatments can make wool shrink resistant. While chlorine chemicals modify the scale structure of wool in order to prevent felting, the process reduces fiber strength. To compensate for this strength loss, synthetic fibers known for their high tenacity, such as nylon, are often blended with the wool.

The chlorine treatments are being replaced by more successful wool shrinkage controls. One type makes use of *resins or cross linking chemicals* that can react with the fiber internally or coat the wool fiber surface in order to prevent felting shrinkage.

Mercerization

Mercerization, combining both chemical and mechnical finishing processes, is a treatment applied primarily to cotton fabrics to *increase fiber strength, luster* and *dye affinity*. The fabric or yarn is immersed in an alkaline solution of caustic soda (NaOH) and held under tension. While under tension in this solution, the fibers' bean-shaped physical structure swells and changes to a rounded, cross sectional shape. After cotton fabric is passed through the NaOH alkaline solution, it must be thoroughly rinsed with water and then neutralized by rinsing in a solution which is mildly acidic. See Figure 5.8.

The reason for keeping cotton fibers under tension during mercerization is to increase the molecular orientation within each fiber. If the stresses are evenly distributed within the molecular chains, the chain is less likely to break. Therefore, the mercerization process contributes to fabric durability by increasing tenacity. It also contributes to a smoother surface on the cotton fiber, and hence more luster, by removing or straightening the convolutions or twists along the fiber length.

QUICK QUIZ 3

1. Sanforized® on the fabric label means that a fabric is guaranteed to shrink less than one percent, which is a good indicator that the fabric will be ____________________ to laundering.
2. A fabric that has been through a compressional shrinkage process will have most of the ____________________ shrinkage removed.
3. Name the two broad categories of shrinkage that should be controlled during fabric finishing, and their causes:
   a) ____________________
   ____
   b) ____________________
   ____
4. Name two types of progressive shrinkage and the fibers involved:
   a) ____________________
   ____
   b) ____________________
   ____

****

ANSWERS QUICK QUIZ 3

1. dimensionally stable
2. relaxation
3. relaxation shrinkage—caused by tensions introduced during yarn or fabric construction; progressive shrinkage—caused by the molecular and physical nature of the fiber
4. felting shrinkage—wool and other animal hair fibers;
   swelling shrinkage—rayon especially, and other cellulosic fibers to some degree

---

Figure 5.8. Fabric mercerization is a part of routine general mill finishing, and a majority of cotton fabrics go through this phase of the finishing line. *(courtesy American Textile Manufacturers Institute, Inc.)*

Slack Mercerization

In slack mercerization, the yarns or fabric remain slack or relaxed during mercerization; this adds comfort stretch to the product. Since the fabric is not under tension, the fibers and yarns shrink in the neutralizing caustic soda solution. The shrunken yarns are easily elongated if stretched, which accounts for the increased stretchability of slack mercerized fibers and yarns. However, this comfort stretch is achieved at the expense of the properties found in mercerized cotton, such as increased tenacity and luster.

REVIEW

1. When proper finishing techniques allow for the adjustment of yarn tensions which cause ____________________ shrinkage, residual shrinkage can be held to within ____________ percent.

2. Fabrics made from fibers with a tendency to swell exhibit ____________________ shrinkage, unless proper care procedures are followed.

3. Name two fibers which are especially subject to fiber or progressive shrinkage, even after relaxation shrinkage has been accounted for: ____________ and ____________.

4. Match the following terms to the types of shrinkage they indicate:

____	1. relaxation	a. process to control relaxation shrinkage
____	2. progressive	b. shrinkage as a result of fiber characteristics
____	3. residual	c. shrinkage as a result of fabric structuring
____	4. compressional	d. guaranteed less than one percent shrinkage
____	5. Sanforized®	e. shrinkage remaining after shrinkage control finishes have been applied
		f. control of progressive shrinkage

Locate Matrix VIII, which includes these durability finishes, and see that it is appropriately completed.

****

REVIEW ANSWERS:

1. relaxation; one or two
2. fiber or progressive
3. wool, rayon
4. 1-c; 2-b; 3-e; 4-a; 5-d

UNIT FIVE

Part III

Finishes Contributing to Comfort of Textile Products

To be comfortable, a fabric must be compatible with an individual's skin. Singeing and calendering are two already mentioned general mill finishes which contribute to comfort by making the fabric surface smooth. However, fabrics are often subjected to more specialized finishing processes when an end use requires special comfort features.

In particular, fabrics made from synthetic fibers may have finishes added to increase comfort by *increasing absorbency*. Also, finishes to *reduce static electricity* are important in synthetic fabrics, since they tend to develop static charges which attract soil to the fabric surface and generate static shocks. They also cause fabrics to cling to the body in an unattractive and uncomfortable way. Specialized finishes added for warm weather comfort frequently involve the addition of some type of chemical to the fiber molecule. Other finishes contribute to cold weather comfort by increasing fabric thermal retention. These *thermal finishes* include chemical and mechanical processes.

The finishes included in this part are listed on Matrix IX. You will need to remove this study matrix from the back of the book and use it as you work through the next section.

Absorbent Finishes

Efforts to increase fabric absorbency by finishing have not been entirely satisfactory. One method is to apply chemicals which cause water coming in contact with the fabric surface to break up into small droplets or vapor. The moisture is dispersed over the fabric and is, therefore, absorbed or evaporated readily; this contributes to comfort by allowing for rapid evaporation of moisture away from the body, eliminating a clammy, damp fabric next to the skin. Since the smaller droplets are evaporated quickly, this finish also contributes to faster drying time.

However, by speeding up the drying time of synthetic fabrics, the problem of dampness is traded for the problem of static electricity associated with these fibers in dry, cool weather. So another method for absorbent finishes has been developed that holds in water molecules to increase drying time instead of breaking them up on the surface and allowing for fast moisture evaporation. This process also keeps moisture away from the body, since the water molecules are held internally rather than passing to the fabric surface.

Regardless of the principles behind absorbent finishes, they have *not proved effective nor durable* and *few are applied to fabrics today*. To ensure adequate absorbency for comfort, the best choice is a fabric made either of natural cellulosic or man-made cellulosic fibers; or, if a product is made of nonabsorbent fibers, select an open structure which allows for air and moisture transfer.

MATRIX IX
FINISHES FOR COMFORT AND CARE

Finish	Process		Typical Fiber or Structure	Purpose
	Mechanical	Chemical		
COMFORT FINISHES				
Absorbent				
Antistatic				
Thermal				
CARE FINISHES				
Stain and Soil Resistant				
Durable Press				
Biological Resistant				
Water Repellent and Waterproof				
Flame Retardant				

Antistatic Finishes

While many antistatic finishes are not permanent, they are more effective and used more frequently than absorbent finishes. Three topical finishing methods are available to help control static electricity; the most effective antistatic finishes use a combination of methods. An antistatic finish may attract moisture to the fabric surface to aid in electrical conductivity; or it may ground electrons, thereby improving electrical conductivity of fibers. Static electricity may also be prevented with a special chemical application which develops an electrical charge opposite to that of the fiber; this neutralizes static charges. Since fibers have varying electrical charges, different antistatic finishes are required.

In addition to these so-called *topical* finishes which are applied to the fiber or fabric surface, much *fiber modification antistatic research* has been conducted on the static-producing synthetic fibers, especially nylon and polyester. They can now have special properties engineered into them to decrease static electricity. In fact, fiber modifications designed to curb static charges are reported to be among the more successful means of decreasing consumer problems associated with static electricity.

Antistatic finishes applied by the converter must frequently be renewed by the consumer to retain their effectiveness. One way to do this is through the addition of *fabric softeners* in the final rinse of a laundry cycle or in automatic dryers to help control static charges. Softeners help ground electrons, thereby contributing both increased electrical conductivity and a lofty, springy hand to fabrics. However, since fabric softeners actually coat fabric yarns and fibers, fabric absorbency tends to decrease with their continued and routine use.

While the finishing industry continues to improve the serviceability of antistatic finishes for synthetics, little mention has been made of protein fibers in regard to antistatic finishes. Yet many have experienced shocks from wool on a cool, dry day. Wool does build up static charges, but not to the

---

QUICK QUIZ 4

1. Name at least two problems created by static charges:
   ____
   ____
2. An antistatic finish that attracts moisture to the surface may reduce static electricity; but it increases drying time and also attracts ______________ stains, which can increase problems related to fabric care.
3. While the accumulation of static charges does not attract moisture, static does attract ______________ particles.

****

ANSWERS QUICK QUIZ 4

1. shocks to the wearer or user; fabric clinging to the body; soil attracted to the surface
2. water borne
3. dirt or soil

---

degree of some synthetic fibers because it is internally absorbent through its hygroscopic characteristics.

With the development of improved antistatic finishes, perhaps the problem of low synthetic fabric absorbency will not be as great, since one method of reducing static charges is to increase the amount of moisture transmited through a fabric. This characteristic is especially important for converters to consider, due to the increased production of synthetic fibers which are not inherently absorvent or comfortable.

Thermal Finishes

Certain finishes, like metallic or plastic coatings, contribute to cold weather comfort by increasing the thermal retention of a fabric. A thin coating of aluminum on the back of a fabric reflects heat and acts as an insulator by preventing air transfer. Aluminum thermal finishes can be recognized by their characteristic gray color. *Milium* is a tradename for one such finish, originally intended for use on coat lining fabrics. However, it is now only occasionally used for this purpose, because the finish has not proven as effective in keeping the body warm as a fabric structure that traps air. Instead, reflective finishes have proven more useful on drapery fabrics where a metallic or plastic coating helps maintain a constant room temperature by reflecting sunlight in summer and retaining heat in winter.

Vinyl or plastic coatings also are used successfully to decrease heat loss by reducing air circulation. They have the added advantage of reducing the amount of soil that can penetrate draperies. Finishes that improve the thermal qualities of drapery fabrics will no doubt increase in demand due to the high cost of energy for heating and cooling.

Napping is a mechanical finish in which fine fiber ends are raised, creating a fabric with increased potential for thermal retention. The raised fiber ends can trap air, which is a good body insulator. Examine fabric 28, a napped tricot, and feel the texture and loft of the fabric surface. In fact, napping is often done as much for esthetic reasons to yield a soft, lofty product as for increased thermal retention. Thus, the process is discussed more fully later, under Finishes Which Enhance Appearance.

UNIT FIVE

Part IV

Finishes Contributing to Fabric Care

Textile products must have proper care to maintain their fresh, new appearance and remain serviceable. Several finishing treatments may be applied to fabrics to make care procedures easier and more effective. These include finishes for *soil and stain resistance, durable press, biological resistance* and *water repellent or water proof.* Fill in Matrix IX on these care-related finishes as you work through this section.

Stain and Soil Resistant Finishes

Soil and stain resistant finishes are designed to simplify care by preventing or retarding the accumulation and penetration of soil and stains into the fabric. Three main mechanisms are used:

1. fluorochemicals or silicones that reduce surface tension of fibers and thereby attract less soil;
2. acrylic and other resins that completely coat surface fibers and form a barrier to stains; and
3. absorbent finishes (polyacrylic acids) that increase the ease of wetting fibers so that water and surfactants can lift off stains and soil.

In Figure 5.9, compare the side that has a stain resistant finish to the unfinished side that absorbed the liquid. Stain resistant fluorochemical finishes like Scotchgard® and Zepel® serve to make fabrics resistant to both oil and water borne stains. A major disadvantage of such finishes is that their effectiveness decreases with laundering. But now, Scotchgard® and Zepel® type finishes are available directly to consumers, enabling them to refurbish stain resistance by spray applications. Such consumer-applied finishes tend to be fairly temporary and not completely effective because commercially applied fluorochemical finishes are cured and better fixed into the fibers.

Stain and soil release finishes have become increasingly important in home furnishing products. Often soil resistant finishes are applied to upholstered furniture to protect fabrics from staining. A fluorochemical finish is the customary type of finish for upholstery; however, the soil release technique used on sample 36A is accomplished by increasing absorbency. In this case, the fabric absorbs moisture to allow cleaning agents to act on stains. The manufacturer of 36A has selected this type of soil release approach because other types of finishes on upholstery are often lost through the heavy abrasion that upholstery fabrics experience.

Durable Press

Perhaps the biggest revolution in the textile industry was the advent of finishes which imparted true no-iron wrinkle resistance to fabrics containing cellulose. Although wash and wear finishes were originally considered big news, they really revolutionized fabric care when they were improved on and evolved into what are now called durable or permanent press finishes.

The chemical resins applied to impart durable press qualities react only with cellulose. Even when a blend

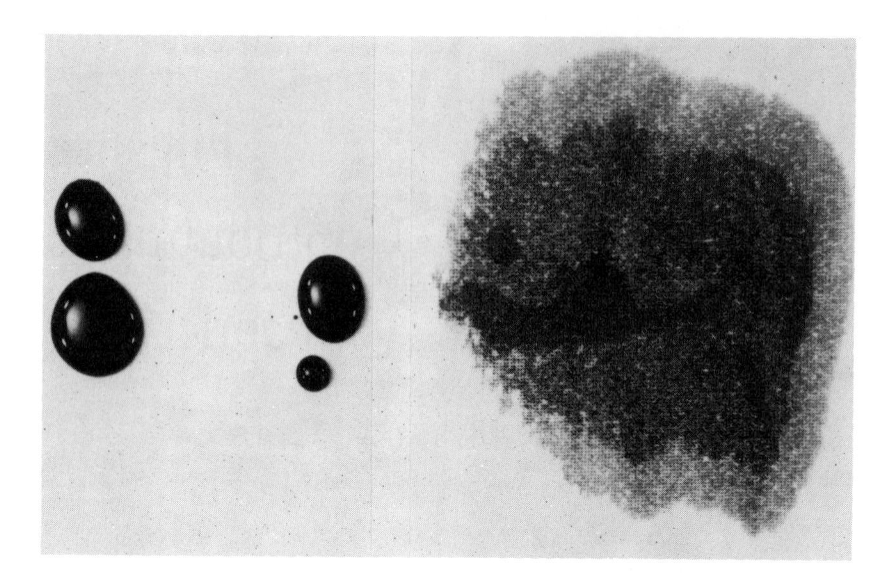

Figure 5.9. Liquids tend to "bead up" on fabrics to which fluorochemical finishes have been applied.

of cotton and polyester is involved, only the cotton accepts the finishing chemicals. These resins were developed specifically to cross link cellulosic molecules to improve the very low resiliency of fabrics made from cellulosic fibers.

Crosslinking between adjacent, cellulosic molecular chains holds the chains together and limits molecular slippage, which aids in wrinkle recovery. The first durable press fabrics were 100 percent cotton. However, the cross linking resins decreased fabric durability to the extent that cotton durable press products split in creased or folded areas after a few launderings and wearings. To deal with this problem, fibers with high tenacity and abrasion

---

QUICK QUIZ 5

1. Samples 11, 14 and 18 have all been treated with Scotchgard®, a ________________ finish that makes fabrics resistant to both ________________ and ________________ borne stains.
2. Drop some coffee, food coloring or other dark fluid on each of the above samples. Describe what happens: ____
3. If stains are allowed to remain on fabric for long periods of time, some soil goes into the fabric despite stain resistant finishes. Therefore, the consumer should remove soil, especially liquid stains ________________________________

****

ANSWERS QUICK QUIZ 5

1. fluorochemical; oil, water
2. The liquid beads up on the fabric and is not absorbed.
3. as promptly as possible

---

resistance were blended with cotton or other cellulosic fibers. This is why many durable press fabrics today are blends of cotton with synthetics, particularly polyester and nylon. Samples 2, 4 and 6 are polyester/cotton blends with a durable press finish. Crease each of them and observe the wrinkle recovery.

Curing Processes

Fabrics can be cured in the flat form—called the precure, or after manufacturing into a textile product—called postcure. In the *precure process,* flat fabric is treated with appropriate crosslinking resins and cured flat by the converter. The cured yard goods are then shipped to apparel manufacturers to be made into garments. To give the garments superior durable press, these garments may be reheated to set the shape. The heat of the presses must be higher than the precure temperature used on the flat fabric in order to set the permanent shape of the garment.

In the *postcure process,* the crosslinking resins are applied in the same way as for precure processing, then the fabric is dried. But the postcure process postpones the final stages of curing until after the fabric has been made into a garment or other end product. Only then is the finish cured and heated to the point where creases and seams are permanently set in as desired.

Despite the many easy care advantages of durable press products, it is almost impossible to alter garments made from these fabrics once their shapes are set. The heat used in curing is higher than available on home irons, so hems and creases cannot be changed and re-set by the consumer.

The durable press yard goods sold at fabric stores are of necessity cured flat, which causes a problem. There is no way for the consumer to set or recure the fabric into its permanent shape; thus seams, creases and hems cannot be pressed in to be flat and sharp.

The *vapor phase* method of durable press curing was developed to provide a better way of finishing 100 percent cotton products, since earlier durable press resins markedly decreased fabric strength and abrasion resistance. Sample 11 is a 100 percent cotton given a vapor phase durable press finish; crease it. In this process, garments treated with nonresin chemicals are put into a vapor chamber and exposed to formaldehyde and sulfur dioxide gas. This treatment also causes crosslinks to form in the cellulosic fibers which contribute to increased wrinkle resistance.

Another durable press development is based on the use of fabric blends of high polyester and low cotton content, such as in sample 16 of 75 percent polyester and 25 percent cotton. No resin chemicals are used in such fabrics. Instead, the durable press garment shape is achieved by heat setting the fabric on specialized equipment called a *hot head press.* The process requires no curing oven, since the high polyester fiber content, rather than durable press chemicals, is responsible for the durable shape and wrinkle resistance of the product. Modified polyester types have been developed especially for use in these kinds of durable press products, and the consumer can expect greater durability than with durable press fabrics made from higher percentages of cellulose or cotton fibers.

Consumer Acceptance and Use

Currently, durable press processes have their limitations: they still cause strength losses within cellulose; further, in cotton/polyester blends, a characteristic called *frosting* may occur when the cellulosic portion of the fabric abrades before the polyester component, giving the fabric a faded appearance. Despite these problems, consumers buy durable press products because they are apparently willing to trade off durability for ease of care.

When cotton prices are relatively low compared to those of synthetics, durable press fabrics tend to have higher proportions of cotton in the blend. These ''mostly cotton'' fabrics have good durable press properties, but not quite as high as those with greater percentages of polyester fiber. However, consumers react favorably to these blends with more cotton because the fabrics are more absorbent and comfortable.

Decreased strength and color loss through the differential abrasion—frosting—of cotton and polyester fibers are not the only problems with durable press fabrics. Stiffness of fabric hand is sometimes observable, and certain resins may impart a ''fishy'' odor when fabrics are exposed to moisture. Unpleasant odors also result when improper techniques are used in heat setting and curing durable press products.

The industry's constant efforts to improve durable press finishing techniques and overcome problems led to the development of *soil release* finishes. They were specifically developed to help curb the problem of soil affinity on blended durable press fabrics, especially those in which polyester fibers were used. Yet soil release finishes have declined in use, partly because of high cost and lack of permanency. Further, according to apparel manufacturers, many items made with soil release finishes, such as men's durable press dress shirts, really do not need such a finish. No research seems to have been done, however, to determine if consumers also believe that soil release finishes are not beneficial.

Technical problems aside, another factor influences the demand for durable press fabrics: fashion. When there is a movement away from woven blends to knits, the need for durable press finishes decreases. This is because knit fibers, do not always require built-in durable press finishes for wrinkle resistance. However, when fashion changes dictate an increased emphasis on woven fabrics, consumers will again benefit from the wide use of durable press finishes on fabrics that have low wrinkle resistance.

The mussed or wrinkled look popular with young people during the past few years has increased demand for products of all cotton without durable press finishes. They may not look very neat, the cotton is strong and durable; its fibers are not weakened by the durable press finishing process. Plus, unfinished cotton products are absorbent and comfortable.

Biological Resistant Finishes

Biological resistant—frequently called *bacteriostatic*—finishes are used for fabrics especially sensitive to biological attack, to retard destruction by *insect pests* and such *microorganisms* as *mildew* and *mold.*

The application of a chemical agent to create a mothproof finish works in the following ways: 1) it can give off an odor which repels moths or carpet beetles; 2) it can give off a gas, such as in moth balls, which is toxic to insects or their larvae; or 3) it can chemically alter the fiber structure so that the fiber does not attract insects. The first two methods are effective in controlling against insect damage, particularly *moths* and *carpet beetles,* by destroying the insect the larval stage when most damage is done.

Due to the casualness of many consumers' care practices, millions of dollars' worth of protein fabrics are unnecessarily lost each year. Simple procedures involving storage of *clean garments in the presence of moth preventatives* usually provide adequate protection. However, to be effective, any mothproofing process used at home should be renewed each season.

In addition to insect repellents, *antiseptic* finishes are useful to retard damage from mildew and mold microorganisms. Such finishes are especially valuable in hospitals to reduce the chance of transmitting organisms that result in illness or skin infections. However, antiseptic finishes tend to lose their effectiveness after laundering and so need to

---

QUICK QUIZ 6

1. Which finish has had the greatest impact on consumer care procedures? ____________
2. Improvements *in the curing processes* for durable press products have increased consumer serviceability. However there are still certain limitations to durable press fabrics. Name two:
   ____
   ____

****

ANSWERS QUICK QUIZ 6

1. durable press
2. any two:
   decreased strength of cellulose; potential color loss from frosting; possible odors; sometimes stiff hand or body

---

be frequently renewed. While mothproofing finishes are not required on nonprotein fibers that do not provide a source of food to insects, antiseptic or bacteriostatic finishes are useful on most fibers. These finishes limit damage from mildew and fungus and reduce the occurrence of infectious bacteria that can be spread by contact.

Treating socks before each wearing with a renewable antiseptic finish is helpful in preventing the spread of athlete's foot, a common fungus problem. Renewable antiseptic finishes applied to T shirts, shoes, or other garments used for high activity are also useful in preventing odors that result from sweating.

Mothproof and antiseptic or bacteriostatic finishes may be combined with other finishes, such as durable press or water repellent finishes. These multipurpose finishing applications provide fabrics with increased consumer serviceability for use under a wide variety of conditions.

Water Repellent and Waterproof Finishes

Water repellent fabrics shed moisture in small amounts, but may be soaked by heavy rains. A *waterproof* finish, on the other hand, completely coats or seals a fabric surface so that no water passes through. Fabric structures such as *bonded vinyls,* (sample 36), serve for all practical purposes as a waterproof finish.

Water repellent finishes may be used on apparel fabrics, home furnishing textiles and on certain industrial products. Although a waterproof finish might seem superior to a water repellent one for rainwear, it does not allow air transmission and tends to be uncomfortable in both hot and cold weather. Thus water repellent finished fabrics are more comfortable for apparel fabrics because they permit the transfer of air through the fabric. The presence of soil, however, causes water droplets to accumulate rather than run off; therefore a soiled garment would be less water repellent.

Urethane coatings provide not only water repellency, but also serve to stabilize a fabric. Sample 27A is a nylon warp knit engineered for use in soft-sided luggage. The nylon fiber provides the strength and abrasion resistance needed in this end use, while the urethane coating provides water repellency and additional stability to the knit structure. During use, luggage is often subjected to stress, abrasion and the weather. The coated nylon knit allows for some flexibility, yet prevents tears, torn seams and water damage.

QUICK QUIZ 7

1. What two major storage precautions are necessary to maintain animal hair products?
____
____
2. Why are antiseptic finishes applied to fabrics? _______________
____
3. When should a consumer choose a waterproof raincoat? _______________
____

   What is the potential disadvantage to the wearer? _______________
____

****

ANSWERS QUICK QUIZ 7

1. store products clean, and with moth preventatives
2. to retard damage from microorganisms such as mildew, mold and bacteria
3. when total protection from rain is desired; decreased comfort due to no air transfer

---

REVIEW

1. In addition to factory-applied textile finishes, there are products designed for home use to improve fabric serviceability. During home laundering __________ are frequently added to whiten fabrics and ____________________ are added to increase fabric loft and decrease static electricity.

2. The consumer can increase the soil and stain resistance of a fabric by spraying a ______________ finish available on the retail market.

3. Why are consumers willing to accept products with durable press finishes despite their limitations? ____

4. Consumers can have mothproof or water repellency finishes renewed during dry cleaning. However, these applied finishes are not ____________________ and are best restored ____________________ to maintain their effectiveness.

****

REVIEW ANSWERS:

1. bleaches; fabric softeners
2. fluorochemical (such as Scotchgard®)
3. for ease of care and wrinkle resistance
4. durable or permanent; at regular intervals

UNIT FIVE

Part V

Finishes Contributing to Appearance of Textile Products

In earlier parts of this unit, two basic classifications of finishes were identified according to how the treatment is applied. Recall that these processes include applications by *chemical* or *mechanical* means, or a combination of both. While esthetic finishes can be categorized under many headings according to the effect produced, mechanical and chemical finishes will be the categories used herein. Study Matrix X to get an overview of finishes included.

Mechanical Finishes Which Enhance Appearance

Shearing and Sculpturing

A mechanical finish known as *shearing* is used on pile fabrics to cut or trim uneven yarns from the pile surface. The amount of shearing varies, depending on the depth of pile desired.

A *sculptured* effect can be achieved by altering the depth of the fabric pile in the shearing process. Such sculptured patterns, often found on carpets, yield a dimensional effect similar to the raised brocade designs created by the jacquard weave pattern.

Sizing

The sizing which serves to stiffen *warp yarn* in preparation for weaving is removed by the preparatory processes of desizing and scouring. However, a sizing finish is sometimes applied to the fabric to give it additional body and stiffness, to add luster, improve abrasion resistance and add smoothness.

Durable resin sizings can be very satisfactory and advantageous when used to stiffen sheer, lightweight fabrics such as curtain fabrics. However, a variety of temporary, starchlike sizings are applied by converters to make less expensive fabrics with low yarn counts appear more firm and desirable; children's Halloween costumes are one example. Laundering soon removes such temporary sizing, leaving a limp fabric that has little body, stiffness or firmness.

To check for sizing durability when buying an inexpensive, low yarn count fabric that seems to have a firm body, gently rub the surface with a fingernail to see if any starchlike substance flakes off. If it does, assume that the fabric has a temporary sizing.

Weighting

Weighting is a special type of sizing applied to silk fabrics which have been degummed in general mill processing. The gummy substance that binds the silk filaments gives the fiber body and acts as a kind of natural sizing. Thus, when the gum is removed, silk is somwhat limp. To reduce this limpness, metallic salts are added to give increased weight and body, or firmness.

MATRIX X
FINISHES FOR ESTHETIC APPEARANCE

FINISH	TYPICAL FIBER OR STRUCTURE	PURPOSE
MECHANICAL FINISHES		
Shearing and Sculpturing		
Sizing		
Weighting		
Brushing		
Napping and Sueding		
Fulling		
Beetling		
Embossing		
Glazing		
Ciréing		
Moiréing		
Schreinerizing		
Flocking		
CHEMICAL FINISHES		
Fabric Softeners		
Acid, as Parchmentizing		
Basic, as Chemical Crepeing		

However, too much weighting has some disadvantages—salts put additional stress on silk filaments, resulting in decreased fabric durability because of a loss of strength and flexibility. Consumers can be misled into thinking that a silk fabric is of much higher quality and firmness if it has excessive weighting from metallic salts added during processing.

To protect the consumer, the Federal Trade Commission passed the Pure Dye Silk Ruling which states that a silk fabric cannot be labeled as *Pure Dye Silk* unless it is made of 10 percent silk with not more than 100 percent weighting. Black silk is allowed 15 percent weighting, since it tends to be more limp than silk dyed other colors. Silk which has more than the prescribed weight of metallic salts must be labeled *weighted silk.*

When manufacturers return some of the degummed material to silk fabric as a part of final finishing procedures, caution is required. This method of stiffening silk tends to increase the likelihood of waterspotting.

QUICK QUIZ 8

1. Name two mechanical finishes used to create esthetic affects on pile fabrics:
 ______________________ and ______________________.
2. Sizing finishes are applied to increase which fabric properties? ______________________
 ____
 In order to be relatively durable, ______________________ sizing should be used.
3. Sample 3 is a high quality 100 percent silk shantung that contains less than 10 percent weighting. Thus, the fabric can be correctly labeled as ______________________.

****

ANSWERS QUICK QUIZ 8

1. shearing, sculpturing
2. body or stiffness, luster, abrasion resistance, smoothness; resin
3. Pure Dye Silk

Brushing

Because pile fabrics are somewhat flattened by various mechanical and chemical finishing treatments, they may be brushed to remove loose fiber ends and to lift the pile. Brushing is also used to raise fiber ends before singeing. This latter use is common when fabrics are made from yarns that tend to be somewhat rough, with considerable fiber ends on the surface.

Sample 12, 13 and 14 are pile weave fabrics which were brushed to raise the fiber ends during finishing. Observe their smooth-finished surfaces.

Napping and Sueding

Napping and sueding finishes are done when a soft, fuzzy surface is desired. Napping is done to fabrics constructed of loosely twisted staple yarns. Figure 5.10 contains a diagram of napping rolls which have short metal wires that raise fiber ends. The sueding process uses rollers covered with sandpaper to raise fiber ends on the surface of the fabric.

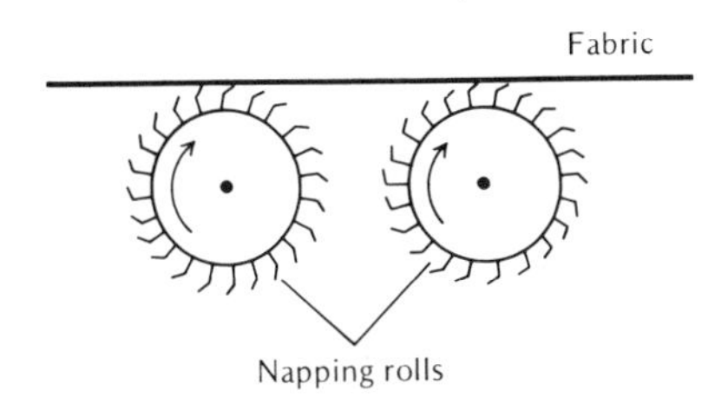

Figure 5.10. Fiber ends are mechanically raised by wirelike napping rollers.

The nap of a fabric should not be confused with a napped surface finish. When a fabric has a definite direction—as with a pile surface, a napped or sueded finish, or a one direction pattern or design—the fabric is said to have a nap. Construction of products from fabrics with such a nap design requires consideration of the direction of pile and the nap finish or fabric design. Any fabric that has a design or finish which requires that all pieces of the item be cut in the same direction has a nap.

Brushing operations also tend to raise a soft nap which is retained on the fabric surface. For example, the denim in sample 8 has a slight nap created by brushing. Manufacturers sometimes prefer brushing to create a nap because it is less likely to weaken yarns.

Teasels (which are natural plant thistles) can be used for creating a soft, fuzzy surface also. Teasels produce a finer, more lustrous nap than napping rollers and their less rigorous action preserves fabric durability characteristics. This process, called *gigging,* is often done on wool fabrics.

A satin fabric with a napped back is shown in Figure 5.11. Satin structures are generally made of filament length fibers; however, to achieve the napped back on the fabric, the filling yarns were made from staple length fibers to allow the fiber ends to be raised. The filament warp yarns which float on the front of the fabric in the illustration impart the characteristic luster or sheen of satin. As a lining fabric, the smooth satin surface would make it easier to slip into a coat. The napped back would add thermal retention to the coat but tend to have less fabric durability because there might be yarn

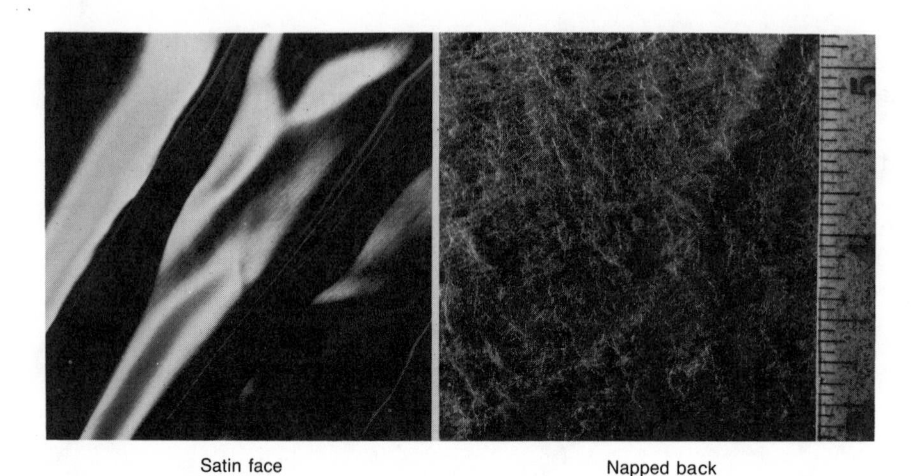

Figure 5.11. SUNBAK® is the trade name of this satin-faced fabric, which was napped on one side.

breakage since napping rolls break, abrade and thus weaken fibers.

A greater nap can be raised on a twill weave than on many plain weaves owing to the longer yarn floats typical of twills, which expose more yarn surface area to the napping rolls. Observe a twill fabric before and after napping in Figure 5.12.

Sueding finishes create fabrics that have a natural, suede look while allowing fabrics to possess the ease of care and durability of manufactured fibers. Samples 25A, 28A and 36A are fabrics that have been sueded. Sample 28A, an apparel fabric, has been manufactured to look like leather through printing techniques and sueding. The consumer thus gets the look of leather with the ease of care, durability and drape of a polyester warp knit. Similarly, sample 36A allows the furniture customer to select an upholstery fabric that is soft and suede-like, yet has the durability of a multicomponent polyester fabric. In both apparel and upholstery, natural suede could be used, but the cost would be much higher and the care procedures more difficult. Sample 25A has been sueded on the wrong side of the fabric to increase comfort during wear.

Fulling

Fulling is a special finish applied to fabric made from wool and other animal hair fibers. It is used to make the resulting fabric appear more closely and firmly structured. The spaces between the yarns are filled in by causing the wool fibers to shrink up and felt together. The process must be carefully controlled to prevent too much felting, resulting in a stiff harsh fabric.

Beetling

Beetling is a mechanical finish specifically used to flatten yarns, notably those made of flax. It is a kind of super-calendering—the fabric is pounded by wooden blocks or metal hammers as it slowly moves over a revolving drum. The finished fabric has a flat, smooth, and lustrous appearance typical of fine, linen damask tablecloths. Cotton is sometimes beetled to provide a product resembling linen damask at lower cost.

Embossing

Additional modifications of the basic calendering process can be achieved by using different types of heated rollers. *Embossed* rollers produce three dimensional designs, ranging from simple designs to those which resemble complex jacquard woven, brocade fabrics. Examples of embossing rollers are shown in Figure 5.13.

To prepare the rollers, a design is stamped onto a smaller metal roller; the smooth large roller, made of a pliable material such as paper or cotton, is moistened so that it is soft. When the metal roll is pressed hard against the large one, it imprints the design onto it. The rollers are run until the larger roll is dry and permanently imprinted. When fabric is passed between the prepared embossing rollers, the raised and lowered design is pressed into it. The design can be heat set for thermoplastic fibers by heating one or both rollers and a resin can be incorporated into the finishing operation for permanence on nonthermoplastic fibers.

Figure 5.12. Notice the different surface appearance of this wool fabric before and after napping.

QUICK QUIZ 9

1. Fabrics which are napped tend to be structured from (staple, filament) yarns which have relatively (high, low) twist.
2. Fulling is used to improve the appearance of ______________________ fibers and beetling is used to flatten ______________________ fibers.
3. Sample 36 is a supported vinyl with the top layer made to look leather-like by ______________________ . Since the vinyl is thermoplastic, the finish can be ______________________ .

****

ANSWERS QUICK QUIZ 9

1. staple; low
2. wool or animal hair; flax
3. embossing; heat set

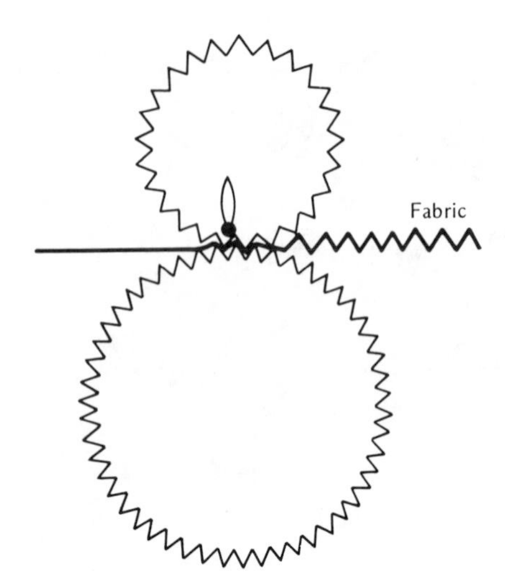

Figure 5.13. In embossing, a raised and lowered design is imparted to a fabric after it is passed between stamped embossing rollers.

Glazing

When chemical resins are used in the calendering process, the very smooth surface typical of polished cotton and chintz results. *Glazing* is the name of this process, which, like most modified calendering finishes, is done at the same time as regular calendering.

For glazing, an extra roller which spins at a faster rate is added to the regular calendaring rolls. The friction created by this fast-spinning roller polishes the fabric to create a glazed surface, which can be made more permanent by adding chemical resins.

Ciréing

Ciré, from the French word meaning ''waxed,'' is a finish which produces a high polish similar to patent leather. A ciréd surface is achieved by impregnating the fabric with wax or some thermoplastic substance, and then passing it through a friction calender. The combination of calendering and wax, or a similar chemical, imparts the ''wet'' look periodically popular in certain apparel.

Moiréing

Moiré, from the French word meaning ''to water,'' is a finishing treatment which creates a fabric that reflects light in a rippling manner resembling the effect of waves on water. The true moiré effect is imparted by pressing together two ribbed fabrics so that the ribs of each impress patterns into the other. The resulting slight difference in the light reflectance angle causes the characteristic wavy or rippling designs associated with these fabrics, which are sometimes called ''watered'' designs.

Schreinerizing

Schreinerizing is another calendering finish, produced by metal rollers which have very fine lines on

---

QUICK QUIZ 10

1. Sample 1 is a 100 percent cotton named chintz. It has been treated with Everglaze®, a trade name for a chemical resin ____________ finish which is (durable, temporary).
2. Another finish used to create a polished surface is called ____________ .
3. Check the following rib fabrics which are suitable for achieving a moiréd effect:
   _____ oxford cloth; _____ faille; _____ denim; _____ taffeta; _____ velveteen
4. Would thermoplastic or nonthermoplastic fibers be best for creating this effect?
   ____________ Why? ____________
5. Sample 5, a 100 percent acetate taffeta, is a ____________ weave fabric with a ____________ finish. Could the finish be heat set? (yes, no)

****

ANSWERS QUICK QUIZ 10

1. glazed; durable
2. ciréing
3. faille, taffeta
4. thermoplastic, because the design could be heat set
5. rib, moiré, yes

---

them. They are similar to embossing rollers except that the almost invisible lines are etched into the rolls so that they impinge on the fabric at an angle of about 20 degrees.

Schreinerizing calenders flatten the yarns to produce compact products which have more surface area, allowing for some increased light reflectance and a more opaque fabric. Schreinerizing gives a softer luster to fabrics than those finishes that result in a polish or shine.

Schreinerizing, used frequently with natural cellulosic fibers to produce a flat, smooth, and compact fabric which has more body than regular calendered fabrics, has also been an advantageous finish for warp knit tricots of acetate or nylon. The flattened knit is highly desirable in lingerie since it retards the characteristic "crawling" of synthetic knits. Nylon tricots tend to be durable to schreinerizing because the nylon can be heat set.

Flocking

Flocking is the last of the basically mechanical finishes used to create esthetic effects. Flocks are composed of very short fibers which are glued to a fabric surface. Flocking a large number of fibers very close together yields a dimensional surface that can be similar to that created by a pile weave, tufted structure, or a napped finish. Texture is the basic visual and tactile difference in the surface qualities of fabrics which have been created by pile weaves, tufted structures, napped, or flocked surfaces.

Fiber flocks are often made from rayon fibers because they are inexpensive; but cotton, wool or other short fibers are also suitable. Either *mechanical* or *electrostatic* methods are used to stand the fiber flock on end prior to gluing it to an adhesive layer on the fabric surface, as diagrammmed in Figure 5.14.

In the mechanical method of flocking, the flock is circulated in a container in order to build up static charges, which cause the fibers to stand upright as they are dropped onto the adhesive layer on the fabric surface.

In the electrostatic method of flocking, the fibers fall through an electrostatic field which forces them into an upright position as they strike the fabric surface. Of the two methods, the electrostatic method produces the more uniform electric field and so a denser flocking is achieved.

In addition to contributing to the tactile and visual pile-like texture of fabrics, flocking contributes to basic design manipulation. Flocked finishes can be used to produce raised, patterned designs on wallpaper and in apparel fabrics. See sample 40, an imitation dotted swiss. However, all flocked designs are vulnerable to abrasive wear.

The *permanence* of mechanical finishes depends on the quality of the *chemical resins* applied to the fabrics, or on the fabric being made from *thermoplastic fibers* which can be heat set. Chemical finishes are usually more durable than mechanical ones that have no resins applied. However, all finishes are subject to reduced serviceability over time, especially as a result of laundering, dry cleaning or whatever method of care is used.

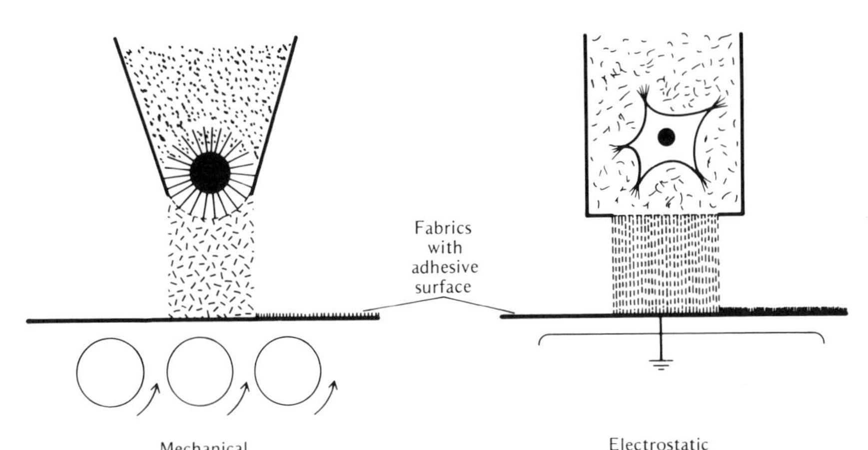

Figure 5.14. Fiber flocks are adhered to a fabric layer which has been treated with an adhesive.

QUICK QUIZ 11

1. What finishing process can be used to achieve a pile effect? ____________
2. Briefly describe the basic process: ____________

____

****

ANSWERS QUICK QUIZ 11

1. flocking
2. short fibers are glued to a fabric service to create a pile-like effect

---

Chemical Finishes Which Enhance Appearance

Chemical finishes added to enhance appearance include those used to reduce harshness through the application of *fabric softeners*; to produce transparent effects, such as in organdy, by the use of an acid *parchmentizing* finish; and to create *chemical crepeing,* such as in plissé, through the application of caustic soda. Also, because chemical fumes in the air may attack dyestuffs in certain fibers, especially nylon and acetate, special chemical inhibitors can be added to the dyestuffs to control the problem of *fume fading*.

Fabric Softeners

Oils, fats, wax emulsions, synthetic detergents, substituted ammonium compounds and silicone compounds are typical finishing chemicals used to soften fabric. While many fabric softeners are not durable and must be replaced at each care period, a few softening finishes retain their characteristics during use and care. However, since most detergents on the market contain some softener, the reapplication of softeners is not difficult.

Acid Finishes

Transparent or parchmentlike effects in cotton goods can be produced by acid finishing. Under carefully controlled conditions, cellulosic fibers are immersed for a short period of time in sulfuric acid, after which the fabric must be thoroughly rinsed to prevent acid damage and may require neutralization with a dilute alkali. This process is used in making organdy.

Acids can be applied to specialized design areas to create etched or "burn out" patterns in fabrics whose fiber content is chosen to react properly to the applied chemical. By blocking from the acid certain portions of the fabric with a resistant substance, the acid will burn out a design in the remaining part. The fabric areas treated with the acid resistant substance are not affected. Another way to create a fabric with "burn out" designs is to use two types of fibers, one sensitive to a chemical, the other resistant to it.

While the durability of burn out or etched fabrics is generally reduced, their serviceability is usually acceptable since such fabrics are usually intended only for occasional dressy uses. Acid finishing tends to markedly reduce the resiliency of fabrics; therefore, additional ironing may be required during care procedures.

Basic Finishes

Acid solutions react with some fibers, and basic or alkaline solutions react with others. Mercerization is an example of an alkaline finish applied to fabrics made of natural cellulose fibers, particularly cotton.

Recall that mercerization requires immersing natural cellulosic yarn or fabric under tension in a solution of NaOH. This finish increases luster, dye affinity, and fabric strength.

Alkaline finishes can also be applied to yield a crinkle or crepe effect, such as in plissé, sample 41. It is a plain weave fabric made by chemical crepeing. Frequently, such crinkling is restricted to striped design areas to imitate seersucker. The resulting fabric is not as durable as genuine seersucker with its woven-in crinkle.

The crinkles in plissé can be heat set if the fabric is made of thermoplastic fibers; generally, however, the finish becomes less evident as the crinkles fade with use and during care procedures. One of the fastest ways to remove the crinkle in plissé is by applying excessive heat and pressure during ironing or pressing.

While this unit offers only a quick consumer overview of fabric finishes, you should be aware that they are very much integrated with the fiber content of a fabric, the type of yarn structure, and the type of fabric construction. Fabric finishes also play a part in the application of color, to be discussed in the next unit.

QUICK QUIZ 12

1. Fabric softeners are (temporary, durable) finishes used for what purposes? ____________ ____________
2. Acids can be applied to fabrics to create transparent or ____________ effects; yet these finishes have a disadvantage in reducing fabric ____________ ____________
3. Alkaline finishes also are used for certain design effects, such as ____________ crepeing, which produces fabric called ____________.

****

ANSWERS QUICK QUIZ 12

1. temporary; to enhance fabric hand, body, texture and loft, and to decrease static electricity
2. parchmentlike; resiliency or ease of care
3. chemical; plissé

UNIT FIVE

Part VI

Finish Summary and Fabric Flammability

Only a few of the numerous finishes available can be reliably identified visually; therefore, consumers must depend on fabric labels or hangtags to give information about finishes. Before a fabric receives any finishing treatments, it is referred to as grey or greige goods. The first finishing treatments to which these greige goods are exposed are the preparatory or general mill finishes which can involve singeing, desizing, scouring, carbonizing for wool, and bleaching. Then the fabrics are thoroughly rinsed to neutralize and remove alkalies and other chemicals that cause fiber damage over time.

Practically all fabrics go through some form of initial general mill cleaning processes, as well as later smoothing and straightening finishes. These final finishes, which are applied to most fabrics, are calendering and tentering or crabbing for wool.

Although some finishes are important in adding desired end use characteristics, increasingly performance properties are achieved through fiber modification rather than finish application. Of course the natural fibers are much less amenable to modification than the manufactured fibers. Among the manufactured fibers, chemical and mechanical alteration can allow for specific performance characteristics that are often permanent in nature. Additives may also be included in the fiber spinning solution of manufactured fibers rather than as a topical treatment, creating a more permanent characteristic. Blending of various fibers is another route to achieving desired properties. The result of increasing fiber modification has been a decline in the use of finishes, especially to achieve durability, comfort and care properties. Fiber modifications and blends generally result in better consumer products than topical applications.

Flame retardant fabrics are an example of this approach. In the early 1950's, the Flammable Fabrics Act was passed to prohibit the sale of ''dangerously flammable'' fabrics. To clarify the Act, it was amended in 1967, 1972, 1975 and 1981 to further specify flammability standards in certain end uses. However, flammability problems are still far from solved, and additional flammability standards are continually being developed. The Consumer Product Safety Commission has been given the authority to establish standards for fabric flammability, to establish labeling and testing methods, and to require fabrics to be categorized as to degrees of flammability permitted. Currently, a number of flammability standards in effect provide mandatory protection from flammable fabrics in some designated end uses. These include *carpet and rugs standards,* which became effective in 1971; *standards for children's sleepwear,* sizes 0–14, established between the late 1960's and mid-1970's; and *standards for mattresses and matress pads,* also developed for each of these products. In addition, local fire codes also influenced the choice of interior textiles, especially for businesses and institutions.

It is important to know the differences among the following terms as they relate to fabric flammability and its control:

Flammable means that the fabric, once ignited, will burn unassisted. It provides its own source for flame.

Nonflammable means that the fabric will not burn or contribute to combustion when exposed to flame.

Flame retardant means that the fabric will burn, usually very slowly, if exposed to a source of flame but will self-extinguish when the source of the flame is removed.

QUICK QUIZ 13

1. What three product categories are covered by federal flammability standards?

   ____

2. Fiber content significantly affects fabric flammability. Match the following combustion characteristics with the applicable fiber or fiber category. More than one characteristic may apply to a specific fiber or category.

___	1. natural cellulosic	a. burns and chars
___	2. acrylic	b. burns and melts
___	3. nylon and polyester	c. burns briefly and chars
___	4. rayon	d. self-extinguishing
___	5. acetate	e. not self-extinguishing
___	6. protein	f. scorches
___	7. mineral	g. flameproof

3. Currently most fabrics are made flame retardant through the use of ______________ .

****

ANSWERS QUICK QUIZ 13

1. carpets and rugs; children's sleepwear sizes 0–14; mattresses and mattress pads
2. 1-a, e, f; 2-a, e; 3-b, d; 4-a, e, f; 5-b, e; 6-c, d, f; 7-g
3. modified fibers

Fiber content and fabric structure both influence the flammability characteristics of a fabric. For example, although acrylic fibers are wool-like and contribute similar properties to blankets at a lower cost than wool, they burn and are not self-extinguishing. If you want a flame resistant blanket, wool is preferred, since it burns only briefly and is self-extinguishing. The level of flame retardance related to fabric structure is based on the ability of air to get through the yarns and fabric, to support combustion of flammable fibers and to increase the speed of burning. A dense twill weave, therefore, tends to have some natural flame retardancy due to its compact structure. In the case of fabrics with fuzzy pile surfaces, the flame will advance rapidly along the surface pile or nap and carry the flame across the surface of the fabric. Flat fabrics do not tend to move the flame as rapidly.

When the Flammable Fabrics Act Amendments were first passed, flame retardant finishes were the typical method of altering fabrics in order to meet guidelines for sleepwear. Sample 28 is an example of a fabric made flame retardant through the use of a finish. Following concerns regarding the possibility of carcinogens in finishes and some undesirable serviceability properties such as stiff hand, however, flame retardant fibers became the common technique for producing flame retardant products. Modified manufactured fibers, such as flame retardant (FR) polyester, or inherently flame retardant fibers, such as vinyl or vinyon, provide flame retardancy without the use of a finish. Serviceability properties such as hand are also more desirable. Without the use of a finish, however, it is impossible to flame retard some natural fibers, especially cotton. The result has been that cotton has disappeared from the children's sleepwear market even though it has other desirable comfort and care properties appropriate for sleepwear.

Using fibers rather than finishes to achieve desired properties, such as flame retardancy, is a common technique. For the most part, using fibers or additives in the fiber spinning solution allows for better performance. Abrasion resistance, for example, is important in upholstery end uses. A fiber that is inherently abrasion resistant, such as nylon, is a better consumer choice than rayon with an abrasion resistant finish. Similarly, the new generation of soil release carpeting achieved through additives in the fiber solution, perform better than surface soil release treatments that can be subject to abrasion. Advanced fiber technology has meant a decline in the use of finishes and an increase in the use of custom designed fibers. Despite improved products through fiber alteration, finishes are still used on many consumer products, however. A review of the finishes you have read about in the previous section follows.

QUICK QUIZ 14

1. General mill finishes involve routine finishing suited to the fiber and structure used. Given the fiber content listed below, to what general mill treatments would each fabric probably be subjected?
   a. wool gabardine ______________________________
   ____
   b. cotton chintz ______________________________
   ____
   c. textured polyester/knit ______________________________
   ____
2. In order to produce the smooth surfaces on samples 1, 4, and 6, which contain staple-length cotton fibers, the fabrics were subjected to ______________________________ ______________________________ general mill finishing treatments.
3. Singeing is sued on fabrics made from staple-length fibers in order to ______________ ______________________________

****

ANSWERS QUICK QUIZ 14

1. a. desizing, singeing, scouring or carbonizing, bleaching, fulling, crabbing
   b. desizing, singeing, scouring, bleaching, calendering and glazing, tentering
   c. desizing, scouring, bleaching or optical brightening, calendering, including heat setting.
2. singeing, calendering
3. remove the fuzz produced by loose, staple fiber ends

Review of Fabric Finishes

Finishing is especially important to control relaxation shrinkage which occurs as a result of tension during yarn and fabric structuring. A label with Sanforized® on it indicates that relaxation shrinkage has been controlled to the extent that the fabric will not shrink over one percent during care.

Mercerization, another finish which increases fabric durability by increasing fiber strength, enhances the esthetic properties of luster and dye affinity. Since mercerization involves immersion under tension of the yarns or fabric in a concentrated solution of NaOH, cotton and other natural cellulosic fibers can be successfully mercerized. They react favorably with alkalies, whereas wool is damaged by them.

Absorbent and antistatic finishes contribute to comfort in textile products, as well as to care procedures. The finishing industry continues to try and improve these finishing techniques to make them more durable or serviceable. Researchers, working to improve the permancence of antistatic finishes, have been successful in modifying the fiber structure of certain nylons and polyesters to decrease their potential for static electricity. These improvements are desirable in order to prevent synthetic fabrics from clinging to and shocking the wearer, and from attracting soil and dirt.

Another way of controlling static electricity includes combining metallic fibers with other fibers to make yarns that will conduct electricity. This method is used primarily in commercial carpeting. However, many antistatic finishes, especially the topical ones, must be renewed periodically. While the consumer can renew antistatic performance by using fabric softeners during the rinse cycle of laundering or during drying, such fabric softeners should be used with some restraint. Over time, softeners can build up on fabrics, coating the yarns, which reduces fabric absorbency and increases flammability.

Minimum care finishes are an asset because they increase the wrinkle resistance of fabrics, resulting in products which require little or no ironing. However, the consumer must be willing to accept decreased strength or durability in return for durable press finishes.

Soil and stain resistant fluorochemical finishes are not visible, but their presence becomes evident when liquid is dropped on the fabric surface and stains ''stand up'' or form beads of liquid on the surface.

Regardless of the presence of a stain resistant finish, a user should not rely on it entirely to protect the fabric. In the event of a spill or stain, the soil should be removed immediately to keep it from permanently staining the fabric.

Numerous finishes are applied solely to improve the esthetic appearance of fabrics. Sizing is often used on inexpensive fabrics with low yarn counts to improve their appearance, body or hand and luster. Durable sizings are chemical resins that stay in a fabric after several launderings; temporary sizings are starch-like substances often removed in the first wash.

Weighting is a finish used to increase the body of fabrics made of silk fibers. Sample 3 could have up to 10 percent weighting and remain within the limits of the Pure Dye Silk Act.

To increase wool body, fulling can be used to cause the fibers to shrink and felt slightly. And if a raised, fuzzy surface is desired on a woven or knitted fabric, it can be napped with wirelike rollers that raise the fiber ends.

Children's sleepwear is often made of flannelette, a plain-woven fabric napped on one side. Such sleepwear is required by law to be flame retardant, which is generally achieved either through the use of flame retardant fibers.

A fabric can have its yarns flattened by a pounding process known as beetling, which is typically used on flax fibers. Another way to make a fabric lustrous is to compress a resin into it by means of a rapidly spinning friction roller which glazes the surface.

A satin or sateen weave structure yields a smooth surfaced fabric that is suitable for glazing because it polishes up better than regular plain weave fabrics due to its long yarn floats. However, plain weave fabrics also polish satisfactorily, as in sample 1.

Finishes applied solely as a design feature include embossing, ciréing, moiréing, flocking, parchmentizing and chemical crepeing. The fabric set has examples of these kinds of finishes applied for design effects.

The taffeta fabric that has been moiréd, sample 5, is structured with a rib weave to produce the wavy, rippling effect. Embossed sample 36 is a nonwoven film structure, finished to create a leather-like fabric. The chemical crepe effect in plissé fabric 41, is not as durable as the creped effect in seersucker fabric 21, which is woven in by the slack tension process.

Flocked sample 40 is generally referred to as dotted swiss, a name originally reserved for fabrics that had their dots structured-in during *spot weaving*. Today, however, much dotted swiss has the dots flocked on in order to increase speed of production and decrease cost.

Mechanically applied finishes are generally temporary unless used on thermoplastic or synthetic fibers which can be heat set. If the fibers are nonthermoplastic, a mechanical finish can be made relatively durable by the application of resins.

QUICK QUIZ 15

1. What fiber must be present in durable press fabrics for the resins to create cross linkages? ______________
2. Why is durable press better than the older wash and wear finish? ______________ ______________
3. Sample 28 is a 100 percent polyester which meets children's sleepwear flame retardant standards. What type of construction was used? ______________
   What finish was used to modify the appearance and hand of the fabric? ______________
4. Which of the pile fabrics in your sample set were probably brushed? ______________
5. Which fabric sample was finished by glazing? ______________

****

ANSWERS QUICK QUIZ 15

1. cotton (or other cellulose)
2. Durable press chemicals allow curing of end products to retain a smooth shape; plus they are more permanent than wash and wear finishes.
3. warp knitting, napping
4. 12, 13
5. 1

The use of appropriate finishes is determined by the wide variety of consumer created demands. Many fabrics would not be serviceable for their end uses, were finishes not available to improve durability, comfort, care and appearance. Because many functional finishes are invisible, the consumer should identify the presence of special fabric finishes with information obtained from the label or hangtag to ensure that serviceability requirements are satisfied.

Identification of Finishes

Take out the Mounting Sheet for Finishes and determine which types are indicated. Many preparatory and functional finishes are invisible, but you should still attach the designated samples to your mounting sheet as representative finishes. On the other hand, finishes for special appearance features are readily identified.

1. For the following samples, give the preparatory or routine finish used to achieve these effects:
   to increase body and stability of sample 2:
   ____
   to remove natural gum and then add body to sample 3: ______________________
   and ______________________
   to remove staple fiber fuzz on sample 4:
   ____
   to increase luster, strength and dye affinity of sample 12: ______________________
   to even up cut pile surface of sample 15:
   ____
   Cut, mount and label swatches from each of these fabrics under Preparatory Finishes.
2. Sample 27A, a luggage fabric, has a urethane coating that adds ____________ and ____________ . Attach and label sample 27A on the Mounting Sheet under Finishes for Durability. Include a notation that the urethane coating also enhances the water repellency of this fabric.
3. Samples 6, 11 and 14 have one or both of these primary finishes that contribute to easy care: __ ______________________________ and ______________________________ .
   Attach and label the samples on the Mounting Sheet under Finishes for Ease of Care. The fiber content and nonvisible finishes are given below so you can label the finishes appropriately:
   6—Durable press blend, 60 percent cotton/40 percent polyester
   11—Durable press, 100 percent cotton; stain resistant, Scotchgard®
   14—Stain resistant, Scotchgard®; 65 percent cotton/35 percent rayon
4. Sample 28 has a functional finish that makes the sleepwear fabric ______________________, as well as a ______________________ finish to improve hand and appearance. Attach a swatch from 28 under Finishes for Appearance, and label both on the Mounting Sheet.
5. Attach swatches from these samples under Finishes for Appearance and label the visible finishes:
   1: ______________________
   5: ______________________
   13: ______________________
   25A: ______________________
   28A: ______________________
   36: ______________________
   36A: ______________________
   40: ______________________
   41: ______________________

****

Identification of Finishes Answers:

1. 2-sizing
   3-degumming, weighting
   4-singeing
   12-mercerizing
   15-shearing
2. water repellency; stability
3. durable press, stain resistant
4. flame retardant; napped
5. 1-glazing, 5-moiréing, 13-brushing (12 is also brushed; add brushing under the mercerizing label), 25A-sueding, 28A-sueding, 36-embossing, 36A-sueding, 40-flocking, 41-chemical crepeing

Posttest

1. Singeing shirt fabrics made of staple length polyester and cotton fibers helps prevent
   a. stretching out of shape.
   b. pilling around the collar.
   c. raveling at the seams.
   d. ring around the collar.
2. Calendering and tentering finishes are used to create ______________________ fabrics.
   a. dimensionally stable
   b. wrinkle resistant
   c. smooth, on-grain

3. As a general mill finish, bleaching serves to remove stains as a result of processing and to whiten
   a. protein and synthetic products.
   b. natural cellulosic products.
   c. man-made cellulosics.

4. Abrasion resistant finishes are used to make synthetic fibers durable.
   a. true
   b. false

5. A couch upholstery fabric made of _________ fibers would benefit from an antislip finish in order to provide maximum serviceability.
   a. man-made cellulosic
   b. natural cellulosic
   c. synthetic
   d. protein

6. A Sanforized® label means the fabric is guaranteed to shrink less than _________ percent.
   a. three
   b. one
   c. two
   d. two and one-half

7. What is the type of shrinkage remaining in a fabric after the converter has applied appropriate shrinkage-control finishes?
   a. progressive
   b. residual
   c. relaxation

8. Which fabric is likely to present the greatest shrinkage problem due to its structure?
   a. plain filling knit
   b. plain weave
   c. double knit
   d. twill weave

9. What advantages do consumers derive from mercerization used on a majority of cotton fabrics?
   a. increased tenacity and cohesiveness
   b. increased dye affinity and luster
   c. increased relaxation shrinkage
   d. decreased progressive shrinkage

10. Absorbent finishes which promote the pickup of moisture on the fabric surface tend to
    a. decrease drying time.
    b. decrease elctrostatic charges.
    c. decrease electrical conductivity.
    d. increase electrostatic charges.

11. With continued routine use, fabric softeners can coat fabric yarns to a degree that fabric absorbency is (increased, decreased); the softener itself attracts moisture, however, which (increases, decreases) static electricity.
    a. decreased, increases
    b. decreased, decreases
    c. increased, increases
    d. increased, decreases

12. Rank the order of the following three-dimensional designs from most to least abrasion resistant, assuming all examples are upholstery fabrics with the same fiber content.
    ___ a. sculptured pile
    ___ b. uncut uniform pile
    ___ c. jacquard brocade
    ___ d. flocked floral design
    ___ e. uniform flocked surface

13. Which product made of acetate would be most susceptible to fume fading?
    a. draperies
    b. evening dress
    c. pillowcase
    d. sports coat

14. Which of the following finishing processes can be used to create three-dimensional designs less expensively than woven-in pile or jacquard designs?
    a. sculpturing
    b. tufting
    c. flocking

15. A soil release finish would be most needed on which product listed below?
    a. olefin carpet
    b. polyester uniform
    c. cotton slacks
    d. acetate drapery

16. Although durable press products are favored by many consumers due to their ease of care, which problem below is a potential result?
    a. decreased heat tolerance
    b. decreased durability
    c. decreased luster
    d. increased static electricity

17. Which fiber content would yield the most serviceable fabric for durable press work slacks?
    a. 100 percent cotton
    b. 65 percent polyester; 35 percent cotton
    c. 80 percent polyester; 20 percent cotton
    d. 65 percent cotton; 35 percent polyester
    e. 80 percent cotton; 20 percent polyester

18. Generally, the ____________________ soiled a raincoat, the ____________________ water repellency it will have.
    a. less, less
    b. more, more
    c. more, less

19. If you were selecting draperies for auditorium windows and safety were a major criterion, which would be the best choice?
    a. cotton, basketweave, flame retardant finish
    b. rayon, leno weave, flame retardant finish
    c. glass, plain weave, silicone finish
    d. acetate, pile weave, flame retardant finish

20. What finish(es) might a fabric designer recommend to improve the body of inexpensive cotton fabrics which were to be used for costumes?
    a. weighting
    b. parchmentizing
    c. sizing
    d. beetling
    e. all of the above

21. A consumer would be smart to avoid purchasing fabrics labeled "weighted silk" because the fabric would
    a. be extra heavy.
    b. have decreased tenacity.
    c. have a metallic sheen.
    d. have decreased strength and increased stiffness.

22. Napping can be used to improve the loft and thermal retention of which of the following blankets?
    a. acrylic needle felt
    b. wool felt
    c. wool twill flannel
    d. acrylic tufted

23. Fulling processes on wool slacks
    a. are generally performed after the garment is structured.
    b. are durable for the lifetime of the fabric.
    c. need renewal to maintain their effectiveness.

24. Beetling makes the woven-in design of a linen damask tablecloth
    a. flat and lustrous.
    b. raised and dull.
    c. flat but weakened.
    d. raised but weakened.

25. Embossing, glazing, ciréing and moiréing can be made relatively durable by
    a. using an appropriately modified calendar.
    b. applying resins to the fabric surface.
    c. using a fabric made of nonthermoplastic fibers.

26. If you were selecting a tufted-pile carpet, match the finish that would be best with each fiber choice for that end use.
    ____ 1. 100 percent wool living room carpeting
    ____ 2. 100 percent acrylic cut-pile playroom carpeting
    ____ 3. 100 percent nylon hall carpeting
    ____ 4. 100 percent polyester family room carpeting
    ____ 5. 100 percent olefin kitchen carpeting
    a. no special finish
    b. flame retardant
    c. mothproof
    d. antistatic
    e. soil resistant

27. Match the fabrics listed to the finishes the converter would automatically apply.
    ____ 1. cotton broadcloth
    ____ 2. silk brocade
    ____ 3. wool flannel
    ____ 4. acetate velvet
    a. shearing
    b. carbonizing
    c. degumming
    d. mercerization

28. Match the finish which could be used to achieve the following esthetic effects:
    ____ 1. transparent appearance
    ____ 2. plissé crinkle
    ____ 3. pilelike texture
    ____ 4. rippling, wavy effect
    ____ 5. patent-leather look
    a. moiréing
    b. flocking
    c. ciréing
    d. parchmentizing
    e. chemical crepeing

29. Match the finish which would most likely have been used on the following fabrics:
    ____ 1. cotton chintz
    ____ 2. cotton flannelette
    ____ 3. noncrawl nylon tricot
    ____ 4. wool gabardine
    ____ 5. cotton corduroy
    a. napping
    b. brushing
    c. glazing
    d. fulling
    e. schreinerizing

Answers for Unit 5 Posttest:

1. b
2. c
3. b
4. b
5. c
6. b
7. b
8. a
9. b
10. b
11. b
12. most—b, a, c, e, d—least
13. a
14. c
15. b
16. b
17. c
18. c
19. c
20. c
21. b
22. c
23. b
24. a
25. b
26. 1-c; 2-b; 3-d; 4-e; 5-a
27. 1-d; 2-c; 3-b; 4-a
28. 1-d; 2-e; 3-b; 4-a; 5-c
29. 1-c; 2-a; 3-e; 4-d; 5-b

UNIT SIX

Color and Design of Textile Products

One good example of the best side of human nature is the eternal effort of people to bring beauty and harmony into their everyday surroundings. From the beginning, individuals have looked for ways to make body coverings and the near environment more attractive by adding color and design to textile products they used.

The earliest expressions of human values in color and design were made possible by natural coloring from plants, insect fluids, and sea life. The gathering and processing of usable quantities of coloring material were arduous and costly, yet people were willing to pay a high price for beauty. The ancient Phoenicians acquired Tyrian dyes from small shellfish off the Eastern Mediterranean coast. One single gram of precious Tyrian purple required 12,000 hand processed shellfish; thus, to this day, purple is associated with richness and royalty, as only kings and queens could afford such colors in earlier times. Today, skill in manipulating chemistry has led to a wide variety of beautiful and economical fabric colorings.

There is no single standard by which one can judge the absolute beauty or attractiveness of color and design added to textile products, or any other product for that matter. Entire disciplines have developed to enable people to better understand color and design appeal.

Why certain colors and designs are chosen, and the significance of these choices, are not considered in this discussion of textiles. Rather, this unit is concerned only with the serviceability of color and design applications.

The basic information presented on the ways in which color and design can be applied and evaluated is important background for pursuing the art of color and design selection. Textile designers, consultants, retailers and interior designers must be concerned with understanding individual needs in the color and design areas. But everyone should acquire some understanding of textile color and design alternatives available for various design applications. Successful application of color and design to textile products depends upon creative ingenuity and the awareness of various alternatives.

Objectives for Unit 6

1. You will be able to describe the techniques involved in the following coloring methods:
   A. Mass pigmentation or solution dyeing
   B. Fiber or stock dyeing
   C. Yarn dyeing
   D. Piece dyeing
      1. Union dyeing
      2. Cross dyeing

2. You will be able to describe the differences among the following design application procedures, and the serviceability implications for each:
   A. Roller or direct printing
   B. Warp printing
   C. Transfer printing
   D. Screen printing
   E. Other resist printing
      1. Batik
      2. Tie-and-dye
      3. Stencil
   F. Discharge printing

3. Using the fabric samples, you will be able to identify the various coloring methods and design application procedures used.

4. Given a particular end use or a method of coloring and/or design application, you will be able to list the applicable criteria relative to the serviceability concepts of:
   A. Durability
   B. Comfort
   C. Care procedures
   D. Esthetic judgments regarding choice of color and design

Pretest

1. Drapery fabric should be especially colorfast to
   a. light.
   b. crocking.
   c. laundry.
   d. drying.

2. Fabrics that are colored with dyes will be more likely to lose color through ____________, while fabrics colored with pigments will be more likely to lose color through ____________.
   a. migration, bleeding
   b. crocking, bleeding
   c. bleeding, crocking
   d. crocking, migration

3. Rate the following methods of coloring fabric from most to least expensive:
   a. piece dyeing
   b. fiber dyeing
   c. yarn dyeing

4. What advantage do printed-on designs have over those that are woven in? Print designs
   a. are less expensive to produce.
   b. are often reversible.
   c. have the design structured with the grain.
   d. have more thorough dye penetration.

5. Which of the following design applications has become increasingly competitive on the retail market?
   a. block printing
   b. stencil printing
   c. rotary screen printing
   d. screen printing

6. The most serviceable and economical method for coloring mass-produced cotton print fabric is ____________ printing.
   a. batik
   b. roller
   c. flat bed screen
   d. block

7. The soft-hazy designs used on certain upholstery fabrics are achieved by
   a. fiber dyeing.
   b. warp printing.
   c. batik printing.
   d. yarn dyeing.

8. Silk screen printing is used when
   a. large yardages of fabric have small designs.
   b. large designs are printing on limited fabric yardages.
   c. small designs are printed on limited fabric yardages.

9. Discharge printing is used on fabrics which have
   a. relatively light backgrounds.
   b. relatively dark backgrounds.
   c. coarse, open textures.
   d. hazy, indistinct designs.

10. Resist printing involves which of the following methods?
    a. roller printing
    b. transfer printing
    c. using a stencil or wax
    d. block printing

11. Which of the following yarns are used primarily for design effect?
    a. core and textured
    b. bouclé and crepe
    c. singles and cord
    d. ply and tweed

12. The design is an integral part of the fabric structure in which of the following?
    a. rotary screen printed batiste
    b. yarn dyed brocade
    c. tie-and-dye broadcloth
    d. piece dyed satin

13. Which of the following fabric finishes are used solely for design effect?
    a. schreinerizing and flocking
    b. moiréing and embossing
    c. parchmentizing and mercerizing
    d. fulling and beetling

14. Which of the following textile products would most likely be fiber dyed?
    a. bouclé yarns in a sweater
    b. tweed yarns in a coat
    c. slub yarns in a drapery
    d. ply yarns in work pants

15. Which fabric is typically yarn dyed?
    a. gingham
    b. corduroy
    c. tweed
    d. faille

16. Cross dyed fabric may appear to be ________ dyed.
    a. union
    b. solution
    c. fiber
    d. piece

17. Match the design application to possible descriptions.
    ____ 1. batik
    ____ 2. discharge
    ____ 3. transfer
    ____ 4. tie-and-dye
    ____ 5. direct
    ____ 6. rotary screen

    a. printing with design on paper
    b. printing with bleach
    c. resist dyeing with thread
    d. resist dyeing with wax
    e. printing process that combines two methods
    f. printing with etched copper rolls

18. Match the most likely color application for each end use.
    ____ 1. wool tweed skirt
    ____ 2. acetate drapery
    ____ 3. cotton plaid shirt
    ____ 4. solid-colored cotton sheet

    a. yarn dyed
    b. piece dyed
    c. solution dyed
    d. fiber dyed

Answers for Unit 6 Pretest:

1. a
2. c
3. most—b, c, a—least
4. a
5. c
6. b
7. b
8. b
9. b
10. c
11. b
12. b
13. b
14. b
15. a
16. c
17. 1-d; 2-b; 3-a; 4-c; 5-f; 6-e
18. 1-d; 2-c; 3-a; 4-b

UNIT SIX

Part I

Color

This unit covers the use of dyes to impart color and design. Matrix XI contains color-related terms used through this part that you will need to learn. Using this Matrix from the back of the book, fill in the definitions and their significance as you proceed.

Color Serviceability

Color can be applied to fabric by numerous methods—before, during or after finishing treatments, as well as at various stages of fabric manufacture. Regardless of method and type of color application, the principal concern is that the color be securely fastened to the fabric so it is durable in use and care. Fabrics with durable coloring are considered *colorfast*. However, some explanation is needed about the environmental factors affecting colorfastness.

Considerable research has been undertaken to make dyestuffs colorfast to *bleeding, crocking, migration, light*, and *fume fading*, which are the major ways color loss or change can occur. Bleeding is the color loss that results from laundering or dry cleaning; crocking is the result of color loss from rubbing or abrasion; and color migration is the movement of one color into another, which can occur as a result of laundering, dry cleaning, and other environmental conditions.

Actually, colorfastness is relative, depending on the end use of the item. A color may be fast to one destructive agent but not to another. A given dyestuff applied to draperies may be fast to laundering, crocking or migration, but not fast to sunlight which causes color to fade. In this case, draperies should not be sold as colorfast for the designated end use. Whether or not fabric is colorfast is usually determined in relation to the designated end use and care procedures.

In addition to problems of color transfer through abrasion and laundering, color loss also can occur as a result of chemical changes in the dyestuff from detergents and other laundry additives. Perspiration can bring about chemical changes in the dye; sunlight, particularly ultraviolet rays, can break down the dye and cause color change; and fumes in the atmosphere, such as gas, smog and other contaminants, may contribute to color loss.

These various types of color loss or fading can bring about changes in color hue, intensity, and chroma over time, which reduce product serviceability. Draperies, in particular, can be damaged by fume fading as well as by sunlight since they are constantly exposed to both. Lining draperies helps prevent color loss associated with sunlight, but that will not have much effect on color loss due to fumes and air contaminants.

Color loss from light and atmospheric gases may also be accompanied by fabric degradation. Sulfur compounds in the air, common in industrial areas, can react with humidity to form sulfuric acid which is harmful to cellulosic fibers. Other compounds in the air can create hydrochloric acid fumes which are harmful to fibers such as nylon. Figure 6.1 illustrates what can happen when a natural cellulosic product is made even more sensitive to atmospheric damage through the use of a sulfur-type dyestuff. Notice the fingers protruding through areas of fabric where it has degraded to the point of breakdown. The sulfur

MATRIX XI
COLOR TERMINOLOGY

COLOR TERMS	DEFINITION	SIGNIFICANCE TO CONSUMER
Colorfast		
Bleeding		
Crocking		
Migration		
Light fading		
Fume fading		
Dyes		
Pigments		
Dye site		
Dye medium (carrier)		

Figure 6.1. With continued exposure to atmospheric contaminants, sulfur dyes can create fabric weakening and eventual disintegration.

QUICK QUIZ 1

Define the following terms by explaining how color loss occurs:

1. bleeding ______________________________
2. crocking ______________________________
3. migration ______________________________

****

ANSWERS QUICK QUIZ 1

1-from laundering or dry cleaning solvents
2-through rubbing or abrading the fabric surface
3-from color shifting to adjacent areas in printed fabrics, which can result from laundering or dry cleaning

---

in the dye reacts with atmospheric and laundering moisture, and sometimes added sulfur in the air, to form sulfuric acid in a concentration sufficiently strong to destroy the fabric. Sulfur dyes are not as likely to damage fibers which already contain sulfur in their chemical structure, such as wool or other animal hair fibers.

The frequency of cleaning required and the need for colorfastness to laundering depends on the end product. For example, consider draperies versus shirts—shirts obviously are cleaned more often than draperies because of differences in end use. Thus, fabrics with fibers that are resistent to laundering and color loss are more important for washable apparel than for home furnishing products.

The infrequency with which many draperies are cleaned, due to the high cost of dry cleaning as well as the inconvenience of taking them down, can contribute to fabric degradation and color loss. The acidic gases present in homes with other than electric central heating are hard on drapery fabric. Consequently, when draperies are returned from the cleaners, consumers are often surprised to find the fabric has faded and split especially when made of rayon and acetate. Rather than blame the dry cleaners, the consumer should realize the cause of damage was likely lengthened exposure of draperies to fumes and sunlight.

The only way to be sure of colorfastness of a textile product is to check the label or hangtag. Yet, labels are seldom of great help because there is no legal requirement that information be provided about colorfastness properties. If information is to be of value to the consumer, it must indicate to what environmental conditions the dye is colorfast. Simply indicating that a product is colorfast tells nothing useful. A tag or label needs to state the product is colorfast to laundering, sunlight, crocking, or whatever, if the consumer is to have information or help in determining the serviceability of the fabric. Increasing numbers of consumer requests for information on colorfastness properties of textile products should encourage manufacturers to provide it.

Just as selecting optimum care intervals to prolong serviceability of a garment or home furnishing fabric is a compromise for the consumer, so is selection of appropriate dyestuffs a compromise for the converter. Of major importance to both is matching fiber and fabric colorfastness properties to the intended end use.

Fiber content determines to a great extent the type of dyestuff the color chemist prepares for a fabric, since fibers are chemically different and therefore have different affinities for dyestuffs. When new fiber modifications are developed, new dyestuffs are also required. With the development of additional types of synthetic fibers, particularly those which have low absorbency and consequently low dye affinity, the job of the dye chemist has become increasingly challenging.

For example, olefin, with its zero absorbency has proven virtually impossible to dye. Therefore, it is colored by adding pigments or dyes to the polymer melt solution prior to spinning.

QUICK QUIZ 2

1. Suppose a designer were asked to specify under what conditions dyes should be colorfast. Considering end uses listed below, match qualities for which each product should be colorfast:

____	1. floral-printed fabric for a skirt	a. bleeding or laundering
____	2. upholstery fabric with color pigments glued to fabric surface	b. crocking
____	3. bright colored wash-and-wear golf shirt	c. migration
____	4. unlined solid-colored drapery fabric	d. perspiration
____	5. solid-color acetate sleeveless cocktail dress	e. sunlight
____	6. striped beach umbrella	f. fume fading

2. What are five types of colorfastness for a textile converter to consider, depending on end use specified?

________________ ________________

________________ ________________

____

****

ANSWERS QUICK QUIZ 2

1. 1-a, b, c; 2-b, c, f; 3-a, b, d, e; 4-b, e, f; 5-b, d, f; 6-b (from sand), e, f
2. any five: bleeding or laundering, crocking, migration, atmospheric gases or fumes, sunlight, perspiration

Coloring Agents

There are two basic types of coloring agents—*dyes* and *pigments.* Dyes for absorbent fibers are usually soluble in either a water bath or a chemical solution. However, dyes for fibers with very low levels of absorbency are frequently of the disperse type—which means the dyes are dispersed and not dissolved into a solution; the dispersed particles then migrate into the fiber during the dyeing operation. The solution into which dyes are either dissolved or dispersed is called the dye bath or the dye medium. *Soluble* dyes have *dispersable* color particles that penetrate absorbent fibers and attach securely to fiber surfaces. Such dyes must be "fixed" by some type of molecular interaction if they are to be colorfast. Fibers with low absorbency, like hard-to-dye polyesters are colored by *insoluble* disperse dyes in water suspensions, while acrylics are dyed with water soluble basic dyes.

On fibers that are very hard to dye, the application of pigments can be less complex and more rapid than many of the processes involved in dyeing. For example, insoluble pigments may be introduced into fiber solution during manufacture of fibers prior to actual spinning of the fibers. This process, called *mass pigmentation,* is used for certain hard-to-dye synthetic fibers, and for modified cellulose such as acetate. When pigments are applied to the fiber solution prior to spinning, color loss by crocking is not as great as when the pigments are applied to the fabric surface by an adhesive.

Pigments, made up of insoluble particles that may not penetrate as deeply into fibers as dyes do, are usually sealed or attached to surfaces with an adhesive. Pigments are primarily used for printed color applications, except for polymer mass pigmentation used in solution dying, as with olefin.

By far, the vast majority of coloring agents used on textiles at the present time are dyes rather than pigments. Dyes are subject to color transfer problems associated with bleeding during laundering or dry cleaning, and with migration; pigments used for prints are subject more to color loss through crocking or abrasion. Both dyes and pigments are subject to problems caused by light and fume fading.

In the past, low absorbency fibers could be dyed in only one of two ways: using disperse dyes for polyester and acetate, and basic dyes for acrylics. The disadvantage to both of these methods was that the process required chemical carriers or special heat fixation to swell the fiber and assure adequate dye penetration. The only other option was to color hydrophobic fibers by adding pigments to the spinning melt, or by printing the color onto the fabric itself. Today, however, there are synthetics which have been modified, either with chemical additions to the spinning melt or with the addition of polymer segments which form copolymers. Either procedure makes low absorbency synthetics more readily dyeable with a wider range of dye classes. However, the improved colorfastness and color selectivity resulting, has to be weighed against the higher costs of these modifications.

Absorbent fibers will generally be colored by soluble dyes, since the dye is easily dissolved in the dye bath and can thoroughly penetrate fibers. Natural cellulosic, rayon, and protein fibers dye very satisfactorily due to their hydrophilic chemical groups and the availability of dye sites in their molecules. Dye sites are positions in fiber molecules that either attract dyes chemically, or are simply spaces in the molecules where dye can be held. The number of dye sites in a fiber determines its ability to accept, as well as retain dye. Hence, determining the quality of the resulting color is a matter of understanding fiber molecular make up and presence of dye sites.

There is one type of dye which does not involve color at all in its usual sense. This is *fluorescent dye* which functions when light invisible to the eye strikes its atoms and excites them. This causes them to give off light waves that are added to reflected, visible light. The effect is to make either white or colored fibers and fabrics seem brighter.

In dyeing a fiber or fabric, fibers are first wet thoroughly and then passed through the dye bath or medium in which the dyes are either dissolved or dispersed in the solution. A padding mangle like that used for finishing fabrics may be used to press the dye into the fabric and maximize color penetration. Look back to Figure 5.7; the dye solution would be where the finishing solution is shown. The dye attaches itself to fiber molecules either by absorption or chemical reaction.

To assist in dye penetration, the dye solution may be heated to a high temperature. Heating causes the fiber molecules to move farther apart, exposing available dye sites, which allows the dye to penetrate further. For some fibers, dyeing can be further assisted by using "carriers," which are special chemicals that help dye penetrate by causing fiber molecules to move apart. This maximizes the probability that dye particles will find available dye sites.

Dyeing after-treatments include removal of excess dye, and most important, the final setting of the dye in order to make fabric as colorfast as possible. During drying of dyed fabric, molecular chains return to original positions, thus trapping dye particles in the fiber and setting the color.

Fiber Affinity to Dyes

Since wool is damaged by alkalies, dye baths are neutral or acidic; otherwise, the wool fibers would be destroyed in the dyeing process. Since wool is absorbent and presents many dye sites to the typically acidic dyes used, wool is described as having high dye affinity. However, given the propensity of wool to felt and shrink, certain precautions must be taken during dyeing. The temperature of the dye solution, concentration of chemicals and dye, and

---

QUICK QUIZ 3

Match the best alternative for coloring fibers listed below:

____	1. olefin	a. soluble dyes
____	2. cotton	b. pigments adhered to fabric surface
____	3. wool	c. dyes that disperse into fibers
____	4. polyester	d. mass pigmentation
____	5. glass	

****

ANSWERS QUICK QUIZ 3

1-d; 2-a; 3-a; 4-c; 5-b or d

---

agitation must be controlled in order to prevent felting.

Most dyes do not combine chemically with cellulosics, but special reactive dyes have been developed recently that combine with their molecules to improve dye affinity and colorfastness. Recall that the molecular structure of certain natural cellulosic fibers, such as cotton, has a somewhat amorphous component. This means that fiber molecules tend to be randomly oriented in some sections of fiber and crystalline in other areas. Thus, availability of dye sites is variable.

The function of the dye bath when coloring cellulosic fibers is twofold: first, it is used as a medium to apply dyestuff; second, it is used to make dyeing more uniform or level across the amorphous and crystalline areas. The latter occurs when all dye sites are reached uniformly. Further, these arrangements prevent dye from clotting or packing into some amorphous areas or voids between fibers and not spreading evenly throughout fiber or fabric. Once dyes have been fixed at dye sites, the dye medium is no longer needed and is rinsed away.

While special dyestuffs are developed for each fiber type, such as the proteins and cellulosics, there have been greater difficulties in developing dyes for fibers with low absorbency or reactivity. However, the problems related to low dye affinity have been overcome to a degree and, currently, most fibers can be successfully dyed. For example, problems of dyeing polyester fibers have been solved by using disperse dyes, combined with high temperatures and sometimes dye carriers, to achieve sufficient dye penetration to yield colorfast products for most end uses.

A discussion of specific types of dyestuffs is not included in this program because dye classification is complex and requires an indepth treatment of chemistry. Besides, labels seldom give any information about types of dyes used, so there is no way consumers can identify dye to help with fabric evaluation. Rather, you have to accept the manufacturer's word in regard to degree of colorfastness, if any information is given at all.

Color application, as in many other stages of textile production, has been improved by the use of computers. Modern dyehouses now use equipment that automatically weighs, mixes and dispenses the correct amoung of color. Color scanning equipment for quality control is also used. The technology is now available to achieve colorfastness for most end uses.

Color Application

Color can be applied during four stages of manufacture, and the dyeing processes are named according to when color is added. The four processes used are *solution, fiber, yarn* and fabric *piece dyeing.*

All of these dyeing processes can result in textiles that are adequately colorfast for a given end use. The *types* of dyes or pigments used on various fibers make more of a difference in colorfastness than the stage at which color was applied. Of course, the more thorough color penetration, the greater the likelihood of a colorfast product.

Solution Dyeing or Mass Pigmentation

Solution dyeing or mass pigmentation, sometimes also referred to as dope dyeing, involves addition of dyes or pigments to the fiber solution before extrusion through the spinnerette. Because the dyeing is accomplished while fibers are still in solution, the fiber filaments are already colored when they are extruded.

One cannot tell from looking at the surface of a fabric if it is made from solution-dyed fibers.

QUICK QUIZ 4

1. Fabric coloring methods have greatly improved and even hard-to-dye low absorbent fibers in the ______________ category can be successfully colored using ______________ dyes or pigments.
2. To help the dispersed dye particles find available dye sites, dyeing temperatures can be ______________ and dye ______________ used.

****

ANSWERS QUICK QUIZ 4

1. synthetic (as olefin, polyester), disperse
2. elevated, carriers

MATRIX XII
COLOR AND DESIGN PROCESSES

COLOR & DESIGN PROCESSES	DEFINITION	USES
Solution dyeing		
Fiber or stock dyeing		
Yarn dyeing		
Piece dyeing		
Union dyeing		
Cross dyeing		
Block printing		
Roller (surface) printing		
Warp printing		
Stencil printing		
Screen printing		
Other resist		
Tie-and-dye		
Batik		
Discharge printing		

However, if uneven color occurs within a yarn, you can assume the fibers were not solution dyed or mass pigmented. To ensure even color penetration, vinyl films are typically solution dyed before being extruded in sheet form. See Figure 6.2.

Fiber Dyeing

Fiber or stock dyeing is a process whereby fibers are dyed prior to being spun into yarns. Flake or tweed is one yarn type likely to be constructed of fibers dyed by this method.

Figure 6.3 shows a fabric with different colored fibers in the same yarn. Solution-dyed fibers can create a similar effect if they are cut into staple form and then spun into yarns with different colored fibers; however, tweed effects are more likely created through fiber dyeing.

Figure 6.2. For solution-dyed vinyl, fiber polymers are colored in solution prior to fabric structuring and finishing.

Figure 6.3. To create tweed effects, fibers are generally colored in the lap stage prior to yarn spinning. The different colored fibers are then spun into one yarn.

Design wise, there is little advantage in dyeing fibers one color and then spinning a solid-colored yarn. Further, the fiber dyeing process is expensive when compared to yarn and piece coloring methods. So, if fiber-dyed stock is used in yarns, tweed or heather effects are generally created.

Yarn Dyeing

When a fabric calls for different, solid-colored yarns, such as in a plaid, the logical coloring method is yarn dyeing. When yarns of different colors are used in a fabric, woven-in designs can be created. Notice in Figure 6.4 how the patterns were formed for the dobby and jaquard weave structures. When plaid and checked gingham fabrics are colored by yarn dyeing, the fabrics are reversible.

Figure 6.4. Yarn-dyed fabrics usually have structured-in colored designs, to take advantage of the different yarn colors.

Piece Dyeing

Piece dyeing involves immersing the whole piece or length of structured fabric in a dye medium to produce solid-colored fabrics. Piece dyeing is the most economical coloring method, and it is used more often than other coloring methods which determine fabric color at an earlier stage of manufacture.

Two variations of piece dyeing have been developed—*union* dyeing and *cross* dyeing—to allow for more versatile and economic piece-dyed fabrics that are made from two or more fiber types. Due to differences in chemical properties of fibers, as well as modifications of a single fiber type, varied effects can be created by different types of dyestuffs. The same dye solution can include two or more types of dyes, and each dye type will attach itself to one fiber type or modification in the fabric.

Figure 6.5. In piece dyeing, already structured fabrics are immersed in dye solutions. Solid-colored fabrics frequently result; but special multicolored effects can be achieved by using different fibers in the fabric.

QUICK QUIZ 5

1. Dyes or pigments can be added to color fabrics during what four stages of manufacturing?
   ____________, ____________,
   ____________, ____________
2. Which is used most often and why? ____________
   ____
3. How might you determine differences between a solid-colored fabric which has been piece dyed and one that has been solution dyed?
   ____
   ____
4. In order to achieve a solid-colored fabric made from two or more fibers, ____________ dyeing would be required because ____________
   ____________.
5. If a combination fabric—one made of different types of single-fiber yarns—were cross dyed, a ____________ dyed effect could be achieved. The typical designs that would result include ____________.

****

ANSWERS QUICK QUIZ 5

1. solution, fiber, yarn, fabric piece
2. piece; beacuse it is the most economical process, allowing color decisions to be made after fabric structuring
3. By removing separate yarns, it is frequently possible to observe that dye penetration is not as thorough on piece-dyed fabrics as with solution-dyed goods.
4. union; different fibers take color differently, and dyestuffs would have to be combined to dye the different fibers uniformly
5. yarn; checks, plaids, stripes

If two dyestuffs are selected to produce the same hue on each fiber type involved—called union dyeing—a solid-colored fabric results. If dyes that produce different colors on different fiber types or modifications are used, a fabric with either the appearance of a yarn-dyed or fiber-dyed product can result. This is called cross dyeing.

If a fabric made of blended single yarns were cross dyed, the resulting fabric would have the appearance of a fiber-dyed item. A tweed or heather appearance can thus be produced relatively inexpensively.

When a solid color is desired in a blended or combination fabric, different types of dyestuffs must be selected that work on each fiber type; to give a solid uniform color, union dyeing must be used. To get the most uniform coloring on a piece-dyed fabric, whether it be union or cross dyed, high temperatures are used in order to ensure maximum penetration of the dyestuffs.

Consumers do not normally need to identify a solid-colored fabric as being either piece, fiber, yarn, or solution dyed. Sometimes the difference is negligible, since piece dyeing methods are sophisticated and dye penetration can be quite uniform. In fact, more solid-colored fabrics are piece dyed than by any other method. Piece dyeing allows the converter to keep structured fabrics on hand and dye them as color orders are received to meet changes in fashions. Fabrics structured from fibers dyed by solution, fiber, or yarn dyeing are already colored; therefore, changes in fashion can mean fabrics might not sell, causing financial hardship to manufacturers.

Solution dyeing or mass pigmentation is not as expensive as some other methods; however, determination of color must be made at a very early stage in the textile manufacturing process. This results in a need to maintain large stocks of fabric in a wide variety of colors. *Vinyon* in film form, *acetate,* and *olefin* are the major synthetics which are solution dyed because of their relatively low dye affinity.

Garment dyeing is a relatively new technique employed on apparel products. In this case, the completed garment is inventoried in the greige stage and is dyed as orders are received from retailers. The principle advantage of dyeing in the garment stage is rapid response to fashion change. Consistency of shade and shrinkage are difficulties that garment dyers face. Garment dyeing has been successfully employed on sweaters and on garments with a worn casual appearance. It is anticipated that the techniques used for garment dyeing will improve so that smoother, neater looks can be achieved. Garment dyeing will likely continue to be used in fashion end uses where quick response to consumer demand is important.

This discussion has included some basics of coloring methods and the application of dyes or pigments. The many dyestuffs on the market and the complexities of coloring procedures have not been discussed, nor are they necessary for a consumer approach to evaluation of serviceability of the textile product. Rather, a consumer should rely on possible information provided on fabric labels or hangtags to determine if a fabric will meet the criteria of colorfastness to various environmental considerations for the proposed end use. However, labels may not provide adequate information, since no laws require information concerning colorfastness properties.

UNIT SIX

Part II

Applied Design

Surface designs may be added to fabrics by a variety of methods. The most important are *printed applications* usually done by one of three ways: *direct, resist,* or *transfer* printing.

Direct printing includes *roller printing, block printing,* seldom used for quantity printing and *transfer printing,* a relatively new method for applying surface designs to fabrics. It is used mostly on synthetic fibers, primarily polyester, for short runs of fabric. Resist methods include *screen, stencil, tie-and-dye,* and *batik* printing; screen printing is the most used commercial resist method. With the exception of roller and screen printing is the most used commercial resist method. With the exception of roller and screen printing, the other methods of resist and direct printing are limited primarily to craft applications. *Discharge printing* is also infrequently used today.

Each of the above methods is described and discussed in this part. Roller printing was the most common method of printing fabrics for a number of years. The time-consuming processes necessary for roller printing and the need for production efficiency have led to a significant decrease in the use of this technique, however. Changes in the technology for rotary screen printing have made it a faster and more cost-effective method of printing fabric. The majority of printed fabrics in use currently are printed by rotary screens. Regardless of the technique used, it is very difficult for the consumer to differentiate between fabrics printed using the various methods. Some clues will be presented in the following section which can assist in distinguishing one type of print from another. For many fabrics, however, you will not be able to determine which printing method was used. As a consumer, the type of dye or pigment used and its colorfastness is more important than the manner in which it was applied. Since this information is often not available to the consumer, you must trust the manufacturer to select the most appropriate coloring agent and method for the fabric and for the end use.

Direct Printing

Direct printing occurs when dyes or pigments are applied directly to the surface of the fabric in a planned pattern. Block and roller printing applications are explained in this section.

Block Printing

Designs have been applied to fabric by block printing for many centuries. Design was first applied to flat fabric by carving patterns into wood blocks and adding color to the raised areas; the design was imprinted by stamping the block onto the fabric surface.

Block printing is a slow and tedious process. The blocks are generally quite small and have to be repeatedly stamped on a fabric to achieve an overall design as shown in Figure 6.6. However, certain high cost specialty fabrics are block printed and are valued for the handcraft involved.

Roller Printing

The principle of carving designs into a surface and printing them on fabric has been adapted to various machine processes which allow for high speed printing. In the roller printing process, designs are etched into copper rolls. Preparation of the copper rolls is time consuming and expensive; therefore, this method is best when *used for printing large amounts* of fabrics. For this reason, the roller printing technique has declined in use and today accounts for a relatively small percentage of printed fabrics.

The design is transferred to the fabric as the roll passes over the fabric. If more than one color is involved in the design, there must be a separate roller for each individual color. Figure 6.7 shows a roller printing machine.

If a fabric has three different colors in a print pattern, three rollers are needed to apply the design and the proper colors. Each roller is etched or indented in

Figure 6.6. The carved block shown was used to achieve the printed-on fabric design.

certain portions to apply color in different areas for the desired design.

A dye trough is at the base of each roller—the roll can either turn directly into the paste solution, or an applicator roller can turn in the paste solution and then roll onto the design roller. A "doctor blade" scrapes off excess dye from the printing roll so that only the etched portion contains color. The roller then rotates over the fabric surface and prints the portion of the pattern that it provides.

The rolls can be cleaned and used for different color combinations so long as the actual design or pattern is not changed. A limitation of direct printing is that the repeat size of the design cannot be any larger than the diameter of the rolls used, so *relatively small repeat designs* are produced.

Figure 6.8 shows an example of a roller-printed fabric. The color difference between the front of the fabric and the back occurs because the dyes are applied to only one side of the fabric; they seldom completely penetrate to the back unless the fabric is extremely sheer.

The wrong or back side of prints is not as intense as the front, unless a modification of roller printing—called *duplex printing*—is used. In this process, the fabric moves between two sets of engraved rollers so that both sides of the fabric are imprinted at the same time. Duplex printing can provide the same design on each side of the fabric, or totally different patterns can be printed simultaneously on the two different sides.

Interesting design effects can be achieved in woven fabrics by roller printing only the warp yarns prior to the actual weaving operation. This process is called *warp printing*. The printed warp yarns are generally interlaced with white or solid-colored filling yarns, which produces a soft, hazy design as shown in Figure 6.9. Warp printing is typically used on upholstery or drapery fabrics of satin weaves, or on rib weave fabrics such as taffeta and faille. The design on the back of a warp printed fabric appears even less distinct than on the front, because the warp yarns are roller printed on one side only.

Figure 6.7. Commercial roller printing allows designs to be applied to large yardages of fabric.

Figure 6.8. This design was applied to the fabric surface by roller printing; a separate roller was used for each color.

One potential problem with high-speed roller printing is that it can result in off-grain designs if the process is not carefully monitored. When this occurs, the design lines are not parallel to the crosswise or lengthwise grain, as pictured in Figure 6.10.

Printed patterns which are most difficult to handle if applied off-grain include large checks, plaids, or geometric designs with specific lines that are especially noticeable when not parallel and perpendicular to warp and filling yarns. This is because the design is more visible than the grain, and products are usually made with the design appearing straight. The customer should examine printed fabrics carefully

Figure 6.9. The warp yarns were printed with a floral design prior to interlacing with white filling yarns.

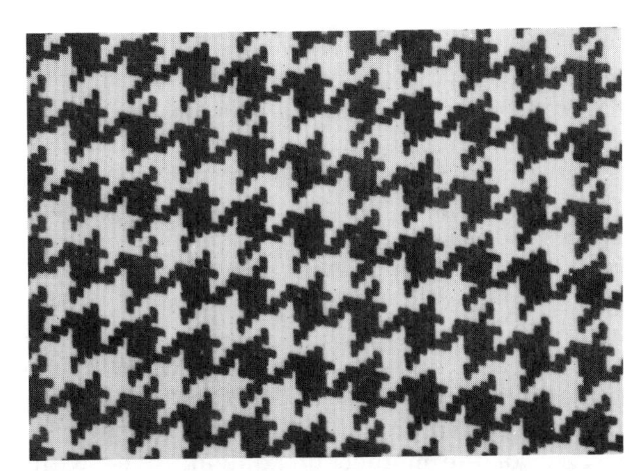

Figure 6.10. To show off-grain printing, the fabric pictured was cut on the crosswise grain. Notice how the design is misaligned with the fabric grain.

before purchase to make certain patterns are on grain or products will not hang straight nor wear well.

Resist and Discharge Printing Methods

In resist printing some material is used to protect the fabric at points where the color is not to be applied; thus, the fabric resists the color at those locations. For each color used, a new resist medium must be reapplied or used to protect the fabric.

In discharge printing, the fabric is dyed a solid color, and then the color is discharged or removed at the points where the design is to occur. In some instances a new color may be added at the discharged design area; however, in many fabrics, the discharge area remains white.

Resist methods discussed in this program include *stencil, screen, tie-and-die,* and *batik* printing. The discharge method described herein is the chemical procedure.

Stencil Printing

In addition to printing the design onto fabric using engraved copper rolls or blocks, a design may be cut into stencil sheets made of cardboard or paper coated with wax, oil or varnish; or from thin sheets of metal. The cutout sheets are placed on the fabric, and thickened dye solutions are applied by brush or spray onto the stencil sheet. The dye penetrates the fabric only in the cutout areas; a separate stencil is generally prepared for each color. The uncut portions of the stencil prevent color penetration. Today, stencil printing is primarily a handcraft and is seldom used in mass production because it is slow and expensive.

One difficulty with stencil printing is that the design areas must be connected to prevent parts of the stencil from falling out. To offset this problem, Japanese stencil artists developed a method of tying the various sections together with silk filament or human hair in order to produce intricate stenciled fabrics as shown as Figure 6.11.

However, most stencil printing yields relatively coarse designs, with obvious connections in the stencil which prevent dye passage at selected points. Because of the method of applying dye—with brushes or sprays—there may be a somewhat raised texture since the dye can build up in the cut out areas. In order to increase production speed, the limited amount of commercial stencil printing is done today with air brushes rather than the hand brushes originally used.

Figure 6.11. This fabric was printed from a hand-cut Japanese stencil.

Screen Printing

A refinement of stencil printing is screen printing. Essentially screen printing is a stencil process—instead of a stencil sheet cut out with background areas left which resist color, a fine screen made of silk, nylon, polyester or metal filaments is used. The screen is covered with some type of nonporous material in all areas except those where the design is to be printed. Just as in stencil printing, the nonporous areas prevent the dye from reaching the fabric by resisting color penetration.

In Figure 6.12, a handcrafted silk screen is shown with the fabric design that results. Both screen and stencil printing are slow and cumbersome when done by hand, but automation has made screen printing feasible for commercial use.

Figure 6.12. A handcrafted screen and the fabric printed from it are shown.

Figure 6.13. Screen printing was a hand process for many years; however, automated machines are now used to produce large designs and short-run yardages. *(courtesy American Textile Manufacturers Institute, Inc.)*

There are two basic commercial methods of screen printing—*flat bed* and *rotary screen*. The rotary screen process is used most frequently at present for producing a variety of designs at high speed. Since rotary screens are larger than the rolls in direct roller printing, design repeats can be considerably larger. Figure 6.13 and 6.14 are illustrations of the flat bed and rotary screen printing processes.

As shown, both rotary and flat bed screen printing call for the fabric to be held taut on a flat surface; whereas in roller printing, the fabric passes at high speed beneath and against the etched out rollers.

Figure 6.14. Certain principles used in automated screen printing and roller printing are combined in rotary screen techniques; circular screens are used to apply design to flat fabric. *(courtesy American Textile Manufacturers Institute, Inc.)*

Since the structure remains flat and stable during screen printing, it is preferred over roller printing for loosely structured or easily distorted fabrics.

If a multicolored design is to be screen printed, there must be a separate screen for each color applied, as in roller printing. The design in roller printing is limited to a repeating pattern whose size is determined by the diameter of the relatively small copper roll. The screen printing process, on the other hand, allows for relatively large designs, especially with flat bed screens. Any limitation is more apt to be determined by the width of the fabric and type of design desired.

Originally, screen printing was considered feasible only for printing relatively small yardages of fabric with large designs. Now with the development of rotary screens, screen printing can be done as fast or faster than roller printing. Plus, rotary screen printing provides increased flexibility in design. Shorter lengths of fabric may be economically printed, since rotary screens are faster and less expensive to prepare than the etched copper rolls required for direct roller printing. The result is that rotary screen is currently the most widely used method of printing fabric.

Tie-and-Dye

A resist method of design was developed years ago by natives of India, Indonesia and Africa. The fabric was tied with waxed thread or some other material to create tightly gathered portions of fabric in certain places. Then, the tied fabric was piece dyed. In Figure 6.15 a fabric before and after tie dyeing is shown.

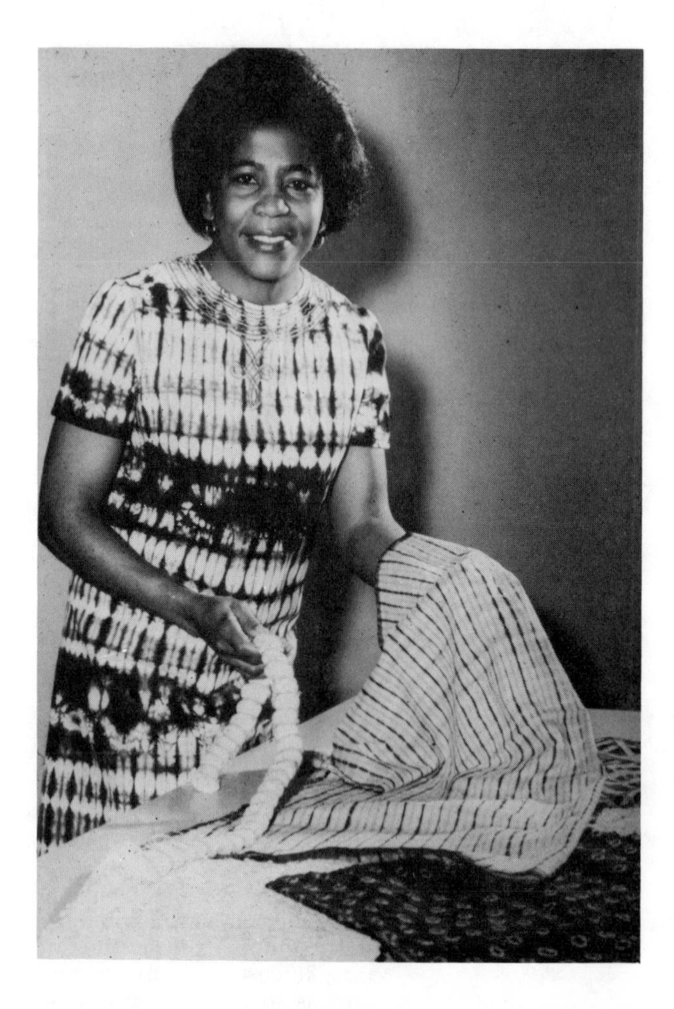

Figure 6.15. This Nigerian woman is wearing an indigo tie-dyed dress she designed; also, she has another tie-dyed pattern that resulted from dyeing the tied sample shown.

QUICK QUIZ 6

1. Although the consumer cannot be absolutely certain whether direct roller printing or resist screen printing methods have been used, what might provide a clue as to which was?
   ____
2. Block, screen and stencil design methods all developed as handcraft processes for printing fabrics. Match which are still primarily craft processes and which are used commercially.
   ____ 1. Block   a. commercial
   ____ 2. Screen   b. craft
   ____ 3. Stencil
3. What consumer problems can the user expect with a fabric that is printed off grain?
   ____

****

ANSWERS QUICK QUIZ 6

1. The size of the design—if there are very large designs, screen printing was probably used.
2. 1-b; 2-a; 3-b
3. Products will not hang straight nor wear well.

Because the dye penetrates unevenly, and in some cases passes through the thread, interesting designs are created.

Different colors and designs can be applied by untying and retying the thread at various places in the fabric. Today tie dyeing is done largely as a craft. Modern tie-and-dye prints are frequently recognized by large splashy areas of color with varying hues of the same color. Since tie dyeing is a hand-coloring technique, the cost of commercially prepared fabric is relatively high. In fact, commercial fabrics which appear to have been tie dyed were likely printed by either roller or screen printing methods.

Batik

Another hand process which has been revived in recent years is batik. Figure 6.16 illustrates a batik fabric whose design was formed by applying wax to certain areas of the fabric. The fabric resists the dye in the areas covered by the wax. If the wax is cracked, the effect resulting may resemble an ancient painting by an "old master." The waxing process is repeated if more than one color is desired. After the wax is set, the fabric is piece dyed, and the process is repeated as many times as necessary to create the planned pattern. All wax is removed prior to final finishing.

Figure 6.16. In batik fabric, wax is used to make the fabric resist dye in certain areas.

Discharge printing

This method yields a printed fabric which looks very similar to a roller printed fabric, except its background is darker than the applied design. Compare the backgrounds and back sides of fabric 1, a roller print, with fabric 13, a discharge print.

A discharge printed fabric is first dyed so that the background color is uniform, and usually dark, on both sides. In a roller printed fabric, all colors are applied to the fabric front. To get the lighter colors to show up on the dark piece dyed background in a discharge print, part of the background color has to be "discharged" or removed before the lighter colors can be printed onto the fabric.

A reducing bleach is applied to the fabric front in the areas where a lighter design is desired. The discharged design areas may be left white, or additional color can be applied either by roller printing methods, or the dyes can be mixed with the discharge chemicals in order to remove the background color and apply additional color in one operation. The added dyes have to be resistant, of course, to the discharge bleaching chemicals.

Likewise, fabric 13 in the sample set is a discharge-printed velveteen, in which the reducing bleach did not remove color throughout the fabric; the back side retained its original piece-dyed color. On the fabric front, however, the background color was discharged and removed in areas where the lighter paisley design was applied.

In Figure 6.17, the reducing bleach was added to the areas where there are spots and wavy lines. The bleach penetrated throughout the fabric and removed most of the black from the background in the design areas. However, in pile fabrics such as the corduroy shown in Figure 6.18, the reducing bleach may not penetrate completely through the fabric; thus, the reverse side will retain most of the dark piece-dyed background color, and only the fabric front will have the discharge design effect.

Compare the roller and discharge printed fabrics shown in Figures 6.8 and 6.17, and note that the piece-dyed background of the discharge print appears the same on both sides. The roller print has more intense color on the fabric front.

Some roller-printed fabrics have darker backgrounds than the design areas, as in fabric 11. Its background color was rolled on the front just as the

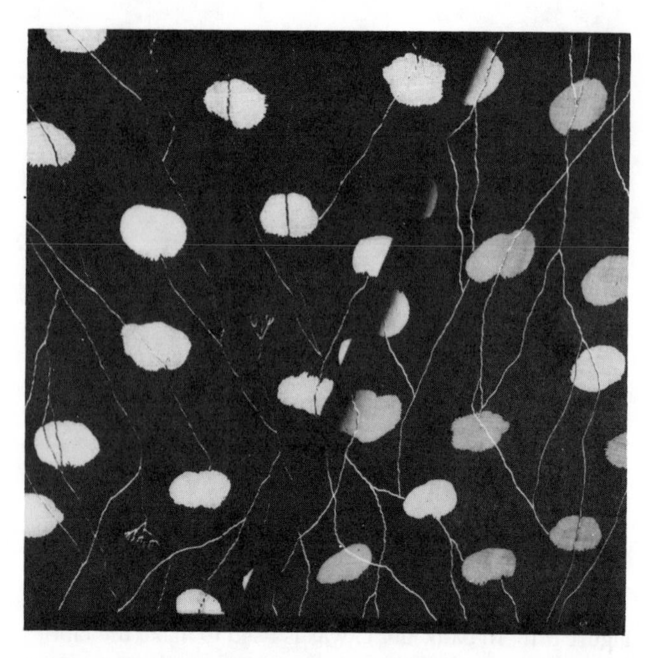

Figure 6.17. When light-colored designs are to be applied to dark, piece-dyed backgrounds, discharge printing can be used.

Figure 6.18. On heavy or thick discharge printed fabrics, the reducing bleach used to remove color on the fabric face may not penetrate to the back of the fabric.

lighter designs were. Compare both sides of samples 11 and 13. The resulting fabric fronts both have backgrounds darker than portions of the applied design, which is characteristic of discharge-printed fabrics. But an examination of their back sides shows the differences—the color was rolled only onto the front of fabric 11; whereas the discharge print in sample 13 was piece dyed, giving a background color which is the same on both sides.

Discharge printing is declining in use because of the required extra steps of piece dyeing, bleaching out, and applying the design colors. Consequently, roller or screen printing methods are often used when a dark background color is desired. There may be some trade-off in that dark dyes do not penetrate the fabric as well when printing on one side only, so there may be somewhat reduced colorfastness.

Transfer Printing

Transfer printing, developed during the 1960's, consists of first printing the design on paper and then transferring it to the fabric by either dry or moist methods. The primary advantages are that it requires less water than wet printing processes, it tends to minimize the likelihood of seconds resulting from shrinkage and smeared designs and it prevents fabric distortion. However, transfer techniques cannot be used on all fabrics; they tend to be more costly than some other methods and disposal of the printed paper can be difficult.

Transfer printing involves two stages:

1. Printing a design onto paper with special dyes that are selected for their sublimation (vapor phase transfer) properties and their attraction to the fibers to which they will be transferred. See Figure 6.19.
2. Feeding the printed paper and fabric to be printed face side together into the heat transfer equipment. The heat vaporizes the dye on the paper, the dye is absorbed by the fabric and the fabric is cooled.

Currently transfer printing accounts for a relatively minor proportion of printed fabrics. The technique is useful, however, for certain specialized esthetic effects. Samples 25A and 28A, for example, are high fashion fabrics with leather looks achieved through transfer printing. Both samples are knitted fabrics that could easily be distorted if other printing techniques were utilized. Transfer printing allows for the detailed leather-like effect without fabric distortion.

QUICK QUIZ 7

1. Upon first examination of the front of sample 42, it appears to be a ______________ print with reducing bleach added to create the ______________ .
2. Yet, you can quickly determine the fabric is not a piece-dyed discharge print by looking on the back side and noting that ______________ .

****

ANSWERS QUICK QUIZ 7

1. discharge, white spots
2. color was applied only to the front surface

The shine on Fabric 25A which contributes to the leather appearance is also a result of transfer printing.

Another application of the transfer printing technique is on garment pieces. T-shirts and children's sleepwear with cartoon characters are often transfer printed. You may have observed transfer printing in T-shirt shops where you can select a paper design which is then transferred through heat and pressure to the front of a T-shirt.

Photographic Printing

One last specialized process in the application of design through photographic printing. This method produces print designs in a manner similar to that used in developing regular photographs. After fabric is impregnated with a solution that will react to photosensitive dyes, engraved rollers print a negative design on the fabric. The fabric is then placed in a developing bath where the photo is developed just as done in regular photographic printing. After the colors are developed, the fabric is thoroughly rinsed and color is set.

This method of printing is reserved for special kinds of designs that required the extreme naturalness and the half tones of actual photographs. It is expensive.

Figure 6.19. Transfer printing involves the transfer of a design from paper to fabrics.

QUICK QUIZ 8

1. Transfer printing was used on samples 22, 25A, 28A, 36A, and 42, all ______________ structures which benefit from transfer printing because ______________________________

2. Name some advantages and disadvantages of transfer printing:
____
____
____
____

****

ANSWERS QUICK QUIZ 8

1. knit; the process does not stretch or distort fabrics
2. It requires less water, prevents shrinkage, fabric distortion and smears, but it cannot be used on all fabrics, is relatively expensive and the paper can be difficult to dispose of.

REVIEW

1. What handcraft printing processes yield relatively small, uniform designs?
______________________________ and ______________________________

2. What method of printing derived from handcrafted processes permits the application of large designs to fabric? ______________________________

3. Indicate when direct roller printing and flat bed screen printing are likely to be used, considering design size and amount of yardage:
Roller: ______________________________
____
Screen: ______________________________
____

4. A more recently developed process combines concepts of roller and screen printing, called ______________________________ printing. Examine samples 34 and 39 printed by this method.

5. Name three advantages of rotary screen printing over roller or flat bed screen printing:
____
____
____

6. Tie-and-dye and batik are generally considered handcraft methods that are classified as ______________________ printing processes.

7. Discharge prints are first piece dyed; then their designs are added after, rather than during, piece dying as in the ______________________ and ______________________ resist methods.

8. To print a design on a dark piece-dyed fabric, part of the background color must be ______________________ so that lighter colored designs can be applied.

9. Fabric thickness determines how the back of a dischare print appears; the thicker the fabric, the (more, less) of the original piece-dyed color likely to remain on the back of the fabric.

10. Discharge printing is used less frequently today because improved methods of applying color to only one side of the fabric can create discharge printed effects, without the time and cost involved in having to first ______________________ dye and later remove color to allow for the application of ______________________ designs.

11. Fabrics with backgrounds darker than design areas can be produced by applying color to the front by ________________, ________________, ________________ printing methods.

12. Fabrics 11 and 42 both have darker backgrounds than the design areas which were created by ______________________ and ______________________ printing respectively.

13. While it is practically impossible to differentiate among roller, screen or transfer printing processes, a true discharge print usually can be identified by examining the fabric back. How would it differ from the other three processes? ______________________
____

14. Samples 1, 32 and 34 all have printed-on designs applied to ______________________ ______________________. Design was applied to sample 1 by roller printing; sample 32 by transfer printing; and sample 34 by rotary screen methods. Can the consumer differentiate among the printing processes? (yes, no)

15. With the many improvements in printing processes, color can be successfully applied to one side of the fabric with adequate dye penetration through the structure for most end uses. Thus, the use of discharge printing will continue to decline, while faster, more versatile and economical methods like ______________________ or ______________________ will increase.

****

REVIEW ANSWERS:

1. stencil, block
2. screen
3. Roller is suited for applying *small designs* on *large amounts of fabric.* Screen is needed for *large designs,* usually applied to *small yardages of fabric.*
4. rotary screen
5. Rotary screens accommodate larger designs than the etched copper rolls; they are less expensive to prepare; printing is faster than flat bed screen printing.

6. resist
7. tie-and-dye, batik
8. discharged or bleached out
9. more
10. piece, lighter colored
11. roller, screen, transfer
12. 11-roller, 42-tranfer
13. The piece-dyed fabric background would appear the same color on both sides; whereas in roller, screen and transfer prints, the color is more apparent on the fabric face.
14. the fabric front or one side only; no.
15. rotary screen or transfer

UNIT SIX

Part III

Design by Fabric Structuring and Finishing

Coloring basics were introduced in Part 1 of this unit, followed by identification of printing methods that add pattern and design in Part 2. This section expands on adding design through certain types of fabric structures and finishes presented in earlier units.

Design by Weaving

Many designs can be produced in fabric through the manipulation of weaving processes. Even simple, basic weave structures can add pattern. For example, a diagonal design can be created by the twill weave and a smooth surface by the satin or sateen weave. More elaborate woven-in designs are produced by combinations of basic weaves or through using the dobby and jacquard attachments. The fabric illustrated in Figure 6.20 is a dobby weave, which can be recognized by its repeating geometric design. The coloring was likely applied by yarn dyeing.

The woven-in design in Figure 6.21 was made with a jacquard attachment, which allows for individual control of the warp yarns. Because the fabric pictured has a dimension effect, you can conclude that the design was achieved by a combination of pile and jacquard weaves, accentuated through yarn dyeing.

In addition to all over designs, both the dobby and the jacquard attachments can be used to produce spot weaves, shown in Figure 6.22, or add border designs in fabric. These attachments can also yield interesting patterns in combination with pile weaving, in which the pile can be uniform or can be sheared in certain areas to achieve a sculptured design effect that is often used in carpeting.

Novelty yarn effects add important aspects of fabric color and texture. Flecks of color are added to a fabric by using fiber-dyed tweed yarns; a crinkled or pebbly surface is created by crepe yarns; a looped, fuzzy surface can be produced by several complex yarns, including chenille, loop, bouclé or ratiné.

But even the most basic textile unit, the fiber, contributes design quality to fabrics. Nontextured filament fibers yield smooth, lustrous surfaces compared with staple or textured fibers, which produce rougher, less uniform surfaces.

Design by Knitting

In addition to woven-in patterns, design effects are created by knitting. The fabric illustrated in Figure 6.23 is a double knit with a repeating geometric design. These types of designs are formed using jacquard attachments on knitting machines to yield complex structured designs.

Many modern knitting machines use computerized controls to form the patterns. While they operate somewhat like the older jacquard controls, computer controls are small and fit into the machine so that each yarn can be knitted in whatever pattern is desired. These computer controls can be changed quickly to make new designs in a matter of minutes.

Figure 6.20. Design by weaving on a dobby loom.

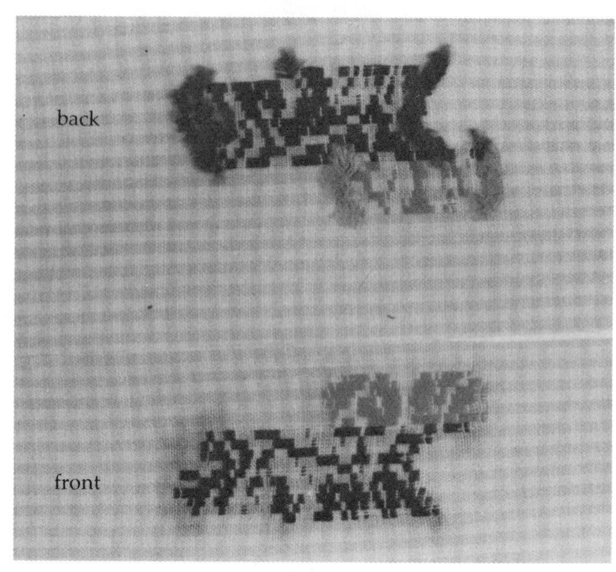

Figure 6.22. Extra warp or filling yarns can be used to "spot weave" varied designs.

Design by Finishing

As discussed in Unit 5 about Finishes, many finishing applications are applied solely for esthetic affects. For example, *flocking* is a finishing process used to create pilelike designs on the fabric surface. *Embossing* and *chemical crepeing* create three-dimensional surface designs, while *moiréing* produces a watered, rippling design on the fabric.

Flocking can create varied pile designs, depending on how the flocked fibers are applied—the fabric surface can be flocked all-over to look like suede, or only in design areas to create floral patterns as in Figure 6.24. A flocked design appears more dense

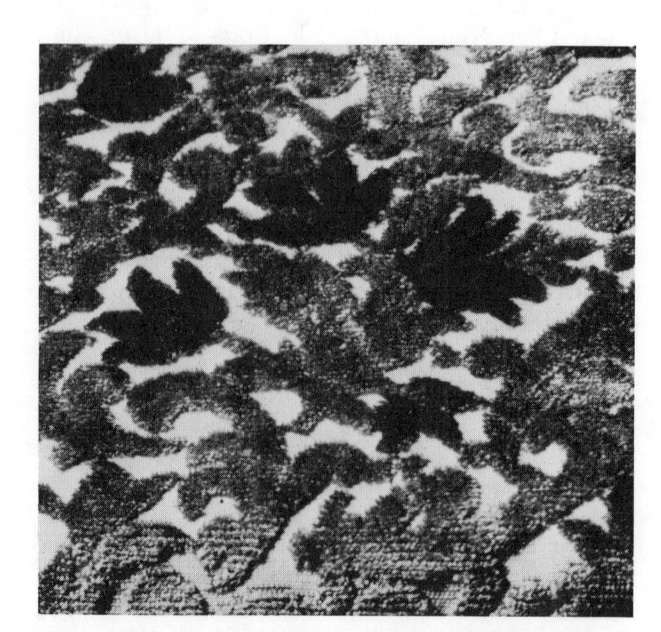

Figure 6.21. Design by weaving with a jacquard loom.

Figure 6.23. Knitted-in complex designs can be created with jacquard attachments or computerized controls.

QUICK QUIZ 9

1. Examine piece-dyed fabrics 16 and 17, which have their design effects created by the ____________.
2. Even greater pattern interests result when dobby or jacquard fabrics are ____________ dyed, as in fabrics 18 and 19 which have their designs created through ____________ attachments.
3. An extra set of warp or filling yarns woven into a fabric structure creates a looped or cut pile effect, which results in dimensional surface designs. Identify the four pile fabrics in the sample packet numbered: ____________, ____________, ____________, and ____________.

****

ANSWERS QUICK QUIZ 9

1. dobby weave
2. yarn; jacquard
3. 12, 13, 14, 15

---

and luxurious the straighter and closer the fibers stand. Colored effects are achieved through the use of fibers that are solution or fiber dyed before they are applied to the fabric; or the floral surface may be roller printed after being glued to the fabric.

Three-dimensional designs also are applied to fabrics through the use of heated embossing rollers or through chemical crepeing processes. These finishes can be made durable in two ways—the fabric can be treated with a resin; or if it is made of thermoplastic fibers, it can be heat set to ensure permanence as long as certain care procedures are followed. Frequently, embossed fabrics have color applied, with results like those pictured in Figure 6.25.

Chemical crepeing produces a crinkled dimensional design, typically applied to solid-colored, sheer woven structures. *Plissé* is the fabric name when a crinkled surface results from this process.

Moiré fabrics are the last to be reviewed in this section on design by finishing. Their watered de-

Figure 6.24. Finishing with flocked fibers created this dimensional floral pattern.

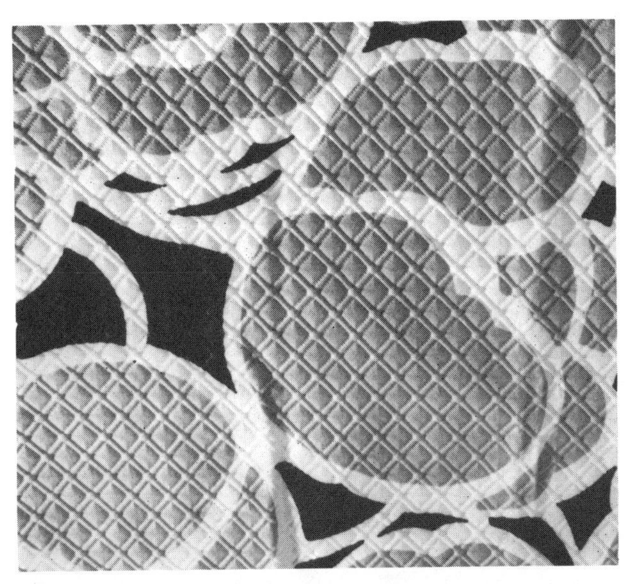

Figure 6.25. A three-dimensional fabric surface was created through embossing and screen printing.

QUICK QUIZ 10

1. The knitted pattern in fabric 25 was created on a double knit machine, which had either a jacquard attachment or a ______________________ control system that works like the jacquard attachment.
2. What coloring method was most likely used in making double knit fabric 25?

____

****

ANSWERS QUICK QUIZ 10

1. computerized
2. yarn dyeing

signs are created on rib weave fabrics through the use of rollers which press the ribs flat in certain areas to create the wavy effects illustrated in Figure 6.26.

Designs Stitched onto the Fabric Structure

Embroidery is an applied design method that can be added to fabrics by hand, or more commonly, machine stitching. Examination of the machine embroidered fabric in Figure 6.27 discloses that the design was created by stitching extra yarns onto the fabric surface.

Popular embroidered designs now on the retail market include the monograms used to personalize shirts, blouses, sheets and towels, and the small animals fastened to name-brand knit shirts. Because design is applied to an already structured fabric, the ground fabric will remain intact if some embroidery yarns are pulled out.

Other stitched-on design applications, also not integral parts of the base fabric, include *appliqué* and *quilted* structures. Appliqué is composed of designs cut from fabric which are sewn or embroidered onto another fabric surface. The process is used primarily on children's clothing and is produced in limited quantities commercially. Appliqué and embroidery are both easily recognized because of the extra dimension added to the fabric by the additional yarns

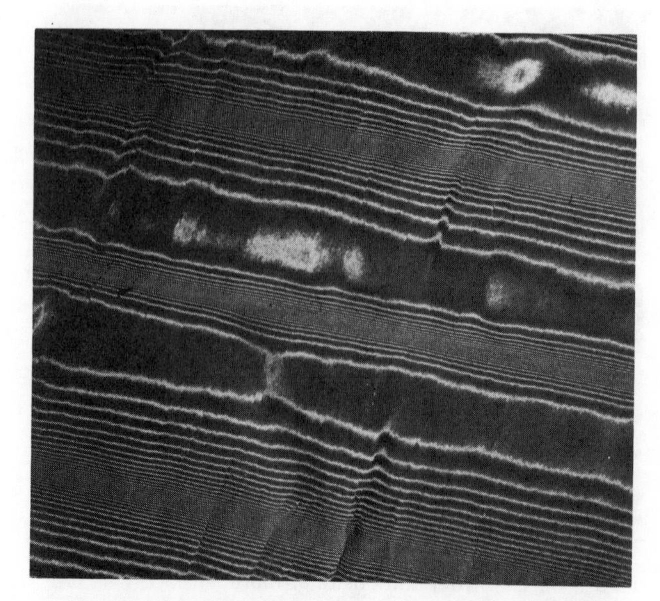

Figure 6.26. A watered effect is created by finishing.

Figure 6.27. Extra yarns are applied to an already-structured ground cloth to create embroidered designs.

QUICK QUIZ 11

1. Which fabric in your packet was flocked? ____________ The flocks were added to create a ____________________ design at considerably less cost than spot weaving.
2. Samples 21 and 41 both have crinkled designs. Give their fabric names and indicate how the designs effects were achieved in each.
   21: ______________________________
   41: ______________________________
3. Finishing rollers can be used to create dimensional design effects on ____________ fabrics and watered design with ____________ finishes.

****

ANSWERS QUICK QUIZ 11

1. 40; dotted swiss
2. 21—seersucker, slack tension weaving; 41—plissé, chemical crepeing
3. embossed; moiré

and/or fabric stitched to the surface, shown in Figure 6.28.

Quilting, a multicomponent fabric discussed earlier, is made by stitching together two fabrics with or without an insulating layer of fibers between. The stitching which joins the two layers together in a quilted structure is often quite intricate and may form geometric or floral design patterns, but these patterns do not classify as woven-in jacquard or dobby designs. Designs produced for commercial sale that involve embroidery or quilting probably have been done by machine, since hand applications are rare and very expensive.

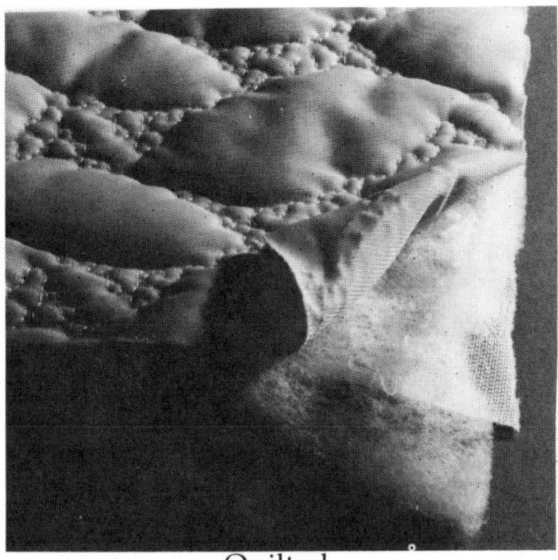

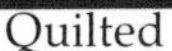

Quilted

Appliqué

Figure 6.28. Quilted and appliqué fabrics require extra stitching on the fabric surface, as does embroidery, to create specialized designs.

UNIT SIX

Part IV

Review of Color and Design Applications

Colorfastness is generally high on the requirement list for textile products. The elements to which a fabric color should be fast are determined by the end use of the product. If fabric 6 were used in an everyday shirt, it should be colorfast to laundering; while an upholstery fabric as sample 14 should be colorfast to crocking. The two major environmental conditions often causing color loss in drapery fabric are sunlight and atmospheric fumes or smog.

Color can be added to textile products at four manufacturing stages: to the fiber spinning solution and to the fiber, yarn or fabric. The most expensive coloring process is fiber dyeing, a process used to create tweed or heather effects. Mass pigmentation or solution dyeing is used with certain manufactured or synthetic fibers which are difficult to dye. It is also an expensive dyeing method in that color must be determined early in the production process, resulting in converters having to maintain large stock inventories.

Color chemists have developed two ways of offsetting the naturally low dye affinity of synthetic fibers such as polyester:

1. the temperature of the dye solution can be raised, causing the fibers to swell and open up dye sites in the molecules;
2. a dye carrier can be used to help the dye penetrate the fiber molecules.

Woven-in colored designs which are reversible can be achieved in yarn-dyed fabrics. Also, single fiber component yarns of different fiber content used in a fabric can be cross dyed to create these effects. But the most economical coloring method is piece dyeing, because the fabric can be structured and left in the grey goods state until orders are received.

Piece dyeing has an advantage over other dyeing methods since the color does not have to be determined before the fabric is structured; thus lower cost fabrics can be produced. However, solution dyeing or mass pigmentation, and even fiber or yarn dyeing, have an advantage when thoroughness of dye penetration is most important.

Dye technology has improved significantly and a majority of solid-colored fabrics are successfully piece dyed to yield adequately colorfast products. Examine fabric samples 2, 3, 4, 9, 12, 14, 15, 16, 17, 21, 26, 27, 28, 31, 38 and 41, which were likely piece dyed. These fabrics represent a variety of fiber contents and fabric structures of varying weights, and the dye penetration appears adequate for normal end uses for which the samples were produced. Many of these piece-dyed fabrics contain blended yarns. Therefore, to assure uniform color of the different fibers present, union dyeing methods were required.

Fabrics woven on dobby or jacquard looms are frequently structured of yarn-dyed yarns in order to create multicolored, woven-in fabric designs. Sample 19 is a yarn-dyed fabric made on a loom with a jacquard attachment. It is a double-cloth structure in which four sets of yarns were used—the fabric name is matelassé.

End uses for matelassé include dressy apparel fabrics and home furnishings. The matelassé sample is obviously designed for use in upholstery. The second set of yarns serving as a supportive layer, adds to the raised design effect. This effect could be a disadvantage in either upholstery or apparel fabrics if there is uneven abrasive wear on the fabric surface which can destroy its appearance. Brocade, sample 18, also has a raised woven-in design created with the jacquard attachment. However, matelassé tends to be more resistant to wear because it has a supportive layer, whereas brocade is a single-layered fabric.

Variations of basic weaves and combinations of weaves also contribute to woven-in design effects. Sample 2 is a variation of the plain weave with striped effects created by varying the warp yarn count. The pattern in sample 43 was formed by combining the plain, twill, and leno weaves. Cut swatches from samples 17, 18, 21 and 43; attach and label them under Design by Weaving on the appropriate Mounting Sheet.

Structural design can also be created through knitting. The design in sample 25 was made using the jacquard attachment on the knitting machine or by using computer controls. The coloring method which contributed to the design in sample 25 was yarn dyeing. The type of knit structure is filling double knit.

Warp knitting machines have great versatility in adding pattern to fabric, particularly the raschel kind, which was used to produce samples 29 and 30. Attach and label swatches from samples 25 and 29 under Design by Knitting.

A discussion of structural design would be incomplete without mentioning the varied complex and novelty yarns that are woven or knit into fabrics for special effects. In your yarn set there are several multi-ply yarns that create surface designs, such as seed, bouclé and ratiné yarns.

Single yarns also can have characteristics that increases the esthetic appeal of fabrics. In sample 3, the thick-and-thin slub yarns contribute to its nubby surface; and in sample 10, the smooth filament-length yarns are woven in a satin weave to create a smooth lusterous surface.

From studying your fabric set, however, you can conclude that a majority of yarns used in fabrics are relatively uniform, simple and basic. This is because varied or loosely twisted complex yarns reduce fabric serviceability through potential snagging, decreased surface abrasion and strength.

In addition to the structural ways that design can be added—by fiber or yarn type and fabric structuring—printed designs are important ways of adding novelty to fabric. Important commercial printing methods are roller, screen or rotary screen, and transfer.

Samples 1, 11, 22, 32 and 34 are examples of the commercial printing methods most frequently used today. You learned earlier in this unit the printing application method used on each, because a consumer cannot distinguish among roller, transfer, and rotary screen printing methods. Recall that samples 1 and 11 are roller printed, so where they are attached on the Finishes Mounting Sheet, label them roller printed. Then cut, mount and label samples 22, 25A, 28A and 36A as transfer, and samples 32 and 34 as rotary screen under Design by Printing on the Color and Design Mounting Sheet.

In sample 13, you can distinguish that discharge printing was used because the fabric was first piece dyed. Mount and label it as discharge printed under Design by Printing. As a further review, examine printed samples 39 and 42. Sample 42 is transfer printed, which requires the use of designs printed on paper. Sample 39 is a rotary screen print which combines the advantages of roller and screen printing.

Two advantages woven or knitted designs have over printed-on designs are that their structured-in designs will be on grain, whereas applied designs might be printed off grain. Further, structured-in designs are frequently yarn dyed, which means that dye penetration will be greater than when the dye is applied to the fabric surface.

Color and Design Identification

1. Given the fiber contents listed, indicate the probable stage at which color was applied and why?
   Sample 3 a 100 percent silk ____________
   ____
   Sample 6 a polyester/cotton blend ____________
   ____
   Sample 7 a 100 percent olefin ____________
   ____
   Sample 10 a 100 percent acetate ____________
   ____
   Sample 31 a 100 percent polyester ____________
   ____

Now, cut, attach and label swatches from samples 3, 6, 7, 10, 28A and 31 to the Color and Design Mounting Sheet as representative of different coloring stages.

2. Dimensional designs are added by structuring an extra set of yarns into pile weaves and knits. Samples 12 and 13 are ______ pile weaves named: ______ and ______. Samples 14 and 15 are ______ pile weaves named: ______ and ______.
3. Design has been added by weaving to samples 16, 17 and 21, which were structured on ______ looms. Weaving contributed to the patterns present, since these fabrics were all ______ dyed.
4. Samples 16 and 17 are classified as dobby weaves, and sample 17 is named ______. Sample 21 also is made on a dobby loom, but by ______ weaving; the fabric name is: ______.
5. Samples 1, 22 and 34 have printed-on rather than structured-in designs. What is a first step in determining the printing method used? ______
6. The above samples all have color applied to ______. Thus you know that ______ printing was not used.
7. Finishing methods play an important role in creating design. List six methods of finishing which are used primarily for design purposes.

   ______ ______
   ______ ______
   ______ ______
8. Match whether the primary design effects were created by structuring or by finishing in the samples listed below:

   ___ 5 a. structuring
   ___ 12 b. finishing
   ___ 16
   ___ 18
   ___ 21
   ___ 30
   ___ 40
   ___ 41
9. Indicate the finishing methods used to create esthetic effects on the following samples:

   1 ______
   5 ______
   28 ______
   36 ______
10. Cut a swatch from sample 40 as representative of design applied by finishing. Mount and label it as a ______ finish on the Color and Design Mounting Sheet.

Color and Design Identification Answers:

1. 3: piece dyed—least expensive way to dye natural fiber silk, which takes color well
   6: yarn dyed—to create fabric design
   7: mass pigmentation/solution dyed—to provide adequate color penetration on nonabsorbent olefin; observe fiber and yarn dyed effects
   10: solution dyed—to assure dye penetration on acetate, which is difficult to dye and make colorfast
   31: piece dyed—least expensive way to dye polyester, which can be adequately colored at this stage
2. filling; corduroy, velveteen; warp; velvet, terry velour
3. dobby; piece
4. waffle cloth; slack tension; seersucker
5. Examine the fabric back to determine if color was applied to the fabric piece or on one side.
6. one side only or the fabric front; discharge
7. any six: embossing, moiréing, ciréing, flocking, chemical crepeing, glazing, acid etching or burn out, parchmentizing
8. 5-b; 12-a; 16-a; 18-a; 21-a; 30-a; 40-b; 41-b
9. 1-glazed; 5-moiréd; 28-napped; 36-embossed
10. flocked

Regardless of the coloring or design application on a fabric, the important consideration is to read the label for any information that might be given concerning colorfast properties and guarantees. Then the consumer must decide if the fabric will be serviceable for the planned end use.

Posttest

1. Match the textile products with each type of color loss that might occur during the product's normal use and care. More than one may apply.
   ____ 1. solution-dyed carpet
   ____ 2. unlined floral print drapery
   ____ 3. roller-printed striped window awning
   ____ 4. piece-dyed sleeveless evening gown
   ____ 5. yarn-dyed plaid sports shirt
   a. bleeding
   b. crocking
   c. perspiration
   d. sunlight
   e. fume fading

2. Draperies can be damaged by the rigors, not to mention the expense, of too frequent cleaning, just as they may be damaged by ____________ if infrequently cleaned.
   a. bleeding
   b. atmospheric contaminants
   c. migration
   d. crocking

3. Generally, a good compromise in frequency of cleaning home furnishing fabrics in order to achieve the optimum serviceability involves cleaning them approximately every __________ months.
   a. 4
   b. 6
   c. 12
   d. 24
   e. 36

4. Although pigment color is susceptible to crocking, it has several advantages. Indicate all that apply. Pigment coloring
   a. is an inexpensive coloring method.
   b. contributes to a soft, pliable fabric hand.
   c. can be added to fabrics with low dye affinity.
   d. can contribute fluorescent effects to fabrics.

5. Solution dyeing has been especially useful in dyeing ____________ fibers.
   a. synthetic and man-made cellulosic
   b. natural and mineral
   c. man-made cellulosic and protein
   d. synthetic and protein

6. Since synthetic fibers typically have low dye affinities, why is solution dyeing not always used to insure thorough dye penetration?
   a. Dyestuffs have not been developed to solution dye many generic classes.
   b. Solution dyeing is an inaccurate process.
   c. The color range of solution dyes is limited.
   d. Solution dyeing decreases production flexibility.

7. What is the most economical way to create a tweed effect?
   a. fiber dyeing
   b. solution dyeing (filament tow)
   c. roller printing
   d. cross dyeing
   e. union dyeing

8. Rank the relative cost of the following dyeing methods from *most* to *least* expensive.
   ____ a. solution
   ____ b. piece
   ____ c. fiber
   ____ d. yarn
   ____ e. union
   ____ f. cross

9. Considering the fact that a manufacturer will select the most economical method to achieve a design effect, match the most likely coloring method to the products listed below.
   ____ 1. solid-colored olefin carpet
   ____ 2. solid-colored rayon drapery fabric
   ____ 3. solid-colored rayon/acetate bedspread
   ____ 4. reversible check gingham
   ____ 5. wool tweed suiting
   ____ 6. wool and acrylic tweed upholstery
   a. solution
   b. piece
   c. fiber
   d. yarn
   e. cross
   f. union

10. Dobby and jacquard designs are often __________________ dyed to achieve woven in, patterned effects.
   a. fiber
   b. piece
   c. yarn
   d. solution
   e. cross

11. The advantage that dobby and jacquard fabric designs have over roller printed designs is that they are
    a. more elaborate.
    b. often reversible.
    c. less expensive to produce.
    d. structured on grain.

12. Which pair of printed fabrics might the consumer expect to be most expensive?
    a. block and stencil
    b. block and screen
    c. screen and transfer
    d. stencil and warp

13. Which handcrafted printing method has been successfully adapted to limited mass production?
    a. block
    b. flat bed screen
    c. stencil
    d. batik

14. When is flat bed screen printing favored over roller printing?
    a. when a small design is desired
    b. when a large design is desired
    c. when short run yardages are produced
    d. when mass yardages are produced
    e. b and c above
    f. b and d above

15. Roller printing was, until recently, the most freqently used printing method because it is
    a. more precise than other methods.
    b. fast and economical
    c. more versatile
    d. an uncomplicated process.

16. What commercial printing method is increasing in use and surpassing roller printing for many applications?
    a. rotary screen
    b. transfer
    c. discharge
    d. flat bed

17. What advantages does rotary screen have over roller printing? Indicate all that apply.
    a. Larger designs can be produced.
    b. Dye penetration is greater.
    c. Rotary screens are less expensive to produce than copper rollers.
    d. Fabric distortion is less because the fabric is printed flat.

18. What are the reasons transfer printing is attractive for certain applications? Indicate all that apply.
    a. paper transfer rolls are inexpensive
    b. transfer printing uses less water than other techniques
    c. fabrics are less likely to become distorted
    d. more vibrant and intricate designs can be produced
    e. the printer can maintain paper inventory rather than fabrics

19. A discharge-printed fabric is selected over a rotary screen or transfer printed fabric when
    a. a large, bold design is desired.
    b. cost is no object.
    c. a dark, vivid background is required.
    d. the fabric is to be bleached.

20. A disadvantage of a discharge-printed fabric which is laundered frequently is that it might
    a. weaken in the design areas.
    b. weaken in the dark background areas.
    c. shrink in the design areas.
    d. stretch in the background areas.

21. In addition to discharge printed fabrics, what other design methods involve piece dyeing?
    a. batik and rotary screen printing
    b. screen printing and tie-and-dye
    c. block printing and stencil printing
    d. tie-and-dye and batik

22. Which of the following novelty yarns would wear most satisfactorily in school clothes?
    a. bouclé
    b. crepe
    c. slub
    d. flake or tweed

23. Which finish creates a wavy design on taffeta?
    a. moiréing
    b. ciréing
    c. embossing
    d. glazing

24. If a consumer wanted a carpet with a dimensional effect, which finishing method would be selected?
    a. brushing
    b. sculpturing
    c. embossing
    d. napping

25. Which finishing process contributes softness and loft to a fabric?
    a. sizing
    b. shearing
    c. napping
    d. singeing

26. Match the coloring or design methods which would be most feasible and serviceable for each of the end uses indicated. The choices may be used more than once or not at all.
    ____ 1. floral glass drapery
    ____ 2. olefin upholstery
    ____ 3. polyester/cotton solid-colored slacks
    ____ 4. wool tweed sports jacket
    ____ 5. solid-colored linen dress
    ____ 6. cotton plaid bedspread
    ____ 7. flowered cotton kitchen curtains
    ____ 8. flax birdseye toweling
    ____ 9. acetate brocade evening jacket
    ____ 10. black-and-white silk print dress
    a. fiber dyed
    b. yarn dyed
    c. piece dyed
    d. solution dyed
    e. cross dyed
    f. union dyed
    g. pigment printed
    h. roller printed
    i. discharge printed

27. Match the design or finishing method which would achieve the following effect. Finishing methods may be used more than once or not at all.
    ____ 1. raised design created with applied yarns
    ____ 2. raised and lowered designs pressed into fabric
    ____ 3. pilelike surface
    ____ 4. crinkled surface
    ____ 5. polished surface
    ____ 6. flat, lustrous surface
    ____ 7. patent leather look
    a. embossing
    b. flocking
    c. chemical crepeing
    d. ciréing
    e. moiréing
    f. glazing
    g. beetling
    h. embroidering

28. In fashion retailing today, quick response to consumer demand is important. Which of the following would be the most efficient method for coloring fashionable junior sportswear?
    a. solution dyeing
    b. yarn dyeing
    c. garment dyeing
    d. screen printing

Answers for Unit 6 Posttest:

1. 1-d; 2-b, d, e; 3-a, b, d; 4-a, c; 5-a, c, d
2. b
3. c
4. c
5. a
6. d
7. d
8. most-c, a, d, e, f, b-least
9. 1-a; 2-b; 3-f; 4-d; 5-c; 6-e
10. c
11. b, d
12. a
13. b
14. e
15. b
16. a
17. a, c, d
18. b, c, e
19. c
20. a
21. d
22. d
23. a
24. b
25. c
26. 1-g; 2-d; 3-f; 4-a; 5-c; 6-b; 7-h; 8-c; 9-d; 10-h or i
27. 1-h; 2-a; 3-b; 4-c; 5-d or f; 6-g; 7-d
28. c

UNIT SEVEN

Textile Decisions: Serviceability Summary

In the preceding units the components of textiles were discussed as separate segments, beginning with a general overview of fundamental terminology associated with the four serviceability concepts of durability, comfort, care, and esthetic appearance. Then textile development from the smallest element of textiles—the fiber—through the yarn and fabric stages was discussed. Finally, fabric finishing and color and design applications were presented.

In each case, the consumer-oriented serviceability concepts were applied to the developing framework of this system of learning textiles. Consumers must learn to deal with all these factors and evaluate them as a totality, because the problems encountered by a textile decisionmaker are not based on isolated component parts of a fabric.

As you have learned, fiber properties can be greatly altered by certain yarn and fabric structures, and fabrics can be considerably changed according to finishes, design applications, and apparel or furnishings manufacturing techniques. Therefore, this final unit provides the opportunity for you to integrate what you have learned and evaluate the many aspects of a textile product for a given end use.

In working through this final unit, consider the responsibilities of such professionals as the interior designer, retailer, manufacturer's representative, consumer consultant and educator, as well as the ultimate consumer. Determine the appropriate responses to the problems presented and make creative textile decisions in the following hypothetical cases. Keep in mind that the simulations presented are over-simplifications. Many other factors also influence real life decisions. For example, company policy, available resources, economics and political considerations play a part from time to time for the textile professional.

Remember that you will not have memory matrixes, fabric packets and review sections while shopping; or while a client is waiting for alert, well thought-out, concise and articulate textile recommendations. Also keep in mind that there are seldom absolute right and wrong judgments!

Objectives for Unit 7

1. You will be asked to recall the specific properties of fibers, yarns, structures, finishes and various color and design alternatives as they relate to the four serviceability concepts.

2. You will be asked to apply the above properties to specific hypothetical situations which are, of course, heavily laden with value judgments. Individual perspectives will vary, and up to a point, you might take issue with certain choices indicated as the most appropriate answers based upon the authors' level of awareness. However, the importance of working through these decision-making activities is to provide experience that will help you make effective consumer decisions, and help consumers understand the rationale behind whatever answer is chosen.

Suppose You are an Interior Designer

Problem A

You have been commissioned to redesign a large living room. A middle income family of four with two children wants an elegant, contemporary interior which will hold up under *constant* use. Indicate which of the following fibers, fabric structures, finishes and design applications you would choose for each end use. *Justify* your selections.

1. Check the best *carpet* alternative:

_____ a. Wool, uncut pile-woven structure, piece dyed, soil resistant
_____ b. Polyester, sculptured tufted-pile structure, piece dyed, soil resistant
_____ c. Nylon, uncut tufted-pile structure, piece dyed, abrasion resistant
_____ d. Wool, uncut tufted-pile structure, fiber-dyed tweed effect, abrasion resistant
_____ e. Olefin, needle-felt structure, rotary-screen printed, soil resistant

****

b

****

Justification: ______________________________

****

The *second* alternative is probably the best choice, considering elegance and durability factors; sculpturing is often associated with elegance, and polyester fibers are durable and do not have the metallic sheen associated with some nylon carpets. Piece dyeing is adequate for good color fidelity, and the soil-resistant finish would retard soil pickup.

****

2. What would be limitations with each of the other carpet choices for the use specified above?

a. ______________________________

c. ______________________________

d. ______________________________

e. ______________________________

****

a. A wool, pile woven carpet is unnecessarily expensive.
c. The nylon uncut, tufted pile produces a utilitarian structure, possibly with a metallic sheen (unless the cross sectional nylon fiber shape is altered or a delustering white pigment is used). Also, a nylon carpet does not need an abrasion resistant finish; an antistatic finish would be more beneficial.
d. A wool, fiber-dyed carpet is expensive, and a tweed appearance is not generally associated with the elegance required.
e. A printed needle felt is a less expensive carpeting, generally used for informal surroundings, i.e., porch, basement, garage.

****

3. Choose the most suitable *drapery* for the living room mentioned in Problem A, by selecting one item in each column for the best combination of alternatives:

____ wool	____ plain	____ yarn dyed	____ shrinkage controlled
____ nylon	____ rib	____ roller printed	____ antislip
____ polyester	____ basket	____ discharge printed	____ abrasion resistant
____ cotton	____ pile	____ fiber dyed	____ mercerized
____ rayon	____ satin	____ block printed	____ biological resistant

****

polyester, plain weave, yarn dyed, abrasion resistant

****

Justification:

fiber ____________________

____

structure ____________________

____

color and design ____________________

____

finish ____________________

____

****

Polyester would have the highest sunlight resistance of the fibers listed.

Plain weaves have uniform surfaces which wear well and contribute to contemporary elegance.

Yarn dyeing penetrates each yarn, yielding durable coloring and a reversible fabric.

An abrasion resistant finish would protect the draperies when they were drawn against the wall or floor, and thus inhibit pilling.

****

4. What are the limitations of the other fibers listed for draperies?

wool ____________________

nylon ____________________

cotton ____________________

rayon ____________________

****

Wool yellows due to its sulfur content.

Nylon loses 50 percent of its strength in sunlight (unless finished with a high luster to reflect light).

Cotton is known for losing color and strength in direct sunlight.

Rayon is not dimensionally stable and exhibits the "elevator property."

****

5. Indicate two acceptable fabric alternatives for couch fabric upholstery for this living room:

____ a. Acetate warp/rayon filling, jacquard brocade, yarn dyed, fume-fading resistant

____ b. Nylon warp/cotton filling, jacquard damask, yarn dyed, antislip

____ c. Acrylic warp/worsted wool filling, dobby, yarn dyed, soil resistant

____ d. Cotton warp/rayon filling, twill, rotary-screen printed, abrasion resistant

____ e. Rayon warp/nylon filling, jacquard brocade, yarn dyed, antislip

****

b,c

****

Justification:
first choice ______________________________

____

____

____

second choice ______________________________

____

____

____

****

b—the nylon warp has high durability and warp yarns would bear the stress; cotton filling contributes to fabric absorbency. Damask is a flat surface, not as subject to abrasion as brocade.
c—the acrylic and wool combination would be durable and attractive; the design, through structuring and yarn dyeing, would wear well.
****

6. Check the most comfortable and durable fabric in the list below for lounge chair upholstery.
   ____ a. Vinyon, unsupported film, solution dyed, embossed to look like leather
   ____ b. Nylon warp/cotton filling, twill, yarn dyed, soil resistant
   ____ c. 100 percent wool, twill, rotary-screen printed, abrasion resistant
   ____ d. Vinyon, supported film, solution dyed, embossed

****

b

****

Justification:

____

____

____

****

The nylon and cotton twill would be durable, comfortable and easy to care for with the soil-resistant finish.
****

7. In addition to esthetic characteristics of vinyl that are not associated with living room elegance, what are two potential disadvantages of the film choices listed above?
   1. ______________________________
   2. ______________________________

****

1. Vinyl films are not comfortable in hot or cold weather.
2. Vinyls tend to fade unless given an appropriate finish.

****

Problem B

You are assigned as a designer consultant to help a consumer meet *low income housing needs*. From the alternatives listed below, suggest the floor covering, wall covering and drapery fabric which would be most economical, durable and yet attractive for furnishing a mobile home. Remember to *justify* all choices.

1. Check the best floor covering:
   ____ a. Polyester, sculptured tufted-pile, piece dyed, soil resistant
   ____ b. Nylon, sculptured tufted-pile, solution dyed, antistatic
   ____ c. Olefin, uncut tufted-pile, rotary-screen printed
   ____ d. Acrylic, cut tufted-pile, fiber dyed, soil resistant

****

c

****

Justification:

____

____

____

____

****

The olefin carpeting is the most economical and durable choice; the uncut pile would wear well and does not involve the extra expense of cutting or shearing. Rotary screen printing is an efficient and adequate method for coloring.

****

2. Check the best wall covering:
   ____ a. conventional wallpaper
   ____ b. flocked wallpaper
   ____ c. embossed vinyl wall covering
   ____ d. smooth vinyl wall covering

****

d

****

Justification:

____

____

****

A smooth vinyl wall covering is inexpensive, durable and easy to clean.

****

3. Why might draperies made from glass filaments be particularly desirable for a mobile home?

____

____

****

They are flameproof and increase the safety of a mobile home.

****

4. Given that the heavy weight of glass filaments can be a limitation, choose the best glass drapery fabric:
   ____ a. Twill weave, pigment printed, antislip
   ____ b. Open plain weave, pigments added to filaments, abrasion resistant
   ____ c. Leno structure, pigment printed, abrasion resistant

****

b

****

Justification:

____

____

____

____

****

The lower yarn count plain weave would weigh less than the twill, and put less stress on the low flex life of glass fibers than the twisted leno structure; the coloring applied to filaments would be more durable than a surface pigment application, which could crock off. Of course, an abrasion resistant finish increases serviceability of brittle glass filaments.

****

Problem C

As a textile consultant for a hospital, select the most appropriate recommendations for the listed end uses.

1. Indicate the best choice for surgical gowns and masks:
   ____ a. Cotton, plain weave, bleached, antiseptic
   ____ b. Cotton/polyester, plain weave, bleached, antistatic
   ____ c. Polyester, spunlaced, bleached, antistatic
   ____ d. Polyester, melt-blown microfiber web, bleached, antiseptic
   ____ e. Polyester, melt-blown microfiber web bonded to spunbonded web, bleached, antistatic

****

e

****

Justification:

____

____

____

____

****

Melt-blown webs are ideally suited for hospital use, but the micro-fiber webs need the added support of spunbonded webs; the result is a viable disposable fabric with good hand and body that has barrier properties, which prevent the transfer of bacteria. (If reusable gowns were desired, the plain weave cotton with an antiseptic finish would be sanitary, create no static electricity problems, and could be boiled after use.)

****

2. Check the preferred hospital bed linens for intensive care units:
   ____ a. Cotton, plain weave, bleached, antiseptic
   ____ b. Polyester/cotton, spunbonded fiber web, bleached, antiseptic
   ____ c. Acetate, bonded fiber web, bleached, antistatic
   ____ d. Polyester, spunlaced, bleached, antiseptic

****

b

****

Justification:

____

____

____

____

****

The disposable polyester and cotton spunbonded fiber web is an advantage, especially where there were contagious diseases, since it can be discarded after use. The cotton allows for absorbency and comfort, and the spunbonded polyester makes the fabric durable enough for sheets. The antiseptic finish also contributes to serviceability for the end use.

****

3. Check the infant blanket which would withstand constant hospital laundering:
   ____ a. Cotton, plain weave, rotary-screen printed, napped
   ____ b. Cotton/polyester, plain weave, piece dyed, napped
   ____ c. Rayon, needle felt, transfer printed, brushed
   ____ d. Wool/nylon, twill weave, yarn dyed, napped

****

b

****

Justification:

____

____

____

****

A plain weave cotton/polyester blend, piece-dyed flannelette would be comfortable and more dimensionally stable to laundering than a 100 percent cotton printed blanket. The other two choices would not withstand laundering as well.

****

Problem D

As a designer consultant for a motel, indicate the best choices for the end uses listed below.

1. Check the most serviceable choice for bed linens:
   ____ a. 50 polyester/50 cotton, plain weave, yarn dyed, soil resistant
   ____ b. 50 polyester/50 cotton, plain weave, rotary-screen printed, antiseptic
   ____ c. 65 polyester/35 cotton, plain weave, bleached, durable press
   ____ d. 100 percent cotton, plain weave, bleached, durable press

****

c

****

Justification:

____

____

____

****

The 65/35 blend has adequate polyester for durability and ease of care; the bleached, nonprinted sheet would be easiest to maintain.

****

2. Check the most serviceable choice for motel blankets:
   ____ a. Rayon, needle felt, piece dyed, brushed
   ____ b. Acrylic, needle felt, piece dyed, brushed
   ____ c. Cotton, plain weave, transfer printed, napped
   ____ d. Cotton, twill weave, yarn dyed, napped
   ____ e. Wool, twill weave, yarn dyed, napped

****

b

****

Justification:

____

____

____

****

Acrylic, needle felts are serviceable, easy care and economical. Since motels are adequately heated, neither the extra warmth nor cost of wool is necessary.

****

3. Check the most serviceable choice for motel towels:
   ____ a. Flax and polyester, uncut terry-pile, yarn dyed, absorbent
   ____ b. Cotton, cut terry-velour, jacquard weave, mercerized
   ____ c. Cotton, uncut terry-pile, bleached, mercerized
   ____ d. Cotton and polyester, uncut terry pile, jacquard weave, absorbent

****

c

****

Justification:

____

____

____

____

****

White, uncut terry-pile towels launder well and 100 percent cotton provides the most absorbency. A jacquard structure adds cost, and an absorbent finish is not needed on either flax or cotton.

****

4. What is an argument for and against the use of glass fiber draperies in a motel?
   for ____

   ____

   ____

   against ____

   ____

   ____

****

for—They are flameproof, relatively inexpensive and especially resistant to sunlight; they do not absorb soil and therefore require infrequent cleaning.

against—When glass draperies do need cleaning, they present problems to commercial laundries. Fabric weight is heavy, requiring heavy duty rods; and glass fabrics are generally loosely structured to reduce weight, which limits privacy.

****

5. Check in each column the most serviceable combination for table linens in a hotel or motel:

____ flax	____ damask	____ piece dyed or bleached	____ stain resistant
____ cotton	____ brocade	____ pigment colored	____ antistatic
____ polyester	____ matelassé	____ roller printed	____ Sanforized®
____ acrylic	____ satin	____ rotary-screen printed	____ schreinerized

****

cotton, damask, piece dyed or bleached, stain resistant

****

Justification:

fiber ____

____

fabric ____

____

coloring ____

____

finish ____

____

****

Cotton is more serviceable than the other fibers listed due to expense of flax, the oleophilic nature of polyester, and the fuzzy texture of acrylic.

Damask is the most suitable fabric—it is a flat, reversible and does not snag easily.

Piece dyeing or bleaching is a good choice for linens which receive frequent laundering.

A *stain-resistant* finish is the most advantageous. (Incidentally, practically all cottons are routinely mercerized for this end use.)

****

Problem E

As an interior designer for a transatlantic airline, indicate the best choices for the following end uses.

1. The best choice for upholstery and window coverings:

____ a. Nylon, twill weave, yarn dyed, antistatic, flame retardant
____ b. Glass, plain weave, pigment printed, abrasion, antislip
____ c. Nylon, plain weave, piece dyed, antistatic, antislip
____ d. Polyester, jacquard weave, yarn dyed, antislip, flame retardant

****

a

****

Justification:

____

____

****

The nylon twill would be durable for the hard use it would receive; plus, flame retardance is extremely important for public conveyance.

****

Indicate limitations of the other three choices:

b. ____

____

c. ____

____

d. ____

____

****

b—While glass is flameproof, its abrasion resistance and comfort are too low for upholstery; and its heaviness would be a disadvantage for curtains, considering the importance of weight in air travel.

c—With no flame-retardant finish, this choice is not acceptable.

d—The polyester flame-retardant jacquard is acceptable, but more expensive and not as durable as the nylon twill.

****

2. Check the best choice for airline carpeting:

____ a. Wool, uncut tufted-pile, yarn dyed, antistatic
____ b. Nylon, uncut tufted-pile, yarn dyed, flame retardant
____ c. Nylon, sculptured tufted-pile, piece dyed, flame retardant
____ d. Polyester, sculptured tufted-pile, transfer printed, flame retardant
____ e. Olefin, needle felt, rotary-screen printed, antistatic, flame retardant

****

b

****

Justification:

____

____

____

****

The uncut nylon tufted-pile is the most durable due to its abrasion resistance and even texture; flame retardance is a key factor. While wool is flame resistant, it is too expensive to be practical. Other choices either do not have as abrasion-resistance a surface; and/or their coloring is not as suitable.

Suppose You are a Manufacturer's Representative

Problem A

As a shipping officer for a textile manufacturing concern, you are responsible for billing and checking the invoices on orders. Each order below has the same yardage of comparable weight and fiber content suitable for table linens.

1. Match the appropriate relative price with each tablecloth fabric:

____	1. Jacquard	a. $550
____	2. Plain	b. $325
____	3. Malimo	c. $200
____	4. Dobby	d. $120

****

1-a; 2-c; 3-d; 4-b

****

2. A consignment of dobby structures was woven with different fiber contents. Which relative price should be on each comparably woven, dobby fabric?

____	1. Cotton	a. $400
____	2. Rayon	b. $350
____	3. Wool	c. $200
____	4. Silk	d. $120

****

1-c; 2-d; 3-b; 4-a

****

Problem B

You are a troubleshooter for a fabric supplier, and you have been asked to help identify an allergy problem which has become prevalent in an institution. Your company sold the following fabrics to the agency for the patients to use in these end uses:

a. Wool/nylon blend, twill weave, to make into slacks
b. Glass malimo for draperies
c. Cotton plain weave to use in shirts
d. Cotton/polyester blend, jacquard weave, for table linens
e. Cotton terry-pile for beach wear

What fabrics might cause the difficulty and why?

____

____

****

a—Glass fabric due to its abrasiveness for those who made the draperies, or came in contact with them.

b—The wool blend slacks could be irritating to sensitive skins.

****

Problem C

As a polyester sales representative, name three *structuring* and *finishing* techniques which offset problems associated with 100 percent polyester garments.

1 ______

2 ______

3 ______

****

Textured yarns help increase bulk and/or absorbency, and decrease pilling.

Loose fabric structures, like knits, can increase air transfer.

Finishes which improve all-polyester fabrics include absorbent, antistatic, antislip, schreinerizing and heat setting.

****

Problem D

High fashion dictates to you as a manufacturer's representative for women's apparel. If large, bright, splashy prints are "in" next year, which of the alternatives listed below are preferred?

Indicate one choice in each column, and justify your choices.

____ cotton	____ plain	____ roller
____ acrylic	____ leno	____ block
____ polyester	____ basket	____ screen
____ olefin	____ pile	____ discharge
____ acetate	____ dobby	____ rotary screen

****

cotton, plain weave, screen printing

****

Justification:

fiber ______

______

structure ______

______

printing ______

______

****

Cotton has the best dye affinity for bright colors.

A plain weave provides a smooth printing surface which would not detract from the large designs.

Screen printing allows for the largest prints.

****

Suppose You are a Retail Store Buyer

1. Indicate the best combination for a versatile suiting fabric for customers who travel extensively. Justify your selection.

____ cotton	____ double knit	____ antistatic
____ polyester	____ plain weave	____ mercerized
____ olefin	____ plain filling knit	____ durable press
____ flax	____ interlock knit	____ Sanforized®
____ rayon	____ raschel knit	____ schreinerized

****

polyester, double knit, antistatic

****

Justification:
fiber ______________________________
structure ______________________________
finish ______________________________

****

Polyester is most versatile for travel because of its resiliency and ease of care.

Double knits contribute to stability and wrinkle recovery.

An antistatic finish would be most needed to prevent fabric clinging.

****

NOTE: Another combination compatible for travel is *plain weave cotton with a durable press finish*. However, the lack of a washer and dryer when travelling prevents using the best laundering method for durable press.

****

2. A large order of shirts has just arrived, and your assignment is to send them to the appropriate departments. More than one shirt may go to the same department.

____ 1. silk, slub yarn, plain weave
____ 2. combed cotton/polyester blend, high yarn count, plain weave, durable press
____ 3. combed cotton, high yarn count, oxford basket weave, durable press
____ 4. cotton, low yarn count, sizing, permanent press
____ 5. worsted wool, twill weave, light napping

a. men's specialty
b. main floor men's department
c. budget basement

****

1-a; 2-b; 3-b; 4-c; 5-a

****

3. You are ordering suits and tuxedos for the rental shop.
Indicate the fiber content you would order for the following end uses:

____ 1. summer wedding	a. 75 wool/25 polyester
____ 2. winter black	b. 65 acetate/35 silk
____ 3. modern dance	c. 50 polyester/50 cotton
____ 4. concert pianist	d. 40 nylon/30 cotton/30 spandex
____ 5. nightclub entertainer	e. 50 cotton/30 silk/20 metallic

****

1-c; 2-a; 3-d; 4-b; 5-e

****

4. You are ordering blankets for a variety of purposes.
   Indicate the most appropriate blanket structure for each use listed below:

   ____ 1. lightweight, spring blanket
   ____ 2. heavyweight, durable winter blanket
   ____ 3. inexpensive blanket for college students
   ____ 4. average weight, durable blanket

   a. rayon, needle felt
   b. cotton, plain weave, napped (flannelette)
   c. nylon twill, laminated
   d. wool twill, napped
   e. wool, double cloth, napped

****

1-b; 2-e; 3-a; 4-d

****

5. You are buying winter coats and are trying to get a variety of styles.
   Match the coat fabric with the *major* reason for ordering it.

   ____ 1. cotton corduroy, fabric foam laminate
   ____ 2. wool tweed, twill
   ____ 3. worsted wool, twill
   ____ 4. acrylic, double knit
   ____ 5. modacrylic pile knit

   a. smooth textured, tailored
   b. bulk, insulative properties
   c. sporty, rough textured
   d. fake fur look
   e. warmth without weight

****

1-b; 2-c; 3-a; 4-e; 5-d

****

Suppose You are a Consumer Consultant or Extension Specialist

1. Explain four methods of making fabric without the use of conventional yarn interlacing or interlooping.
   a. ____
   ____
   b. ____
   ____
   c. ____
   ____
   d. ____
   ____

****

Any four:

Needle felt: mechanical entangling of fibers with each other or about a center scrim.

Bonded fiber web: a web of fibers adhered by heat, chemicals, or adhesives; generally lightweight.

Film: plasticlike substance formed by extruding a liquid spinning solution into a flat sheet.

Stitch bond: filling yarns, warp and filling yarns, or a fiber web adhered with a chain stitch.

Spunlaced: a web of fibers entangled through water jet action.

****

2. Match the end uses listed to their adaptations in the following nonwoven structures:

____ 1. disposable wipes	a. spunbonded web
____ 2. upholstery	b. needle felt
____ 3. blankets	c. film
____ 4. interfacing	d. stitch bond
____ 5. hospital gowns and masks	e. spunlaced
____ 6. tablecloths	f. melt-blown web
____ 7. apparel	
____ 8. rainwear	
____ 9. carpets	
____10. draperies, curtains	

****

1-a; 2-c; 3-b; 4-a; 5-f; 6-d; 7-e; 8-c; 9-b; 10-d,e

****

3. Discuss the durability features of these structures:

spunbonded web ______________________________

needle felt ______________________________

film ______________________________

stitch bond ______________________________

spunlaced ______________________________

****

Bonded webs and needle felts depend heavily on fiber content and planned end use; at best, they are not as durable as comparably woven structures.

Films have high abrasion resistance and are waterproof; they are durable for uses which do not require absorbency or extreme flexibility.

Stitch-bond fabrics can be relatively durable structures when made from yarns such as malimo. The stitch-bond fiber webs are not highly durable.

Spunlaced structures are durable for many purposes; they are flexible and strong for their weight.

****

4. What key information would you give to help consumers evaluate the fiber serviceability of:

nylon ______________________________

polyester ______________________________

acrylic ______________________________

olefin ______________________________

vinyon ______________________________

spandex ______________________________

glass ______________________________

****

Nylon is known for high tenacity and abrasion resistance; it has low sun resistance and absorbency.

Polyester is known for resiliency and easy care; it is not absorbent and attracts oily stains.

Acrylic is created to be woollike; it is soft, fuzzy and has bulk density. Pilling can be a problem.

Olefin has outstanding stain resistance due to zero absorbency; it is inexpensive, but has low heat tolerance.

Vinyon is used in the popular vinyls; it is waterproof in film form. It also has low heat tolerance.

Spandex is known for its high stretch potential and recovery; it has holding power without weight.

Glass has low flexibility and abrasion resistance; it is flameproof and sun resistant, so can be used for draperies if hung and cared for properly.

****

5. What advantages do the newer nonwoven structures have for the limited income consumer?

____

____

____

____

****

Fiber-formed structures can be produced rapidly and inexpensively for disposable or short-term use items.

Needle felts are suitable for inexpensive blankets and carpeting.

Stitch bond and spunlaced structures appear to have great potential as rapidly produced, inexpensive apparel and home furnishing fabrics, which can appear similar to woven or knit structures.

****

6. Which synthetic fiber classification, often used in needle-felt carpeting, might be recommended to a low income consumer? ____________ Why? ____

____

****

olefin; because it is inexpensive and nonabsorbent

****

Suppose You are an Educator

1. A student asks you to differentiate among the terms *hydrophilic, hydrophobic, hygroscopic* and *wicking*. Include fiber examples of each.

hydrophilic ____

____

____

hydrophobic ____

____

____

hygroscopic ____

____

____

wicking ____

____

____

****

Hydrophilic literally means "water loving"—moisture is readily absorbed, as in cotton, flax and rayon.

Hydrophobic fibers are not absorbent and do not readily accept moisture internally, as the synthetic and mineral fibers.

Hygroscropic fibers absorb moisture vapor without feeling wet to touch; moisture is trapped internally, as in silk, wool, and other animal hair fibers.

Wicking is the ability of a fiber to transmit moisture along its length or around its width. Some hydrophobic synthetic fibers can wick moisture along the outside of the fiber, thereby increasing comfort. Also, flax wicks moisture vertically through its lumen.

****

2. Explain how a hygroscopic fiber can absorb moisture and not feel wet.

____

____

****

Protein fibers exhibit hygroscopic properties: water molecules are held under the scales of animal hair fibers, so that the surface feels dry until the product is saturated with water. In silk the *vapor* is accepted into the molecular structure.

****

3. Discuss the major significance of the TFPIA. ____________________

____

Give three items which the Act requires on a fabric label to benefit consumers.

1 ____________________

2 ____________________

3 ____________________

****

Generic classifications of fibers were created, thereby grouping common fibers in a market filled with trade names. Requirements include:

1) generic classification of all fibers present in amounts over 5 percent (2 percent for elastomers);
2) fiber percentage by weight; and
3) manufacturer's name.

NOTE: The TFPIA does not require the label to be sewn in. A more recent FTC Care Labeling Ruling requires permanently attached care information on apparel products.

****

4. What other labeling law which requires fiber content labeling, should be explained?

____

****

Wool Products Labeling Act (WPL)

****

5. Give the three requirements of the WPL, and any applicable subdivisions under them.

1 ____________________

2 ____________________

3 ____________________

****

The WPL requires: l) fiber percentage by weight; 2) manufacturer's name; 3) type of wool—specifying *new* or *virgin*, *recycled*, *reused* or *reprocessed*.

****

6. What is the value of the WPL to the consumer?

____

____

****

The consumer can evaluate the serviceability of a wool garment based on the percentage by weight and type of wool; for example, a product of recycled wool is less soft and durable than one of new wool.

****

7. What advantage does the FTC Care Labeling Ruling give the consumer that previous textile acts did not provide?

____

____

****

Direct care information, *permanently attached*, without consumers having to know demands of a given fiber.

****

Suppose You are a Consumer

1. You are interested in buying shirts for school or work, and you want them to require no ironing. Indicate which you would buy and why.
   ____ a. all cotton, durable press
   ____ b. 65 polyester/35 cotton, durable press
   ____ c. 20 polyester/80 cotton, wash-and-wear
   ____ d. 65 cotton/35 polyester, durable press

****

b

****

Why? ____

____

****

The *durable press blend with the most polyester* contributes to the highest wrinkle resistance and recovery after laundering.

****

2. Which fabric listed below would you select for a shower curtain liner?
   ____ a. cotton, high yarn count, plain weave, water repellent
   ____ b. glass, high yarn count, plain weave, abrasion resistant
   ____ c. vinyon, unsupported film, fade resistant
   ____ d. vinyon, supported film, abrasion resistant

****

c

****

Why? ____

____

****

The unsupported film is waterproof and lightweight.

****

The list of hypothetical situations could go on and on! Hopefully, you are now prepared to apply the concepts of textile serviceability to real situations you will face many times in the future as a consumer, if not as a textile professional.

Where Do You Go From Here?

The intent of this consumer-oriented textile text is to give you a systematic and general overview of the textile industry, and to provide an appreciation of the complexities involved in producing textile products for the varied consumer market. Beginning with the basic concepts and terminology shared by professionals in the textile field, you were introduced to the fundamental considerations of fibers, yarns, fabric construction, finishing and finally coloring and design applications. What might you do with this background information?

If you have a scientific or engineering bent, you may find the implications of a technical career in textiles an exciting possibility. For information on schools which specialize in the *technical training* necessary for entry into the textile industry at the *scientific research* level, contact:

American Association of Textile Chemists
and Colorists
P.O. Box 12215
Research Triangle Park, NC 27709

American Association of Textile Technology
295 Fifth Ave., Rm. 621
New York, NY 10016

American Textile Manufacturers Institute, Inc.
1801 K St. N.W., Suite 900
Washington, DC 20006

From a different point of view, you may have been challenged to look into the retail side of the industry and to explore career opportunities in textile marketing, sales or management. For information on *business-oriented programs,* write:

American Apparel Manufacturers Association
2500 Wilson Blvd., Suite 301
Arlington, VA 22201

American Textile Manufacturers Institute, Inc.
1801 K St. N.W., Suite 900
Washington, DC 20006

National Retail Merchants Association
100 West 31st Street
New York, NY 10001

In addition, there are textile majors in Colleges of Home Economics, Human Ecology, or comparably named units, which may serve the need for technical training in textile sciences. These programs provide general textile education in preparation for teaching or work within Extension, government agencies, business, or the textile industry. For addresses of colleges and universities that have such programs, which vary constantly from school to school, write:

American Home Economics Association
1555 King St.
Alexandria, VA 22314

However, if your career is undecided, you might consider the challenge of pioneering a new discipline of *consumer consultants.* These individuals are needed to mediate directly between the consumer and industry or government. Such consultants should have no vested interest in either industry or government; they could function as a branch of a university, or as a private agency.

Although various groups engage in some of these activities to a degree, there is still a missing link and need for professionals who can provide consumers with advice on purchasing, and feedback performance information to the textile-related industries.

Consider the consumer, the textile manufacturing industry (with its many parts), and the retailer, who is the direct link between the basic industry and the ultimate consumer. In addition to these elements, there are agencies whose function it is to coordinate the affairs of industry, and there are groups to intercede and represent the interests of the consumer. All these parts fit together into an extremely complex relationship which is simplistically represented in Figure 7.1.

The diagram illustrates the direct link between the consumer and the retailer, which might be mediated by the influence of professional consumer consultants. While the various groups input information into the system, a formal consumer consultant link is not readily identified. There are efforts to fill the gap, as can be observed in the following listings of selected *government agencies.* Their purposes, wholly or in part, are to advise, represent and *protect the consumer.*

Office of Consumer Affairs
Department of Health and Human Services
200 Independence Ave. S.W.
Washington, DC 20201

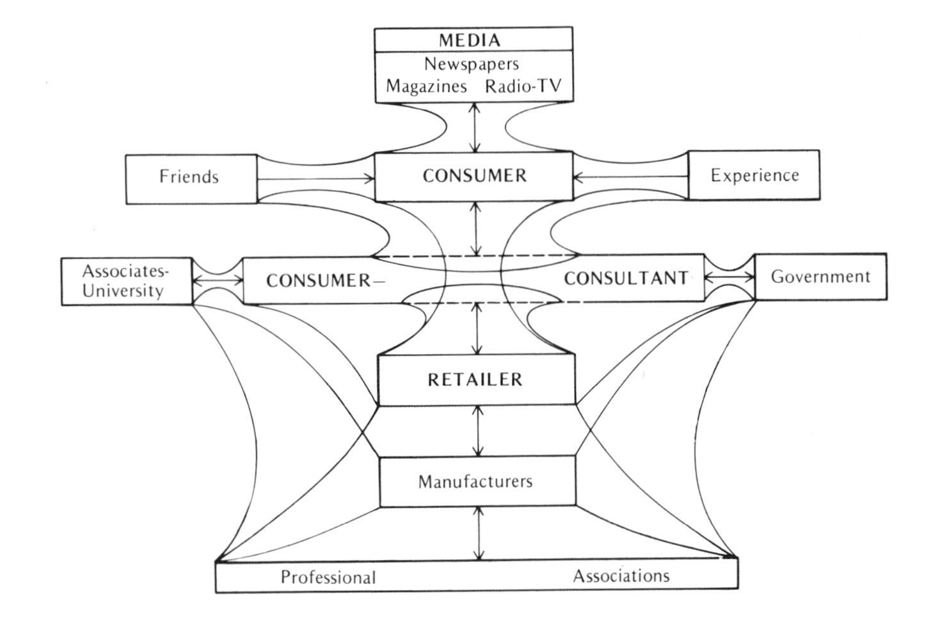

Figure 7.1. Consumer information flow.

Federal Trade Commission
Pennsylvania Avenue at Sixth, NW
Washington, DC 20580

U.S. Department of Agriculture
Fourteenth Street and Independence
Avenue, SW
Washington, DC 20250

Federal Cooperative Extension Service
U.S. Department of Agriculture
Fourteenth Street and Independence
Avenue, SW
Washington, DC 20250

National Bureau of Standards
Department of Commerce
Fourteenth Street between Constitution Ave.
and E. Street, N.W.
Washington, DC 20230

Consumer Product Safety Commission
5401 Westbard Ave.
Bethesda, MD 20207

Various Congressional Committees
Washington, DC

Also, consider a list of some of the *public organizations* whose purpose is to represent and supply pertinent information on behalf of the consumer:

American Council on Consumer Interests
162 Stanley Hall
University of Missouri
Columbia, MO 65201

American Home Economics Association
1555 King St.
Alexandria, VA 22314

Chamber of Commerce of the United States
Consumer Issues Committee
1615 H. Street, NW, Suite 1401
Washington, DC 20062

Consumer Federation of America
1424 16th St. N.W., Suite 604
Washington, DC 20036

Consumer Research, Inc.
800 Maryland Ave. N.E.
Washington, DC 20002

Consumers Union of United States, Inc.
256 Washington Street
Mount Vernon, NY 10553

Council of Better Business Bureaus
4200 Wilson Blvd., Suite 800
Arlington, VA 22203

Federal Trade Commission[1]
Sixth Street & Pennsylvania
Avenue, NW
Washington, DC 20580

National Consumers League
815 15th St. N.W., Suite 516
Washington, DC 20005

There are also industry sponsored self-regulative associations which establish voluntary standards for the benefit of consumer education and protection. Some of the organizations within the textile industry which promote and develop standards and test methods are:

American Association of Textile Chemists
and Colorists
P.O. Box 12215
Research Triangle Park, NC 27709

American Institute of Laundering
12251 Tech Road
Silver Springs, MD 20904

American National Standards Institute
1430 Broadway
New York, NY 10018

American Society for Testing and
Materials (ASTM)
1916 Race Street
Philadelphia, PA 19103

Association of Home Appliance
Manufacturers
20 N. Wacker Driver
Chicago, IL 60606

Carpet & Rug Institute
P.O. Box 2048
Dalton, GA 30720

Industrial Fabrics Assoc. International
345 Cedar Bldg., #450
St. Paul, MN 55101

Upholstered Furniture Action Council
(UFAC)
P.O. Box 2436
High Point, NC 27261

National Institute of Dry Cleaning
12251 Tech Road
Silver Springs, MD 20904

National Retail Merchants Association
100 W. 31st Street
New York, NY 10001

Soap and Detergent Association
475 Park Avenue, S
New York, NY 10016

Textile testing laboratories also fit into this complex scheme of consumer/industry information. There are private and public laboratories which do contractual product testing for manufacturers, retailers, or other interested parties. Also, there are industry-supported laboratories which test the manufacturers' or retailers' own products. A selected listing of some of the *testing laboratories* follows:

Private: Better Fabrics Testing Bureau
101 - 31st Street
New York, NY 10001

[1]The FTC is the enforcing agency for rulings. Violations of labeling requirements should be reported to it on forms provided for this purpose.

International Fabricare Institute
12251 Tech. Road
Silver Springs, MD 20904

Industry: J. C. Penny Merchandise
Testing Center
Quality Assurance Ctr.
1505 Wallace Dr., Suite 102
Carollton, TX 75006

Sears, Roebuck and Company
Testing Laboratory
Sears Tower
Chicago, IL 60684

Celanese Corporation
4000 Barclay Downs Drive
P.O. Box 32414
Charlotte, NC 28232

Monsanto Chemical Company
800 N. Lindbergh Blvd.
St. Louis, MO 63167

Burlington Industries
1345 Avenue of the Americas
New York, NY 10105

E. I. duPont de Nemours & Co.
1007 Market St.
Wilmington, DE 19898

Government: Consumer Product Safety
Commission
5401 Westbard Ave.
Bethesda, MD 20207

Some laboratories give *certified seals of approval or quality* when products measure up to standards. In addition, there are textile manufacturers' guarantee programs which specify a certain level of performance for products. *Licensed trademark programs,* such as those developed by Celanese Corporation, are becoming more common. In these programs, certain manufacturers require that a standard of quality be maintained in products carrying their trade names.

Apart from sales and production, the textile industry annually spends millions of dollars for quality control, testing, evaluation and consumer education. Industry associations try to cope with the problem of technical information transfer within and between the textile related industries. Government legislates against glaring deficiencies in trade practices. But who serves to digest and synthesize and then transfer information among industry, government and consumers? Regrettably, many are poorly informed and equipped to advise on consumer purchases, and the feedback to industry often stops with the salesclerk.

Perhaps you could be a pioneering individual who, after meeting various employee problems, might find crative ways of serving as an *ombudsman* by providing textile knowledge to both the industry and the consumer. This would require commitment to intensive studies and further exploration of the content of textile information and associated areas of marketing and retailing. Energies directed toward this interrelationship should find some reward in the heartfelt appreciation of consumers. Industry, too, should appreciate such efforts, since they would certainly help all aspects of industry as well as the consumer.

However, if career interests lead you in directions other than those of consumer consultant, textile researcher, marketing, or mechandising specialist, then, at least, you should have gained increased understanding of the selection, use and care of textile products. Consumers must be aware of their own responsibilities, as well as recognize that industry and retail establishments have responsibilities. Those who have experienced consumer disappointments should make certain that the information gets back to all individuals and companies who are involved. That means that the salesclerk, retail store, and the manufacturers need to know the problem and recognize their part in preventing it from happening in the future.

FABRIC REVIEW SUMMARY

Below is a description of the characteristics of each of the fabric samples in your packet. You may use this information to test yourself and to review.

Fabric Name & Content	Yarn	Structure	Finish	Color Application
1. Chintz 100% cotton, Everglaze®	singles	plain weave	glazing, mercerization	roller print
2. Batiste (Shadow Voile) 65% Kodel® polyester 35% combed cotton	singles	unbalanced plain weave	sizing, durable press	piece dyed
3. Shantung 100% silk	slub	plain weave	degumming, weighting	piece dyed
4. Racquet poplin 65% Kodel® polyester 35% cotton	singles, combed	rib weave	singeing, durable press	piece dyed
5. Taffeta 100% acetate	multi-filament singles	rib weave	moiréing	solution dyed
6. Oxford 60% cotton 40% polyester Permaset, Dan Crest®	singles	2/1 basket weave	durable press blend	yarn dyed
7. Upholstery Fabric 100% olefin	multi-filament singles	basket weave	*bonded* to fiber web	solution dyed (mass pigmentation)
8. Stretch Denim 75% cotton 25% spun polyester (warp) 100% textured polyester filling	textured filament crosswise singles lengthwise	twill weave	napping	yarn dyed
9. Ultressa® (stretch woven) 100% Dacron® polyester	textured filament	twill weave	—	piece dyed
10. Satin 100% acetate	multi-filament singles	satin weave	sizing	solution dyed
11. Sateen 100% cotton	singles	twill weave	durable press, stain resistant "Scotchgard"®	roller print
12. Pinwale Corduroy 83% cotton 17% polyester	singles	filling pile weave	mercerization, brushing, shearing	piece dyed
13. Velveteen 100% combed cotton	singles	filling pile weave	brushing, shearing	discharge printed
14. Velvet Upholstery 65% cotton 35% rayon	singles	warp pile weave	Schotchgard® stain resistant	piece dyed
15. Terry cloth 88% cotton 12% polyester	singles	warp pile weave	shearing, brushing	piece dyed
16. Patrician Fancies (Dobby) 75% polyester 25% combed cotton	singles—lengthwise multi-filament singles—crosswise	dobby weave	durable press blend	piece dyed
17. Waffle cloth 65% Kodel® polyester 35% cotton	singles	dobby weave	—	piece dyed
18. Brocade upholstery 56% olefin 44% nylon	singles	jacquard weave	Scotchgard® stain resistant	yarn dyed

Fabric Name & Content	Yarn	Structure	Finish	Color Application
19. Matelassé upholstery 35% olefin 35% polyester 30% rayon	lengthwise—singles crosswise—ply & singles	jacquard double cloth	—	yarn dyed
20. Crepe 100% polyester	textured filament	crepe weave	—	solution dyed
21. Seersucker 50% Kodel® polyester 50% combed cotton	lengthwise—singles & ply crosswise—ply	slack tension weave	—	piece dyed
22. Single knit jersey 100% spun Fortrel® polyester	singles	plain filling knit	—	transfer printed
23. Warp knit crepe 75% Arnel triacetate 25% polyester	textured filament	warp knit (crepe effect)	—	piece dyed
24. Interlock 100% polyester	textured filament	interlock filling knit	—	solution dyed
25. Double Knit 100% polyester	textured filament	jacquard double knit	—	yarn dyed
25A. Leather Look 100% polyester	filament	warp knit	—	transfer print
26. Pile Knit 65% acetate by AVTEX 35% nylon	textured filament	pile filling knit	—	piece dyed
27. Tricot 100% Zefran® Nylon	textured filament	warp knit	schreinerized	piece dyed
27A. Guil-Tuf® **luggage fabric** 100% Caprolan® nylon	filament singles	warp knit	urethane coating	piece dyed
28. Brushed Tricot 100% Celanese Fortel®	filament singles	warp knit	flame resistant, napping	piece dyed
28A. Guil Suede Plus **Leather-look apparel** 100% polyester	filament singles	warp knit	sueding	transfer printed
29. Raschel 100% polyester	filament singles	warp knit	—	—
30. Raschel Lace 100% nylon	filament	raschel warp knit	—	
31. Phun Phelt™ 100% Kodel® polyester	—	chemically bonded staple fiber web	—	piece dyed
32. Spun bonded Handi wipe® Rayon/polyester blend	—	spunbonded web	—	rotary screen printing
33. Nexus® 100% polyester	—	unsupported spunlaced	—	
34. Nexus® 100% polyester	—	supported spunlaced backed by 100% crushed acrylic foam	—	rotary screen printed
35. Needle punch 100% olefin	—	needle felt	—	piece dyed

Fabric Name & Content	Yarn	Structure	Finish	Color Application
36. Supported Film	singles—backing	supported film— filling knit backing	embossing	solution dyed
36A. Polyester Chamois 100% polyester	filament singles face	multicomponent—warp knit bonded to fiber web	sueded	transfer printed
37. Malimo Stitch Bond 70% cotton 19% polyester 11% rayon	2 ply	stitch bond malimo top layer with maliwalt backing	—	yarn dyed
38. Tufted 100% rayon	filament singles	tufted	brushing, shearing	piece dyed
39. Quilted 50% Kodel® polyester 50% cotton face 100% polyester filling 100% nylon backing	singles—top layer	quilted structure plain weave top layer center fiber web layer warp knit backing	—	piece dyed and rotary screen printed
40. Dotted Swiss 65% Dacron® polyester 35% Avril® rayon	singles	plain weave	flocking	piece dyed prior to flocking
41. Plissé 65% Kodel® polyester 35% combed cotton	singles	plain weave	chemical creping	piece dyed
42. Interlock 100% polyester	textured filament	interlock filling knit	—	transfer printed
43. Drapery Fabric 100% acrylic	singles and cord (two ply lengthwise twisted together)	combination of weaves plain, twill, leno	—	yarn dyed

FABRIC REVIEW SUMMARY

Fabric Name & Content	Yarn	Structure	Finish	Color Application
1. **Chintz** 100% cotton, Everglaze®				
2. **Batiste (Shadow Voile)** 65% Kodel® polyester 35% combed cotton				
3. **Shantung** 100% silk				
4. **Racquet poplin** 65% Kodel® polyester 35% cotton				
5. **Taffeta** 100% acetate				
6. **Oxford** 60% cotton 40% polyester Permaset, Dan Crest®				
7. **Upholstery Fabric** 100% olefin				
8. **Stretch Denim** 75% cotton 25% spun polyester (warp) 100% textured polyester filling				
9. **Ultressa® (stretch woven)** 100% Dacron® polyester				
10. **Satin** 100% acetate				
11. **Sateen** 100% cotton				
12. **Pinwale Corduroy** 83% cotton 17% polyester				
13. **Velveteen** 100% combed cotton				
14. **Velvet Upholstery** 65% cotton 35% rayon				
15. **Terry cloth** 88% cotton 12% polyester				
16. **Patrician Fancies (Dobby)** 75% polyester 25% combed cotton				
17. **Waffle cloth** 65% Kodel® polyester 35% cotton				
18. **Brocade upholstery** 56% olefin 44% nylon				

Fabric Name & Content	Yarn	Structure	Finish	Color Application
19. Matelassé upholstery 35% olefin 35% polyester 30% rayon				
20. Crepe 100% polyester				
21. Seersucker 50% Kodel® polyester 50% combed cotton				
22. Single knit jersey 100% spun Fortrel® polyester				
23. Warp knit crepe 75% Arnel triacetate 25% polyester				
24. Interlock 100% polyester				
25. Double Knit 100% polyester				
25A. Leather Look 100% polyester				
26. Pile Knit 65% acetate by AVTEX 35% nylon				
27. Tricot 100% Zefran® Nylon				
27A. Guil-Tuf® luggage fabric 100% Caprolan® nylon				
28. Brushed Tricot 100% Celanese Fortel®				
28A. Guil Suede Plus Leather-look apparel 100% polyester				
29. Raschel 100% polyester				
30. Raschel Lace 100% nylon				
31. Phun Phelt™ 100% Kodel® polyester				
32. Spun bonded Handi wipe® Rayon/polyester blend				
33. Nexus® 100% polyester				
34. Nexus® 100% polyester				
35. Needle punch 100% olefin				

Fabric Name & Content	Yarn	Structure	Finish	Color Application
36. Supported Film				
36A. Polyester Chamois 100% polyester				
37. Malimo Stitch Bond 70% cotton 19% polyester 11% rayon				
38. Tufted 100% rayon				
39. Quilted 50% Kodel® polyester 50% cotton face 100% polyester filling 100% nylon backing				
40. Dotted Swiss 65% Dacron® polyester 35% Avril® rayon				
41. Plissé 65% Kodel® polyester 35% combed cotton				
42. Interlock 100% polyester				
43. Drapery Fabric 100% acrylic				

Selected Bibliography

REFERENCE BOOKS

American Association of Textile Chemists and Colorists—Technical Manual. Published annually.

Cohen, Allen. *Beyond Basic Textiles.* New York: Fairchild Publications, 1982.

Corbman, Bernard. *Textiles: Fiber to Fabric.* New York: McGraw-Hill Book Company, 1983.

Gioello, Debbie. *Understanding Fabrics: From Fiber to Finished Cloth.* New York: Fairchild Publications, 1982.

Hollen, Norma and Jane Saddler. *Textiles,* 6th ed. New York: The Macmillan Company, 1988.

Joseph, Marjory. *Essentials of Textiles,* 4th ed. New York: Holt, Rinehart and Winston, 1988.

______. *Introductory Textile Science,* 5th ed. New York: Holt, Rinehart and Winston, Inc., 1986.

Larsen, Jack and Jeanne Weeks. *Fabrics for Interiors.* New York: Van Nostrand Reinhold, 1975.

Lyle, Dorothy. *Modern Textiles,* 2nd ed. New York: McMillan, 1982.

Price, Arthur and Allen Cohen. *Fabric Science.* 5th ed. New York: Fairchild Publications, 1987.

Smith, Betty and Ira Block. *Textiles in Perspective.* New York: Prentice-Hill, 1982.

Totora, Phyllis. *Understanding Textiles,* 3rd ed. New York: Macmillan, 1987.

Wingate, Isabel B. and J. F. Mohler, *Textile Fabrics and Their Selection,* 8th ed. Englewood Cliffs, New Jersey: Prentice-Hall, Inc., 1984.

______., editor. *Fairchild's Dictionary of Textiles.* New York: Fairchild Publications, Inc., 1967.

Yeager, Jan. *Textiles for Residential and Commercial Interiors.* New York: Harper-Row, 1987.

SUGGESTED PERIODICALS FOR SUPPLEMENTARY READING

American Dyestuff Reporter
American Fabrics and Fashions
America's Textiles International
Apparel
Carpet and Rug Industry
Ciba Review
Clothing and Textiles Research Journal
Home Economics
Research Journal
Textile Chemists & Colorist
Textile Hi-Lights
Textile Horizons
Textile Organon
Textile Research Journal
Textile World
Textiles

TEXTILE ENCYCLOPEDIAS,
ABSTRACTS AND INDEXES

American Fabrics Magazine. *AF Encyclopedia of Textiles,* 2nd ed. 1972.

Kirk-Othmer. *Encyclopedia of Chemical Technology,* 3rd ed. 1984.

Klapper, Marvin. *Fabric Almanac.* New York: Fairchild Publications, 1966.

Applied Science and Technology Index
Business Periodicals Index
Chemical Abstracts
World Textile Abstracts

Index

MATRIX I
OVERVIEW OF CONSUMER TEXTILES CONCEPTS

Terms	Fibers	Yarns	Structures	Finishes	Color & Design
	Mineral		Nonwoven		
	Man-made noncellulosic		Multi component		Structuring
	Man-made cellulosic	Novelty	Complex weaves	Chemical	Finishing
	Protein	Ply	Basic weaves	Mechanical	Printing
	Natural cellulosic	Singles	Knits	General	Dyeing

Serviceability Concepts	Developmental Processes					
	Terms	Fibers	Yarns	Structures	Finishes	Color & Design
1 Durability						
Properties:[1] strength or tenacity .... abrasion resistance .... cohesiveness or spinning quality .... elongation .... elastic recovery .... flexibility or pliability ...... dimensional stability						
2 Comfort						
Properties: absorbency .... hydrophobic .... hydrophilic .... hygroscopic .... wicking .... electric conductivity .... allergenic potential .... heat or thermal conductivity heat or thermal retention .. density or specific gravity ..						
3 Care or Maintenance						
Properties: resiliency .... dimensional stability ...... flammability .... chemical reactivity .... heat tolerance .... biological resistance .... light resistance .... age resistance						
4 Esthetic Appearance						
Properties: color .... dye affinity .... luster .... translucence .... drape .... texture .... body or hand .... loft or bulk						

[1]For review you may wish to record property definitions on this matrix to the right of the term.

MATRIX IV
FIBER DURABILITY

DURABILITY PROPERTIES	FIBER CATEGORIES				
	Natural Cellulosic	Protein	Man-made Cellulosic	Man-made Noncellulosic	Mineral
OVERALL					
Strength or tenacity					
Abrasion resistance					
Cohesiveness or spinning quality					
Elongation					
Elastic recovery					
Flexibility or pliability					
Dimensional stability					

MATRIX V
FIBER COMFORT

COMFORT PROPERTIES	FIBER CATEGORIES				
	Natural Cellulosic	Protein	Man-made Cellulosic	Man-made Noncellulosic	Mineral
OVERALL					
Absorbency					
Wicking					
Electric conductivity					
Heat or thermal conductivity					
Heat or thermal retention					
Allergenic potential					
Density or specific gravity					

MATRIX VI
FIBER CARE

CARE PROPERTIES	FIBER CATEGORIES				
	Natural Cellulosic	Protein	Man-made Cellulosic	Man-made Noncellulosic	Mineral
General care or maintenance					
Resiliency					
Dimensional stability					
Chemical reactivity					
Detergents					
Bleaches					
Dry Cleaning					
Heat tolerance					
Flammability (recall burn test)					
Biological resistance					
Light resistance					
Age resistance					

MATRIX VII
FIBER ESTHETIC APPEARANCE

ESTHETIC PROPERTIES	FIBER CATEGORIES				
	Natural Cellulosic	Protein	Man-made Cellulosic	Man-made Noncellulosic	Mineral
Color or dye affinity					
Luster					
Translucence					
Drape					
Texture					
Body or hand					
Loft or bulk					

MATRIX VIII
GENERAL MILL AND DURABILITY FINISHES

Finish	Process		Typical Fiber or Structure	Purpose
	Mechanical	Chemical		
GENERAL MILL FINISHES				
Initial:				
Singeing				
Desizing				
Scouring				
Carbonizing				
Degumming				
Bleaching				
Final:				
Calendering				
Tentering				
Crabbing				
DURABILITY FINISHES				
Abrasion Resistant				
Antislip				
Relaxation Shrinkage				
Progressive or Fiber Shrinkage				
Mercerization				
Slack Mercerization				

MATRIX IX
FINISHES FOR COMFORT AND CARE

Finish	Process		Typical Fiber or Structure	Purpose
	Mechanical	Chemical		
COMFORT FINISHES				
Absorbent				
Antistatic				
Thermal				
CARE FINISHES				
Stain and Soil Resistant				
Durable Press				
Biological Resistant				
Water Repellent and Waterproof				

MATRIX X
FINISHES FOR ESTHETIC APPEARANCE

FINISH	TYPICAL FIBER OR STRUCTURE	PURPOSE
MECHANICAL FINISHES		
Shearing and Sculpturing		
Sizing		
Weighting		
Brushing		
Napping and Sueding		
Fulling		
Beetling		
Embossing		
Glazing		
Ciréing		
Moiréing		
Schreinerizing		
Flocking		
CHEMICAL FINISHES		
Fabric Softeners		
Acid, as Parchmentizing		
Basic, as Chemical Crepeing		

MATRIX XI
COLOR TERMNINOLOGY

COLOR TERMS	DEFINITION	SIGNIFICANCE TO CONSUMER
Colorfast		
Bleeding		
Crocking		
Migration		
Light fading		
Fume fading		
Dyes		
Pigments		
Dye site		
Dye medium (carrier)		

MATRIX XII
COLOR AND DESIGN PROCESSES

COLOR & DESIGN PROCESSES	DEFINITION	USES
Solution dyeing		
Fiber or stock dyeing		
Yarn dyeing		
Piece dyeing		
Union dyeing		
Cross dyeing		
Block printing		
Roller (surface) printing		
Warp printing		
Stencil printing		
Screen printing		
Discharge printing		
Other resist		
Tie-and-dye		
Batik		

Unit III
YARN MOUNTING SHEET

I SIMPLE YARNS: Singles

100% Staple*	Blended Staple	Blended Staple	100% Filament (Multifilament)	Blended Filament (Multifilament)

II SIMPLE YARNS: Ply and Cord

III COMPLEX

Two Ply-Crosswise	Two Ply-Crosswise	Two Ply	Cord	Slub

IV TEXTURED YARNS

Stretch Denim	Stretch Woven	Textured Crepe	Pile Knit	Textured Knit

*The blanks have been filled in on this mounting sheet. On future sheets, label the samples in the blanks provided.

Unit IV
BASIC WEAVES MOUNTING SHEET

I PLAIN WEAVES

Chintz*

Plain Weave Variations: Rib

Basket

II TWILL WEAVES

III COTTON SATEEN (STEEP TWILL)

Satin Weave

*You should write the remaining fabric names in the blanks, where applicable.

Unit IV
COMPLEX OR DECORATIVE WEAVES MOUNTING SHEET

I PILE WEAVES: Filling Pile

Warp Pile

II DOBBY WEAVES

III JACQUARD WEAVES

IV CREPE EFFECTS

Crepe Weave

Slack Tension Weave

V COMBINATION OF WEAVES

Unit IV
KNITTED FABRICS MOUNTING SHEET

I FILLING KNITS: Single

FILLING KNITS: Double

II WARP KNITS: Raschel

Tricot

II WARP KNITS: Tricot continued

Unit IV
NONWOVEN FABRICS MOUNTING SHEET

I BONDED FIBER WEBS

II SPUNLACED FABRICS

III NEEDLE PUNCH WEB

IV FABRIC FROM SOLUTION

Unit IV
SPECIAL STRUCTURES MADE FROM YARNS MOUNTING SHEET

I RASCHEL LACE

II TUFTED

III STITCH BOND

MULTICOMPONENT FABRICS MOUNTING SHEET

IV FABRICS SUPPORTED BY FIBER BACKING LAYERS

V FOAM BACKING

VI FABRICS SUPPORTED BY KNIT BACKING LAYERS

Unit V
FINISHES MOUNTING SHEET

I PREPARATORY OR ROUTINE FINISHES

for body

*

II DURABILITY

Also H_2O Repellency

III EASE OF CARE

*

IV FINISHES FOR APPEARANCE

*

APPEARANCE continued

*Blocks with two label lines underneath will have two samples.

Unit V
COLOR AND DESIGN MOUNTING SHEET

I COLORING STAGES

II DESIGN BY WEAVING

*

DESIGN BY WEAVING continued

III DESIGN BY KNITTING

IV DESIGN BY PRINTING

V BY FINISHING

* * *

*Blocks with two label lines underneath will have two samples.